96
Structure and Bonding

W0042482

Springer-Verlag Berlin Heidelberg GmbH

Molecular Self-Assembly
Organic Versus Inorganic
Approaches

Volume Editor: M. Fujita

With contributions by
I. Bernt, H. Bögge, A.J. Carr, P.S. Corbin,
B. Demleitner, M. Fujita, A.D. Hamilton, T. Kato,
P. Kögerler, M.J. Krische, J.-M. Lehn, B.R. Linton,
R.E. Meléndez, A. Müller, R.W. Saalfrank, E. Uller,
S.C. Zimmerman

Springer

The series *Structure and Bonding* publishes critical reviews on topics of research concerned with chemical structure and bonding. The scope of the series spans the entire Periodic Table. It focuses attention on new and developing areas of modern structural and theoretical chemistry such as nanostructures, molecular electronics, designed molecular solids, surfaces, metal clusters and supramolecular structures. Physical and spectroscopic techniques used to determine, examine and model structures fall within the purview of Structure and Bonding to the extent that the focus is on the scientific results obtained and not on specialist information concerning the techniques themselves. Issues associated with the development of bonding models and generalizations that illuminate the reactivity pathways and rates of chemical processes are also relevant.

As a rule, contributions are specially commissioned. The editors and publishers will, however, always be pleased to receive suggestions and supplementary information. Papers are accepted for *Structure and Bonding* in English.

In references *Structure and Bonding* is abbreviated Struct. Bond. and is cited as a journal.

Springer WWW home page: http://www.springer.de

ISBN 978-3-662-14306-3   ISBN 978-3-540-46591-1 (eBook)
DOI 10.1007/978-3-540-46591-1

CIP Data applied for

© Springer-Verlag Berlin Heidelberg 2000
Originally published by Springer-Verlag Berlin Heidelberg New York in 2000
Softcover reprint of the hardcover 1st edition 2000

Typesetting: Scientific Publishing Services (P) Ltd, Madras
Production editor: Christiane Messerschmidt, Rheinau
Cover: Medio V. Leins, Berlin
Printed on acid-free paper   SPIN: 10702272   02/3020 – 5 4 3 2 1 0

Volume Editor

Prof. Makoto Fujita

Department of Applied Chemistry
Graduate School of Engineering
Nagoya University
Chikusa, Nagoya 464-8603
Japan
E-mail: mfujita@apchem.nagoya-u.ac.jp

Editorial Board

Preface

Self-assembly has been undoubtedly a topic of special interest in current chemistry and is related to very wide scientific areas, including molecular recognition, material science, crystal engineering, nanotechnology, supramolecular chemistry, etc. Recent progress in this field seems to feature the construction of well-defined discrete systems exploiting complementary hydrogen bonding as well as coordination bonding. The purpose of the volume is to introduce the current topics in this very interesting field, focusing on two major subjects: i.e., organic assemblies (Part I) and inorganic assemblies (Part II). Seven international leading experts in this field deal with up-to-date topics in molecular self-assembly.

Part I begins with a review of hydrogen-bonded assemblies by Krische and Lehn (Louis Pasteur University, Chapter 1). Recent development in the work of their group is emphasized. This field has shown impressive progress in the development of novel multi H-bonding donor-acceptor systems, which is thoroughly discussed by Hamilton and co-workers (Yale University, Chapter 2) and Zimmerman and co-workers (University of Illinois, Chapter 3). In Chapter 4, Kato (University of Tokyo) deals with function through assemblies in hydrogen-bonded liquid crystalline compounds.

Part II of this volume focuses on inorganic assemblies involving coordination and metal-oxide linkages. Coordination assemblies are most discussed by Saalfrank (Universität Erlangen), who has been developing unique metal-assembled cage compounds since the late 1980s, and his co-workers (Chapter 5). Fujita (Nagoya Univeristy) also introduces a unique concept of metal-directed molecular paneling (Chapter 6). Finally, the inorganic part is concluded by Müller and his co-workers (Universität Bielfeld) with novel inorganic giant compounds consisting of metal-oxide linkages (Chapter 7).

I hope that chemists who are interested in molecular recognition, material science, nanotechnology, and supramolecular chemistry will welcome this book as an inspiring source for creative research ideas.

Nagoya, February 2000   Makoto Fujita

Contents

Contents of Topics in Current Chemistry, Vol. 198

Design of Organic Solids
Volume Editor: E. Weber

Part I
Organic Assemblies

The Utilization of Persistent H-Bonding Motifs in the Self-Assembly of Supramolecular Architectures

Michael J. Krische[1], Jean-Marie Lehn[2]

[1] Department of Chemistry and Biochemistry, University of Texas at Austin, Austin, TX 78712, USA

[2] Laboratoire de Chimie Supramoléculaire, ESA 7006 of the CNRS, ISIS, Université Louis Pasteur, 4 rue Blaise Pascal, 67000 Strasbourg, France
E-mail: lehn@chimie.u-strasbg.fr

The present account describes some selected examples of the generation of supramolecular architectures by recognition-directed self-assembly of components containing complementary arrays of hydrogen-bonding sites. Specific groups yield one-, two- or three-dimensional motifs. Supramolecular materials such as polymeric arrays and liquid crystals may be obtained.

Keywords: Supramolecular chemistry, H-bonding, Self-assembly, Molecular recognition, Supramolecular materials

1
Introduction

As early as 1920 Latimer and Rodebush suggested that "a free pair of electrons on one water molecule might be able to exert sufficient force on a hydrogen held by a pair of electrons on another water molecule to bind the two molecules together" [1]. Although the term "Übermoleküle" was first coined in 1937 to describe entities of higher organization resulting from the association of coordinatively saturated species such as the H-bonded dimer of acetic acid [2], it was only in the 1970s that a field was defined, termed supramolecular chemistry [3], that specifically dealt with entities held together by non-covalent bonds, in particular hydrogen bonds, donor–acceptor interactions and metal ion coordination. Within the last two decades, hydrogen bonding has become one of the most important non-covalent binding interactions employed in the de novo design and self-assembly of supramolecular architectures from molecular building blocks [4–8].

In this account, an outline of just some of the major advances in H-bond mediated self-assembly in non-biological systems are presented. The focus will be on a description of a few robust non-covalent synthons expressing well-defined, persistent H-bonding motifs and on strategies for their modular use in the design of novel supramolecular architectures and materials, specifically non-covalent polymers and liquid crystals. Such synthons for supramolecular synthesis [3a, 5–7] may be termed *tectons* [8], if the stress is on their structural characteristics, or *codons* if their information content is to be highlighted. The vast literature on synthetic molecular receptors operating through the action of H-bonds will not be discussed.

2
Expression of Persistent H-Bonding Motifs

That molecules possess information manifested in their geometric (shape, size) and interactional features, which is expressed through their intermolecular interaction with other molecules, is a fundamental precept of supramolecular chemistry and, hence, of molecular recognition-directed self-assembly [3, 5, 7]. Quantification of this principle is equivalent to defining a paradigm for H-bond mediated self-assembly and requires identification of the dominant supramolecular interactions and an understanding of their relative importance. Ideally, the information embodied in the molecular components should be a read-out following a well-defined interactional algorithm to selectively yield one of several possible non-covalent objects [3, 7]. Such fidelity requires that the integrity of structural information expressed by the molecular components should also withstand high degrees of chemical, structural, and thermal perturbation. As the integrity of the structural information may be characterized by the strength, directionality and selectivity of the encoded H-bond interactions, the multiplicity, cooperativity and complementarity of the H-bonds are key elements in the design of a persistent motif. Nevertheless, even when the synergistic action of numerous H-bonds are brought into play, the stability of a superstructure may still be merely nominal. Reliable molecular programming depends on the identification and orchestration of the *collectively* dominant supramolecular interactions (steric, van der Waals, π-molecular interactions, H-bonding and solid state packing) such that all operate in consonance.

The requirement of synergy between the dominant supramolecular interactions leads to the heart of the challenge of H-bond mediated self-assembly, as it is as a consequence of this fact that the intermolecular binding interactions one would like to exploit are generally weak (*vide supra*). While the use of weak interactions is advantageous owing to their reversible formation and, hence, the capacity of supramolecular systems to correct errors that may occur in the course of their self-assembly, the outcome may be difficult to control. Reduced predictability becomes especially evident in the solid state, where crystal packing forces may be of the same order of magnitude as the predesigned interactions of the component molecules [9]. Even for very simple systems, kinetic effects or energetically similar crystal packing modes may exist resulting in the formation of crystalline polymorphs [10]. Indeed, the interplay of symmetry constraints and the many attractive and repulsive interactions in the solid state is sufficiently complex that a priori prediction of crystal packing motifs is only viable for rather simple systems and remains a worthy endeavor [11]. Consequently, although X-ray crystal-lographic analysis is a very valuable means of characterizing supramolecular species, the solid state does not necessarily reflect the corresponding population of supramolecular species in solution. The importance of controlled crystal engineering [12] is, nevertheless, underscored by the fact that non-linear optical properties [13], ferromagnetic behavior [14], elecrical conductivity [15] and solid state reactivity [16] all critically depend on the relative orientation of the composite molecules in the solid state.

The design of robust tectons for the construction of supramolecular architectures requires selection of intermolecular interactions that are strong and possess a well-defined directionality. To this end, H-bonds are not ideal. They are usually rather weak, with binding energies in the range of 1–5 kcal/mol [17]; greater values are obtained for ionic H-bonds which contain a large charge interaction contribution [18], while the so-called low-barrier H-bonds believed to be employed in enzymatic catalysis have been estimated as having binding energies as high as 30 kcal/mol in the gas phase [19]. They are also of rather soft directionality, being amenable to appreciable bending with little loss in binding energy [20]. Finally, H-bonding is strongly solvent dependent. The influential role of differential solvation by the surrounding medium undoubtedly accounts for the large disparity in many reported H-bond energies. For example, the presence of only 1% methanol in a CCl_4 solution may lead to a decrease in association constants by a factor of 25 [24].

The designed expression of specific H-bonding motifs requires that robust molecular codons/tectons be devised which contain information strong enough to enforce the generation of a specific supramolecular architecture despite changes in secondary structural features and/or in environmental conditions. In view of these features, the design of H-bonding codons/tectons must rely on the accumulation of multiple H-bond donating (D) and accepting (A) sites in order to ensure that the total interaction between them is both strong enough and geometrically well defined. Suitably selected multi-site subunits may take advantage of synergy and cooperativity [21] in the establishment of multiple H-bonds. Whereas reversibility allows full explora-tion of the energy hypersurface of the system, positive cooperativity aids in directing the selective formation of one of several possible species.

To achieve synergy and cooperativity, it is desirable to employ preorganized systems that possess multiple complementary H-bond arrays defining molecular recognition patterns. Thus, for example, compounds 1–4 contain tridentate ADA/DAD and ADD/DDA arrays. In chloroform such triply H-bonded systems have association constants that vary between $K_a = 10^2$–10^3 M^{-1} for DAD-ADA pairs and $K_a = 10^5$–10^6 M^{-1} for DDD-AAA [22]. These differences in K_a may be interpreted as resulting from secondary electrostatic interactions [23]. On the other hand, for spatially fixed arrays of multiply H-bonded systems in chloroform and CCl$_4$, a simple empirical free energy relationship may be derived in which each H-bond is attributed 7.9 ± 2.9 kJ/mol for each attractive or repulsive secondary interaction [24]. While it is possible to predict the relative stabilities of multiparticle H-bonded aggregates [25], for most purposes tridentate H-bond arrays are sufficiently robust, and even stronger quadruple H-bonding arrays have been described [26]. The anticipated assemblies corresponding to the association of molecules 1–4 represent discrete objects and are indicated in Fig. 1.

The aforementioned discrete associations through molecular recognition patterns may be combined to yield 1-dimensional H-bonding motifs following two distinct strategies. H-Bonding recognition groups may be combined in a modular sense via covalent linkers or by "fusion" of the heterocyclic recognition groups to yield heterocycles possessing two H-bonding faces, termed Janus molecules (Fig. 2) [8]. Similarly, 1-dimensional motifs may be combined to yield 2-dimensional objects and 2-dimensional motifs may be extended to yield 3-dimensional objects.

It must be stressed that programming the self-assembly of supramolecular architectures on the basis of H-bonding recognition patterns supposes that all H-bonds that may form are used in an intra-supramolecular fashion.

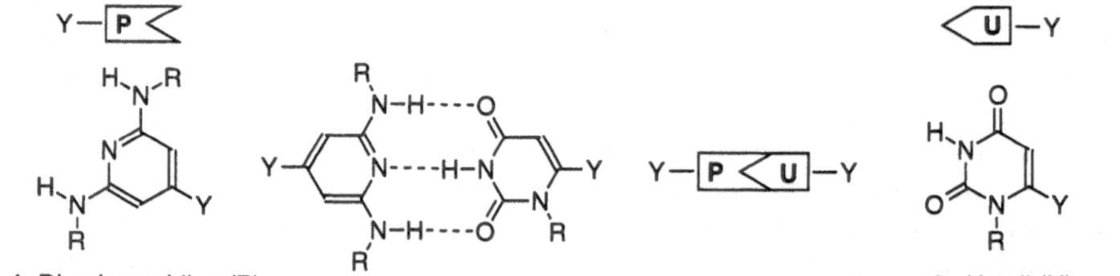

1, Diaminopyridine (P)
ADA H-Bond Array

2, Uracil (U)
DAD H-Bond Array

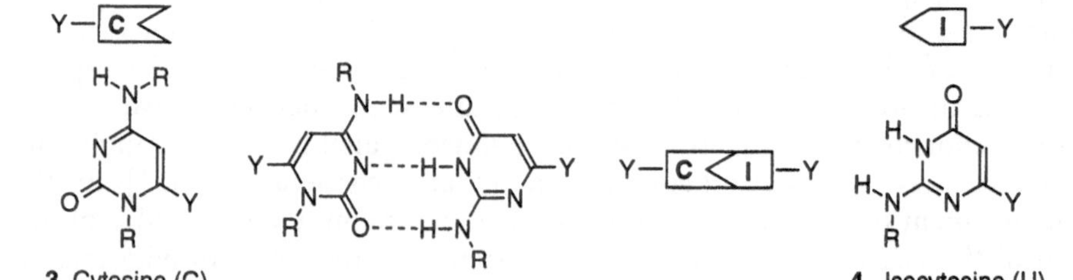

3, Cytosine (C)
AAD H-Bond Array

4, Isocytosine (U)
DDA H-Bond Array

Fig. 1. Associations derived from DAD/ADA and DDA/AAD H-bond arrays

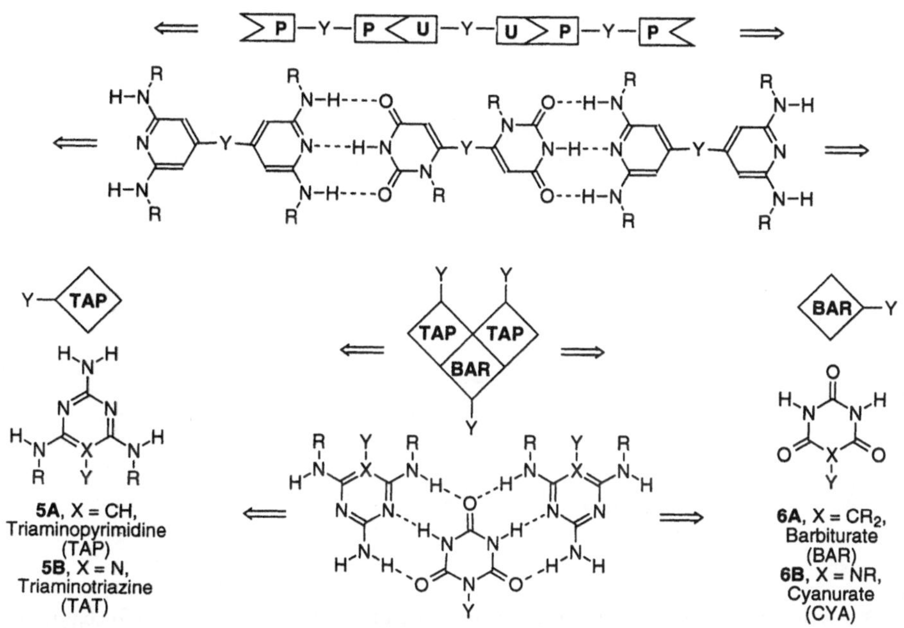

Fig. 2. Combination of molecular recognition patterns to create 1-dimensional arrays

Deviations from the expected arrangement result from the establishment of exogenous H-bonds due to the operation of other intermolecular effects.

3
Discrete Associations

The numerous efforts directed toward the generation of cyclic objects via self-assembly from acyclic precursors are instructive as they illustrate a variety of general design principles employed in H-bond mediated self-assembly. As H-bonded macrocycles may be viewed as arising from the folding of 1-dimensional motifs, the discrimination between linear and corresponding cyclic aggregates becomes a strategic point. One tactic for the selective promotion of one among several modes of self-assembly involves the use of steric and geometric constraints. For example, angular orientation between the H-bonding faces of Janus molecules 5a/5b and 6a/6b permits formation of either a linear tape motif (see Fig. 2) or a cyclic hexamer (Fig. 3). In addition, an intermediate combination of these two limiting arrangements yields a crinkled tape (Fig. 12). These three modes of self-assembly are degenerate with respect to the H-bonding pattern, so that further discrimination requires the operation of other factors such as steric constraints. Thus, for the assembly of 6a (X = CEt$_2$) and 5b (R = p-tert-butylphenyl), non-bonded interactions between the p-tert-butylphenyl moieties, which are more pronounced in the linear array, are relieved in the cyclic aggregate and promote its formation [27]. Alternatively, the steric influence of porphyrin substitution facilitates

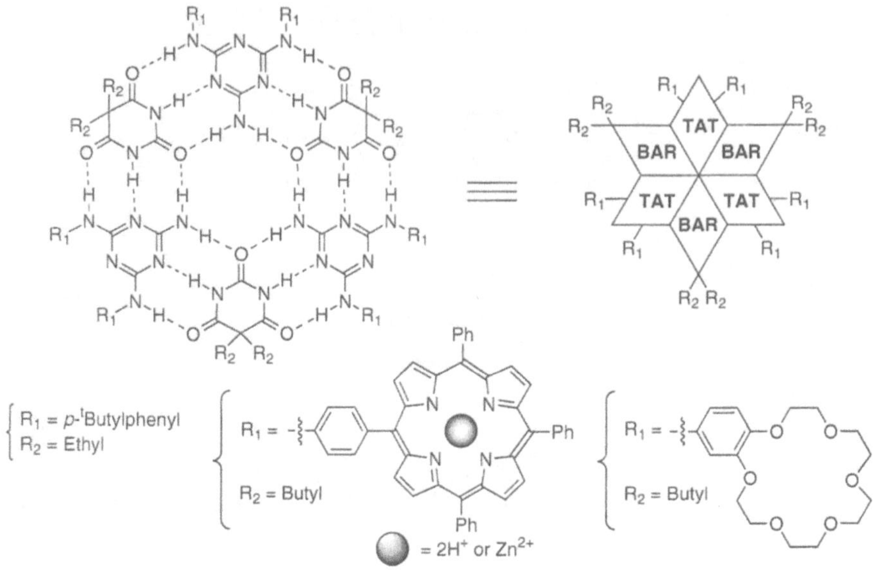

Fig. 3. Steric control in cyclic hexamer formation from Janus-type DAD/ADA recognition groups **5** and **6**

generation of a H-bonded macrocycle in which the relative orientation of six porphyrins is controlled [28]. Finally, substitution of **5** with crown ether residues promotes formation of a cyclic aggregate which can be observed by electrospray mass spectroscopy via ion labeling [29] (Fig. 3).

In addition to steric information, a second strategy towards the construction of cyclic arrays involves preorganization, as in the potassium-templated cyclotetramerization of guanosine monophosphate and related compounds [30]. In the case of Janus recognition groups **5a/5b** and **6a/6b**, the tethering of three molecules of **5b** to a central core facilitates geometric predisposition towards formation of the cyclic hexamer by attenuating the entropy loss associated with the self-assembly event (Fig. 4) [31]. This strategy has been implemented for the self-assembly of homologous doubly and triply stacked systems [32].

More recently, systems exploiting both steric factors and preorganization in the assembly of doubly and quadruply stacked H-bonded macrocycles have been reported. This has been accomplished by appending the Janus H-bonding recognition groups **5** and **6** on a calix[4]arene scaffold as in tectons **7** and **8** (Fig. 5) [33]. Isophthalic acid derivatives also assemble to yield H-bonded macrocycles [34]. Tecton **9**, derived via appendage of isophthalic acid residues to a dendron-substituted acridine scaffold, generates a doubly stacked H-bonded macrocycle (Fig. 6) [35].

However, in order to achieve an unambiguous programming of supramolecular macrocycle formation based exclusively on H-bonding information, it is necessary to devise a suitable pattern of H-bond donor/acceptor sites. The

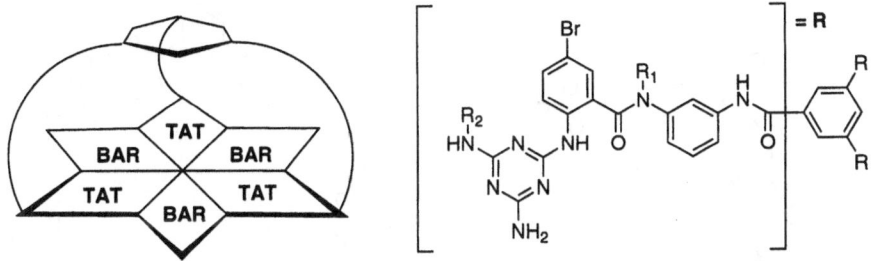

Fig. 4. A capped cyclic hexamer derived from Janus-type DAD/ADA tectons **5** and **6**

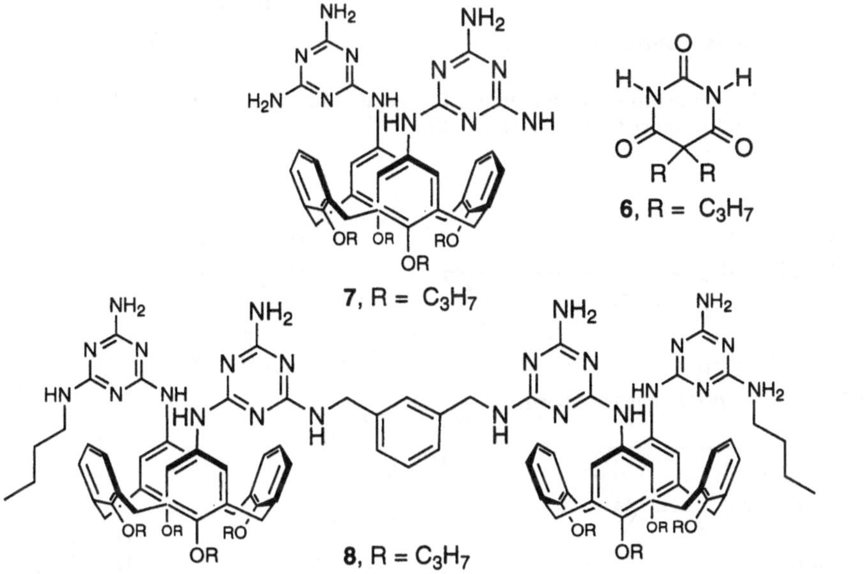

Fig. 5. Calix[4]arene-based tectons **7** and **8** for the generation of doubly and quadruply stacked H-bonded macrocycles

expanded Janus molecule **10A**, possessing DDA/AAD H-bonding faces, can only form a fully interconnected H-bond array in the cyclic arrangement. Thus, the H-bonding information unequivocally determines the outcome of the self-assembly process (Fig. 7) [36]. This strategy has also been successfully applied for the assembly of the related monomers **10B** [37] and **11** [38].

As part of an ongoing program focused on chirality in supramolecular systems [8, 39], it was noted that stereochemical aspects of H-bonding information may also be used to direct self-assembly processes. Racemic **12** produces a pleated ribbon comprised of doubly H-bonded components of alternating chirality, motif **A**. In contrast, optically pure **12** is predicted to form the cyclic hexamer **B**, provided the available H-bonding capacity is fully used

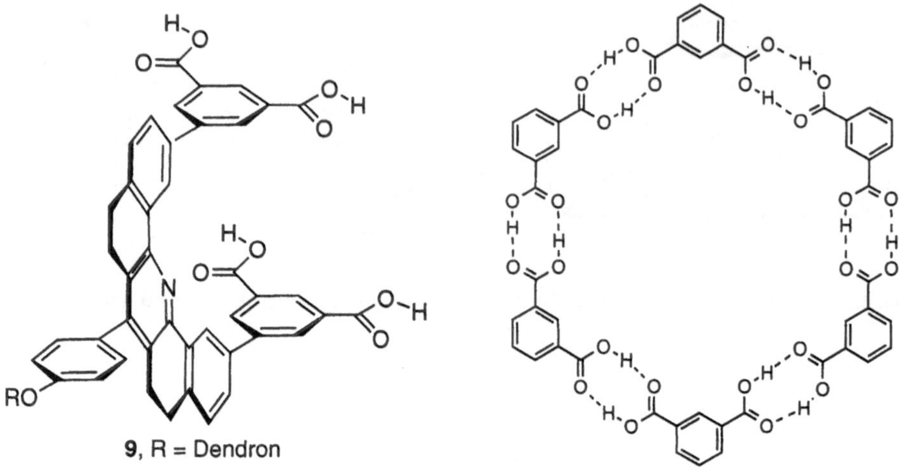

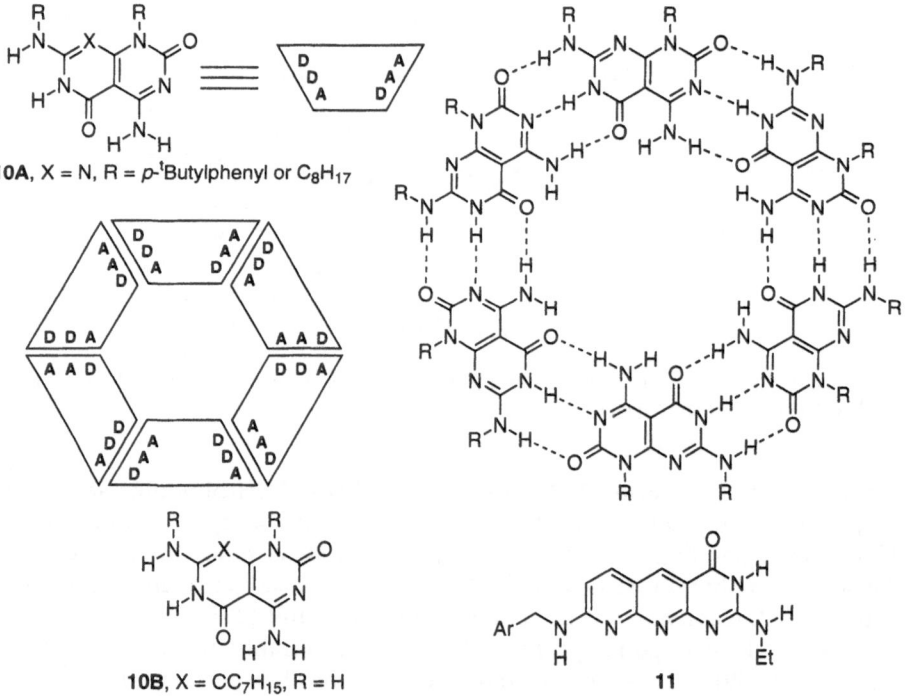

9, R = Dendron

Fig. 6. Dendron-substituted tecton **9** for the generation of a doubly stacked H-bonded macrocycle and proposed mode of H-bonding

10A, X = N, R = *p*-ᵗButylphenyl or C₈H₁₇

10B, X = CC₇H₁₅, R = H

11

Fig. 7. Self-assembly of H-bonded macrocycle enforced by well-defined complementary arrays of H-bonding sites

within the discrete supramolecular assembly. X-ray crystallographic analysis reveals that a truncated version of the macrocycle arises, the cyclic tetramer C, due probably to the fact that it allows the formation of externally directed H-bonds (Fig. 8) [40] and tighter packing [41].

Another case of a supramolecular cyclic structure is illustrated in the formation of the four-component porphyrin cage 13 (Fig. 9) [42]. In the absence of triaminopyrimidine, the uracil moieties appended to the porphyrin exist as a mixture of *syn-* and *anti-*atropisomers. Upon introduction of triaminopyrimidine, the presence of the *anti* isomer decreases to less than 20%, suggesting formation of 13. The presence of 13 is further supported by electrospray mass spectroscopy and vapor pressure osmometry analyses. Similarly, rigid scaffolds bearing complementary DNA bases form closed ensembles, rather than linear polymers, as evidenced by the solution-state dimerization of 14 (Fig. 10) [43].

Non-covalent closed objects are also represented by capsules, cages and spheres assembled through the action of H-bonds. Many beautiful examples of such systems exist and this topic has recently been reviewed [44]. The design principles that have emerged indicate that high degrees of preorganization are required to overcome entropic factors. Thus, tectons 15, 16 and 17 possess

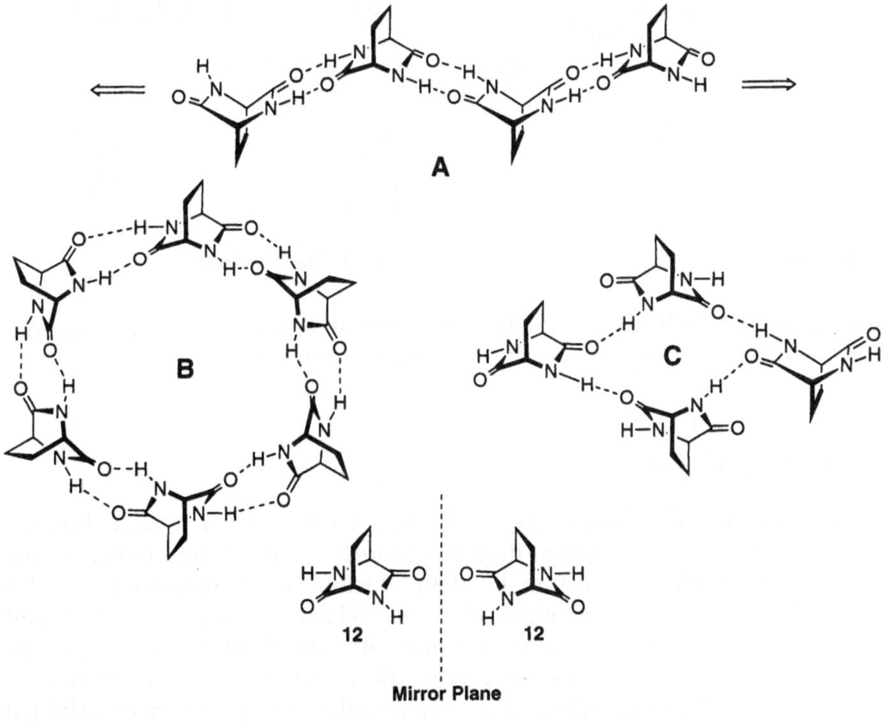

Fig. 8. Self-assembly of H-bonded superstructures from enantiomeric components 12: *A* heterochiral strand; *B* homochiral macrocycle; *C* observed homochiral assembly

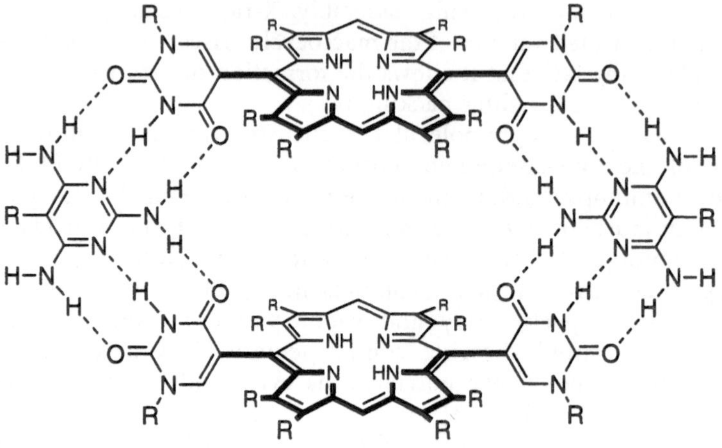

Fig. 9. Self-assembly of a macrocyclic porphyrin cage

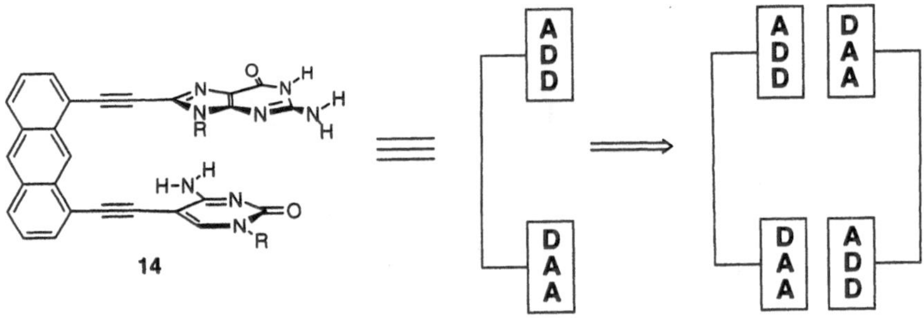

Fig. 10. Formation of a cyclic structure from an artificial dinucleotide **14**

rigid scaffolds which fix the relative orientation of their H-bonding arrays, predisposing the tectons toward homo-dimerization (Fig. 11) [45].

4
1-Dimensional Motifs

As previously noted, discrete assemblies based on 1-dimensional H-bonded motifs, also termed α-networks, may be obtained by covalently tethering the H-bonding recognition arrays or by their "fusion" to yield Janus molecules (Fig. 2). The information manifested in the relative orientation (120°) and juxtaposition of H-bonding donor/acceptor sites (DAD/ADA), in the specific case of **5** and **6**, may generate three types of association: a linear ribbon or tape, a H-bonded macrocycle and an intermediate wavy ribbon or crinkled tape [46]. These structures have been characterized in the solid state by X-ray crystallographic analysis. Homochiral crinkled tapes are formed in the course

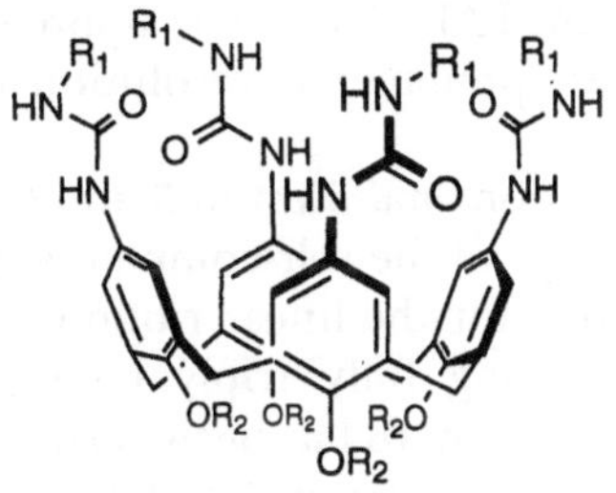

Fig. 11. Representative tectons 15, 16 and 17 for the generation of capsules via dimerization

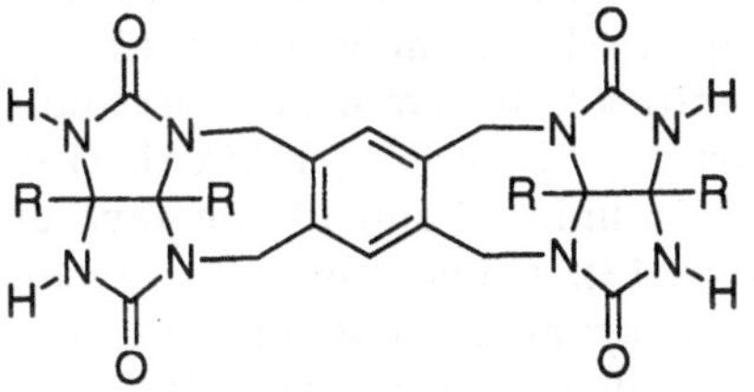

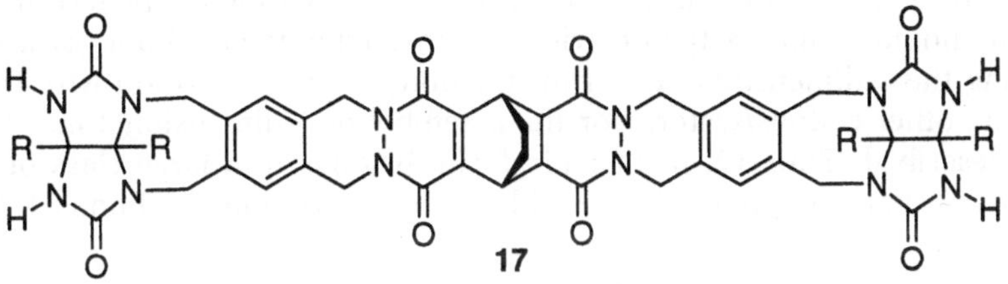

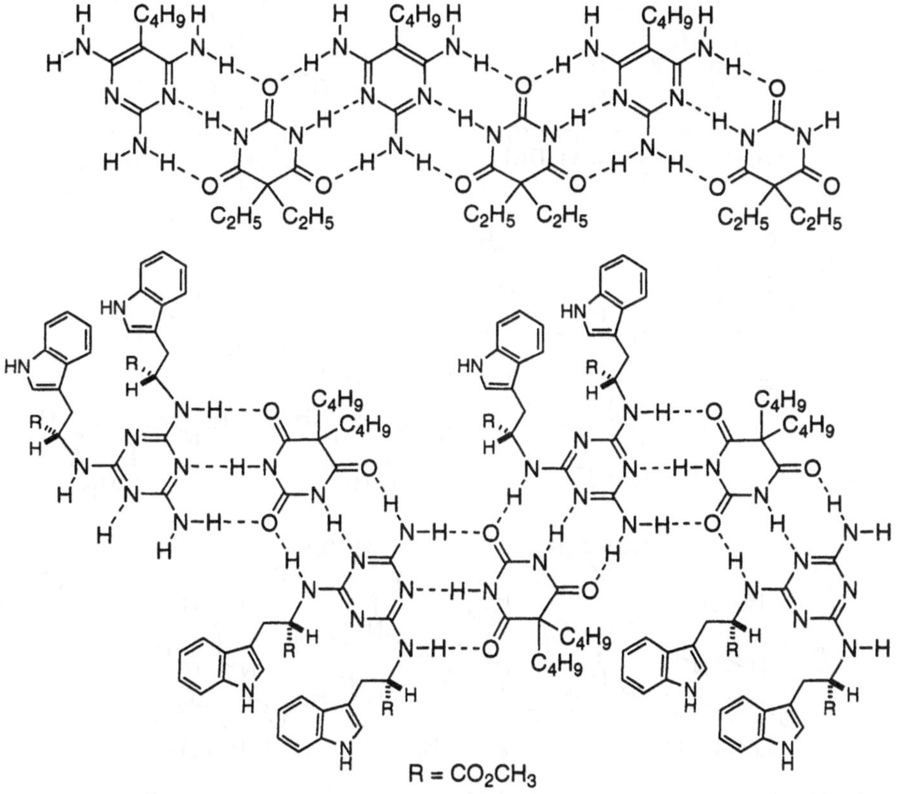

R = CO₂CH₃

Fig. 12. Representation of the crystal structures of Janus tectons 5 and 6 illustrating linear (*top*) and "crinkled" (*bottom*) 1-dimensional H-bonding motifs; the latter is homochiral

of the association of a racemic bis-tryptophan TAP derivative upon self-assembly with dibutyl BAR, demonstrating that spontaneous resolution may occur upon self-assembly (Fig. 12) [47].

In contrast to the ambiguous assembly instructions manifest in 5 and 6, the geometry and plerotopic DDA/AAD H-bond arrays of the self-complementary recognition units 18 and 19 unequivocally encode for the linear ribbon motif only (Fig. 13). Other modes of assembly involving fully triply connected H-bond arrays are not possible. The motif expressed by 18 has been verified by X-ray crystallographic analysis [48]. Interestingly, the unequivalent Janus faces of 18 and 19 are tautomerically related. Thus, protonation at one terminus of a ribbon should induce a tautomeric shift throughout the ribbon as a whole, allowing the 1-dimensional motif to function as a single proton wire [49].

Many other codons/tectons for the assembly of 1-dimensional motifs have been described. Thus 12, in racemic form, belongs to a large class of cyclic secondary diamides that self-assemble to form H-bonded ribbons [35, 50].

X = CH, Diaminopyrimidinone (DAP)   X = CH, Aminopyrimidine Dione (APD)
X = N, Diaminotriazinone (DAT)   X = N, Aminotriazine Dione (ATD)
          18   19

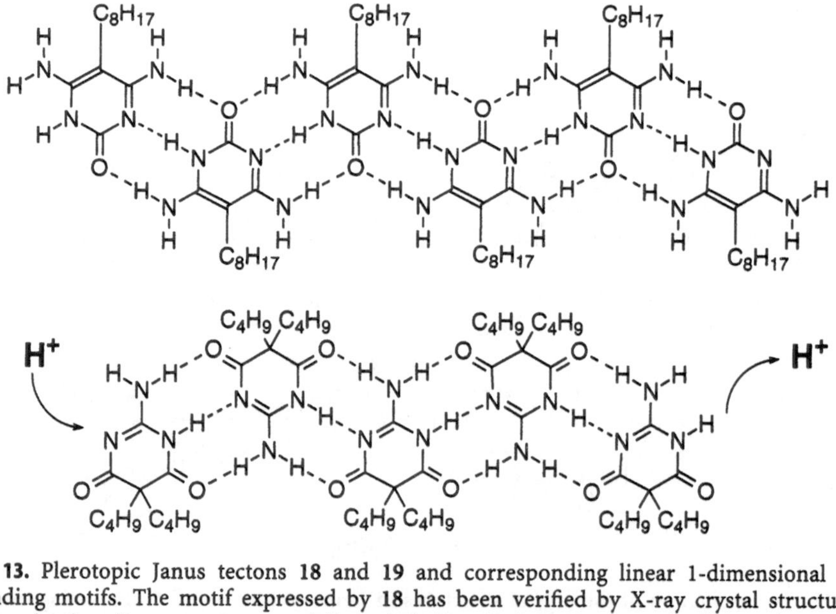

Fig. 13. Plerotopic Janus tectons 18 and 19 and corresponding linear 1-dimensional H-bonding motifs. The motif expressed by 18 has been verified by X-ray crystal structural determination; it represents potentially a chain for single proton transfer (*bottom*)

2-Aminopyrimidines [51], diacids [52] and acyclic secondary diamides [53] have long been known to form H-bonded tapes, and cocrystals of compounds incorporating these functionalities also form 1-dimensional objects [54]. Cyclic [55] and acyclic [56] *N,N'*-disubstituted ureas, the Janus analogs of secondary diamides, and bisamidine-dicarboxylic acid cocrystals [57] also generate 1-dimensional motifs.

5
2-Dimensional Motifs

The combination of like or unlike 1-dimensional motifs represent two distinct strategies for the generation of 2-dimensional motifs, or "β-networks". Consequently, the challenge in creating a 2-dimensional network lies in the requirement that the composite 1-dimensional motifs be sufficiently robust for their integrity to be maintained.

This "modular motif" paradigm is nicely illustrated in the self-assembly of primary amides and diamides, in which the former gives an α-network and the latter a β-network (Fig. 14) [58]. Caution, however, must be exercised for exceptions to this principle abound. For example the motif expressed by urea, the Janus version of a primary diamide, is truncated in one dimension to yield a helical channel structure [59] Combination of unlike motifs, specifically cyclic secondary diamide and diacid motifs, is demonstrated in the solid-state structure of diketopiperazine **20** (Fig. 15) [60].

Janus recognition groups expressing 1-dimensional motifs may also be combined to form β-networks (Fig. 16). Modular covalent combination of

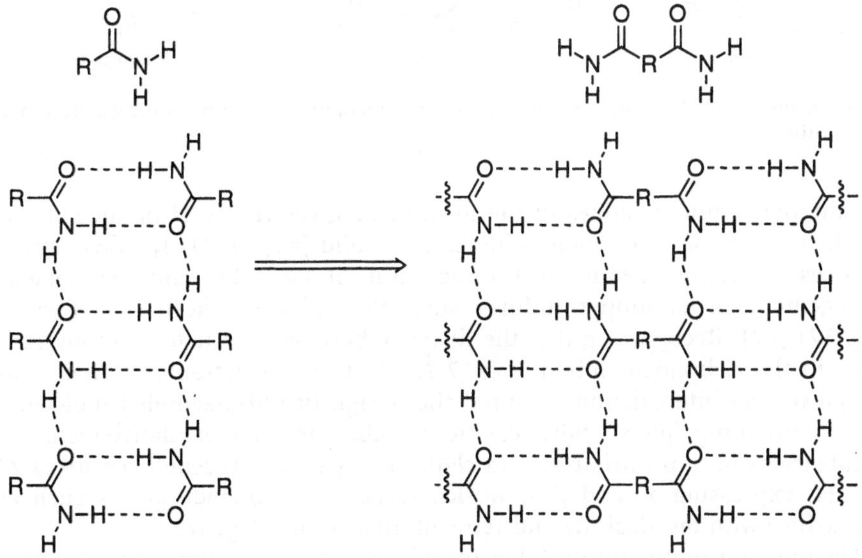

Fig. 14. Combining like 1-dimensional motifs to yield a 2-dimensional pattern

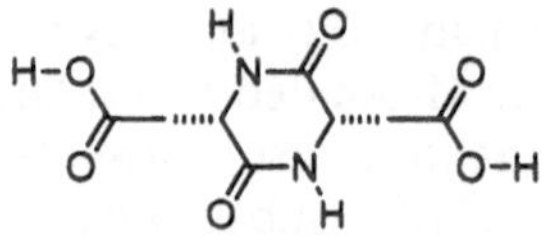

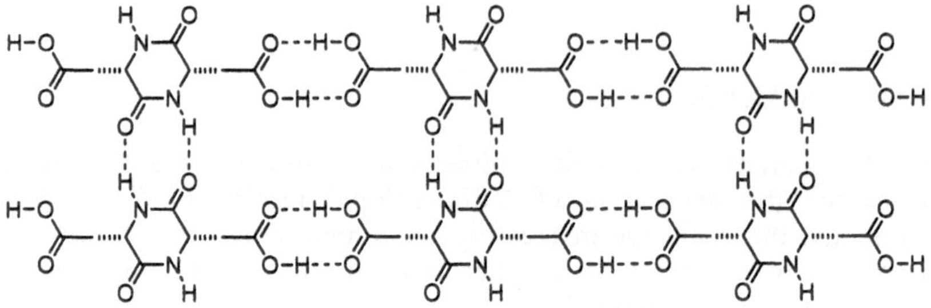

Fig. 15. Combining unlike 1-dimensional motifs to yield a 2-dimensional pattern

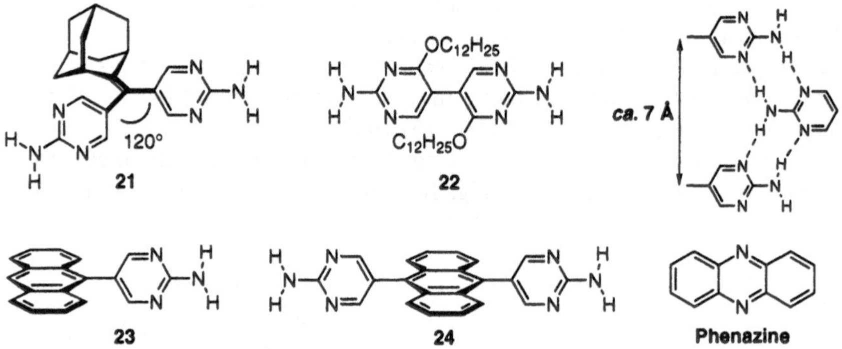

Fig. 16. Components bearing homotopic Janus-type recognition groups for the generation of tape motifs

2-aminopyrimidine residues in the form of **22** gives rise to H-bonded sheets which stack to form an amphiphilic layered solid [61]. In **21**, 1,3-allyic strain enforces a relative geometrical orientation of ca. 120° and syn-cofacial arrangement of aminopyrimidines such that pleated sheets are derived (Fig. 17) [62]. Recognizing that the distance between alternate aminopyrimidines in the H-bonded ribbon is ca. 7 Å, ideal for formation of π-molecular complexes via intercalation, inspired the design of self-assembled molecular arrays containing "pigeon hole" cavities suitable for the intercalative inclusion of substrates of appropriate size in clathrate style. Anthracene derivatives **23** and **24**, expressing α- and β-networks, respectively, include guests such as phenazine (with an ideal fit) and tetrachloroethylene (Fig. 18).

The modular motif approach has fueled a generation of numerous examples of β-networks. Linear molecular tapes derived from tectons **5** and **6** have been

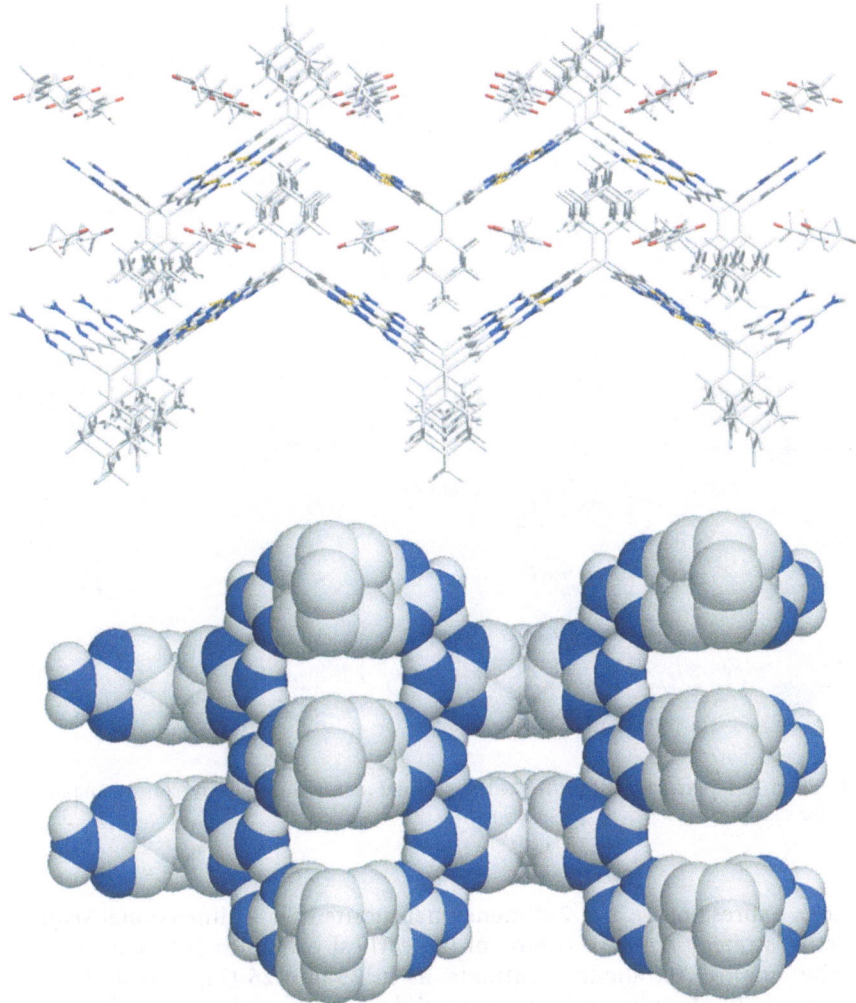

Fig. 17. Pleated sheet motif in the crystal structure of **21**

linked by ammonium carboxylate salt bridges [63]. Oxalimides and ureas possessing terminal carboxylic acid residues [64], sulfamides [65], and guanidinium alkyl- or arylsulfonate salts [66] provide robust 2-dimensional H-bonded networks.

6
3-Dimensional Motifs

Three-dimensional H-bonded networks, or "γ-networks", may be derived via combination of 2-dimensional motifs or via appendage of non-covalent

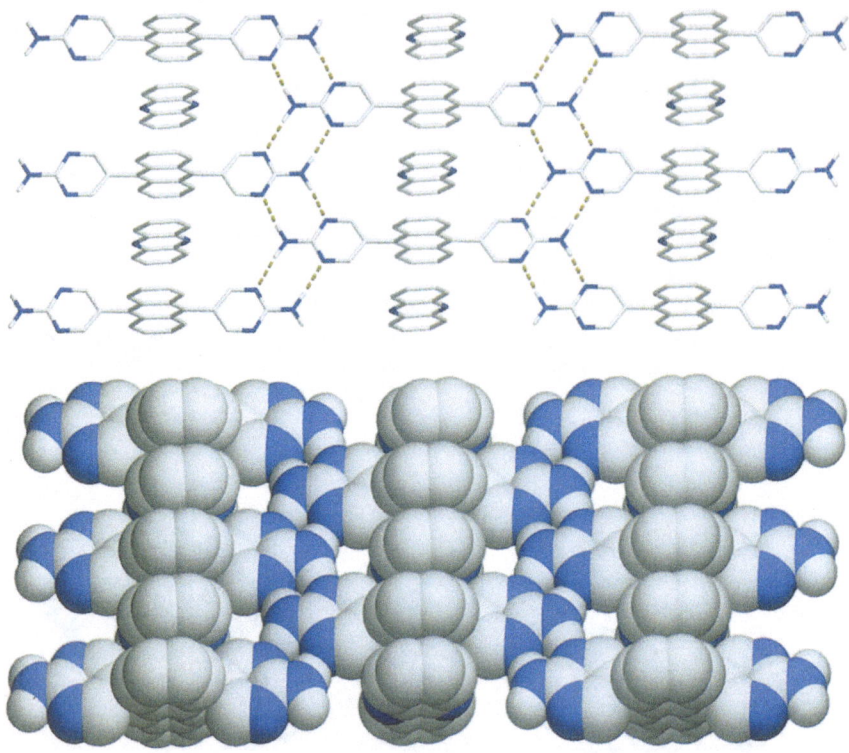

Fig. 18. Solid-state structure of the 2-dimensional array formed by **24** · phenazine π-molecular complex

synthons expressing 1- and 2-dimensional motifs to a 3-dimensional scaffold. The latter strategy embodies some of the earliest examples of γ-networks and typically employs tetrahedral scaffolds, as in **25** and **26** (Fig. 19) and related structures [67]. More recently, this principle has been extended to tectons **27**, **28** and **29** (Fig. 19) [68]. β-Networks derived via the guanidinium alkyl- or arylsulfonate system (*vide infra*) may be combined by the utilization of disulfonates to provide γ-networks (Fig. 20) [69].

An alternate means of accessing 3-dimensional arrays takes advantage of stereochemical information. Appending two 2-aminopyrimidines to the chiral cyclohexyldiamine scaffold (as in **30**) promotes generation of a helical ribbon comprised of H-bonded dimers. The helical interior defines a channel and consumes half the available H-bonds. The remaining H-bonds are satisfied by connecting the helical ribbon to adjacent, parallel helices. A honeycomb-like network results with a channel diameter of 7.65 Å from corner to corner and 6.83 Å from side to side. Taking into account the van der Waals radii, the central void of the channel is slightly under 4 Å in diameter (Fig. 21) [70].

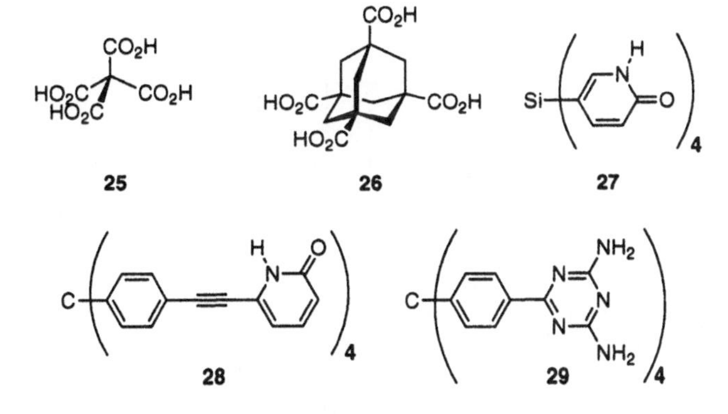

Fig. 19. H-bond recognition groups appended to 3-dimensional scaffolds

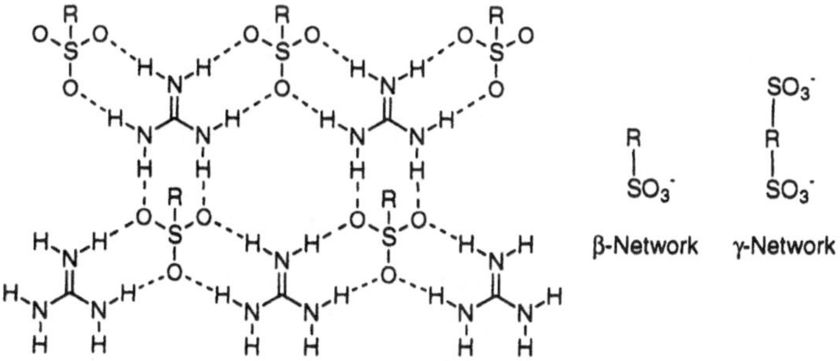

Fig. 20. γ-Networks via modular combination of 2-dimensional motifs of guanidinium and sulfonate components

7
H-Bonded Supramolecular Materials

Recognition processess occurring at the molecular level may lead to changes at the level of the material, thus expressing molecular information on the macroscopic scale. Such "bottom-up" strategies [71] have the advantage over "top-down" procedures [72] in that they give access to nano-size material devices of more or less well-defined structure through self-assembly.

Since liquid crystalline materials may themselves be viewed as extended supramolecular entities, it is natural that the use of H-bonds in mesophase design, in both discrete and polymeric assemblies, has garnered much success [71a, 73]. A common type of molecular species that form thermotropic liquid crystals presents an axial rigid core fitted with flexible chains at each end [74]. Splitting the core into two complementary halves which may associate through H-bond interactions, as in **31**, then sets the stage for the generation of mesogenic supermolecules. While the pure components are isotropic, the 1:1 complexes **31** display mesomorphic phases (Fig. 22) [75]. Indications to this end have also been found in the formation of mesomophic phases derived

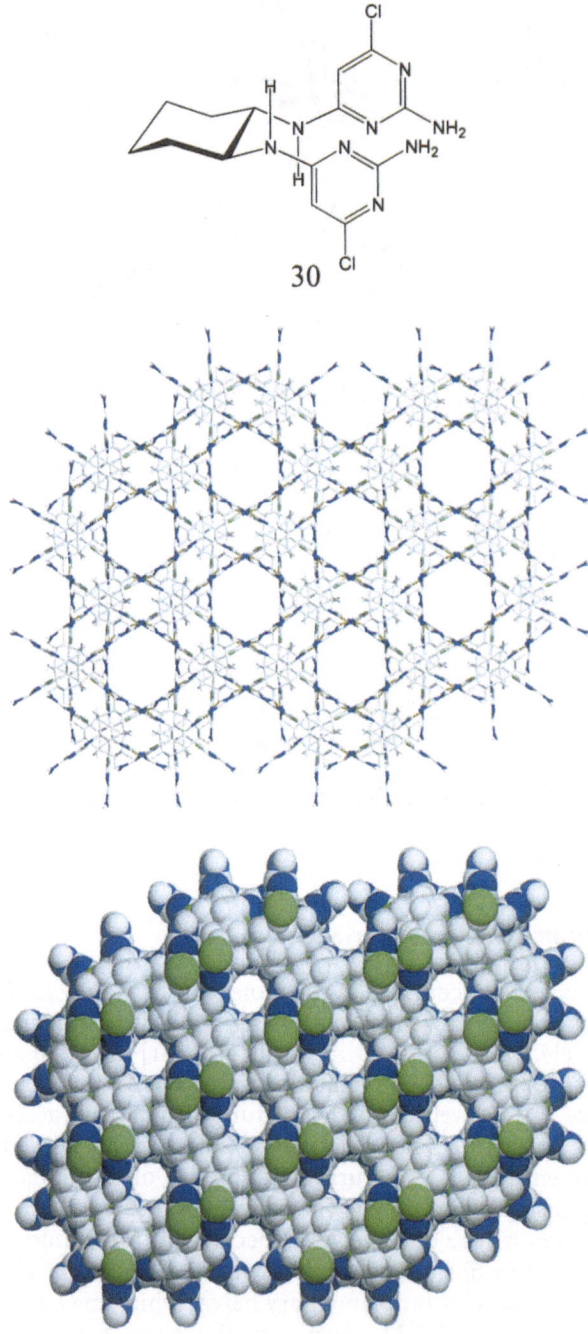

Fig. 21. Tecton **30**: Interwoven helices result in a network of unoccupied channels

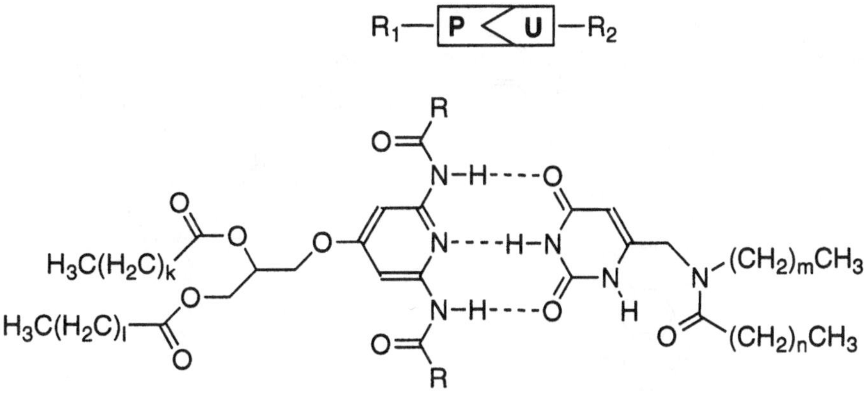

$$R_1 - \boxed{P \diagdown U} - R_2$$

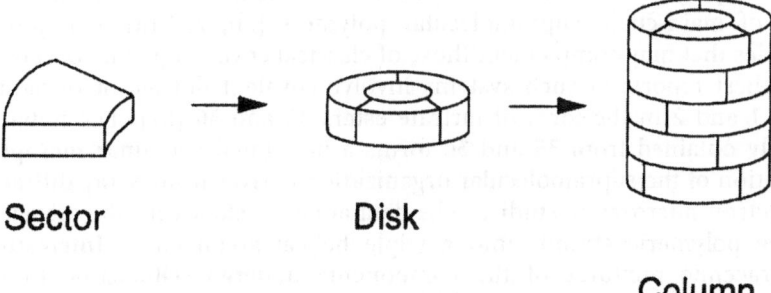

Fig. 22. Assemblies **31** displaying thermotropic mesophases

through the H-bond mediated dimerization of *p*-alkoxybenzoic acids [76a] and 2-pyridone-5-carboxylic esters [76b], and the association of carboxylic acids with various pyridine derivatives [77].

Discotic columnar mesophases may be derived from discrete H-bond assemblies possessing disk-like morphology in a sequential *hierarchical self-assembly* process involving the assembly of components into H-bonded disks which thereafter assemble to form columnar stacks (Fig. 23). Components may organize around a complementary central core, as illustrated in the assembly **32** (Fig. 24) [78]. Alternatively, in the case of the phthalhydrazide cyclotrimer **33**, an H-bonded macrocyclic core may be utilized [79]. A novel aspect in the formation of **33** involves managing the H-bond donor/acceptor information through the control of tautomeric equilibria. The bias toward the requisite lactim-lactam tautomer is reinforced through cooperative formation of the cyclotrimer which thereafter forms discotic columnar stacks. Finally, H-bonds may be used as secondary structural control elements for the rigidification of side chains appended to a high symmetry core, as in **34** (Fig. 24) [80]. It has recently been noted that columnar stacking of the now classical barbituric

Sector   Disk

Column

Fig. 23. Sequential hierarchical self-assembly process resulting in the formation of discotic columns via the assembly of sector molecules into a disk

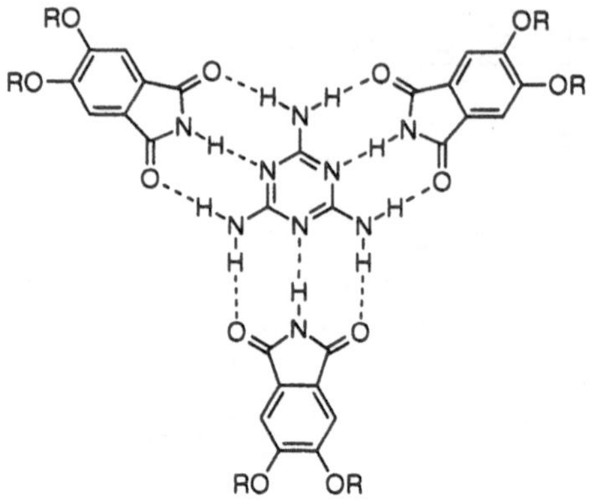

Fig. 24. Discrete assemblies 32 and 33 for the generation of discotic columnar architectures

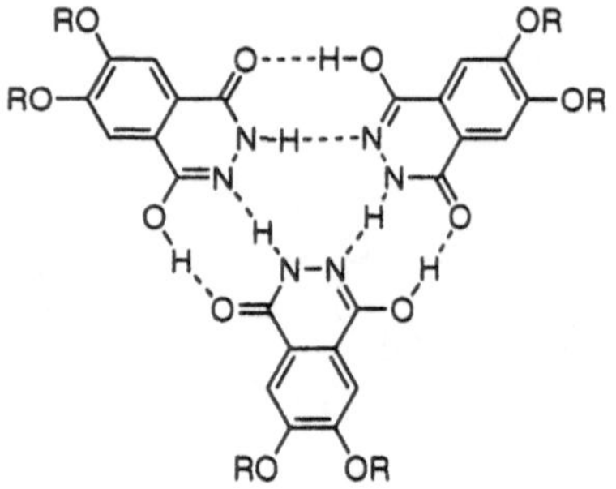

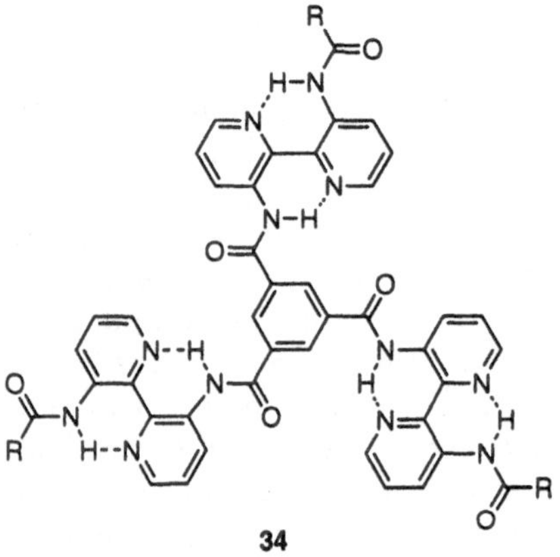

acid-melamine H-bonded macrocycles results in the formation of mesoscopic fibers [81].

The ubiquity of robust 1-dimensional H-bonding motifs makes possible the design of main-chain supramolecular polymers [3b, 7, 71a] with physical properties that may complement those of classical covalent polymers (Fig. 25). The earliest reports of such systems involve covalent linkage of recognition groups 1 and 2 in the form of tartrate esters 35 and 36 [82]. The polymeric assembly obtained from 35 and 36 forms a hexagonal columnar mesophase. Description of the supramolecular organization derives from X-ray diffraction and electron microscopy studies. The data are consistent with the association of three polymeric strands into a triple helical architecture. Interestingly, chiral racemic mixtures of the components undergo self-sorting to form optically pure left- and right-hand helices (Fig. 26). The rigid components 37 and 38 associate into supramolecular rigid rod polymers that form a lyotropic phase (Fig. 25) [83].

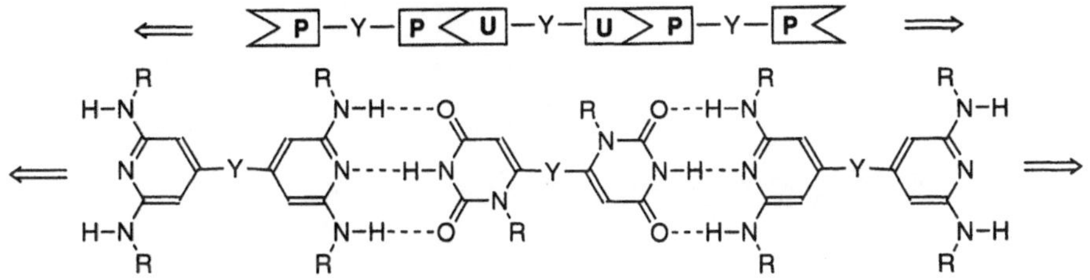

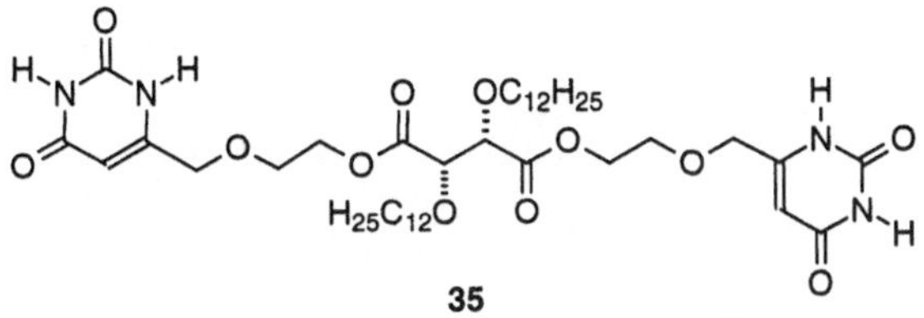

Fig. 25. H-bonded main-chain polymers incorporating recognition groups 1 and 2 linked to a flexible chiral tartaric acid spacer (35, 36) or to a rigid anthracene spacer (37, 38)

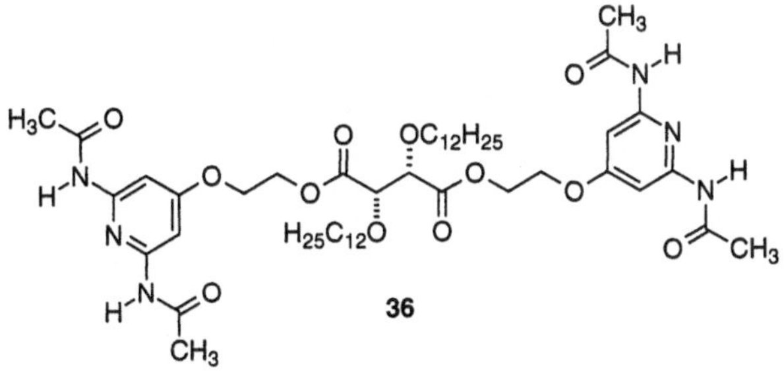

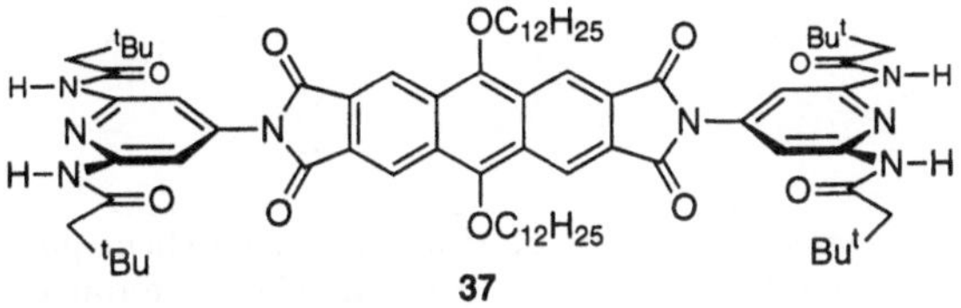

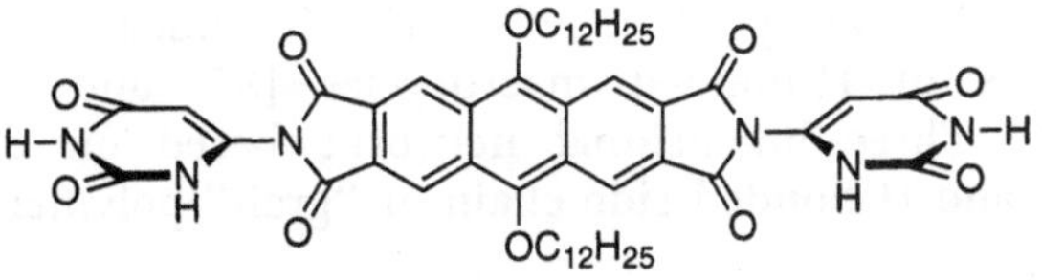

The incorporation of Janus recognition moieties **5** and **6** on a tartrate scaffold in the form of "double-Janus" tectons **39** and **40** permits the design of infinite 2-dimensional polymeric assemblies (Fig. 27) [84]. The assembly of **39** and **40** with "single-Janus" tectons **5** and **6**, respectively, would then result in a truncated 2-dimensional assembly, a double H-bonded ribbon. More recently, 1-dimensional main-chain H-bonded polymers have been reported based on

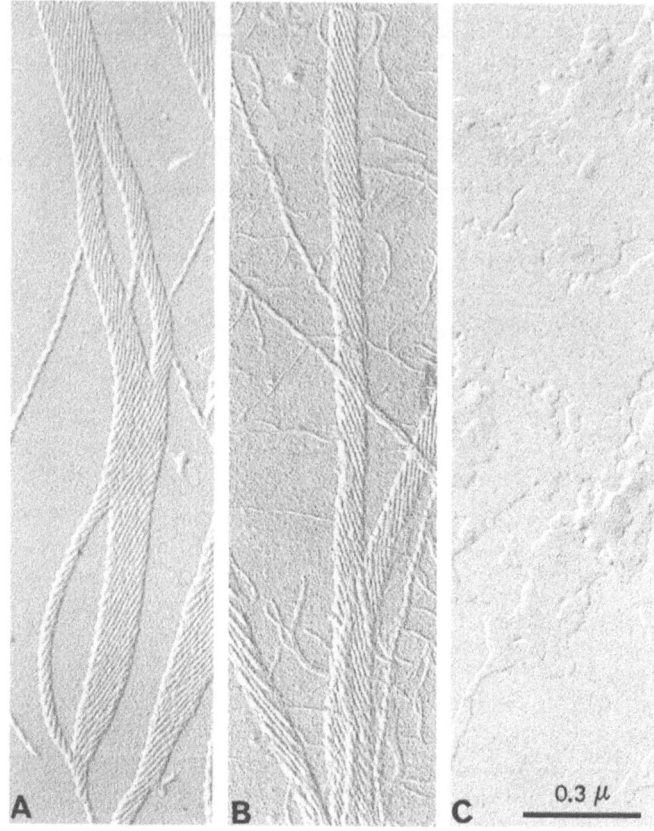

Fig. 26A–C. Electron microscope images of H-bonded main-chain polymers obtained via association of **35** and **36** derived from *A* L-, *B* D- and *C* meso-forms of tartaric acid

quadruple H-bonding arrays [85], polymeric H-bonded capsule motifs [86], polymeric stacks of H-bonded macrocycles [87] and diacid-bipyridine aggregates [88]. Three-dimensional networks based on diacid-bipyridine aggregates [89] and H-bonded side chain or "graft" polymers [90] have also been reported.

8
Conclusions

The design of persistent and selective H-bonding arrays toward the development of a library of modular recognition groups is a goal which, in turn, permits the design of recognition-directed "smart materials". For example, modular attachment of complementary H-bonding recognition units to vesicles promotes their specific aggregation and fusion [91]. Similarly, selective association of nanoparticles with various electrochemically active partners may be mediated via H-bonding recognition groups [92]. Of great

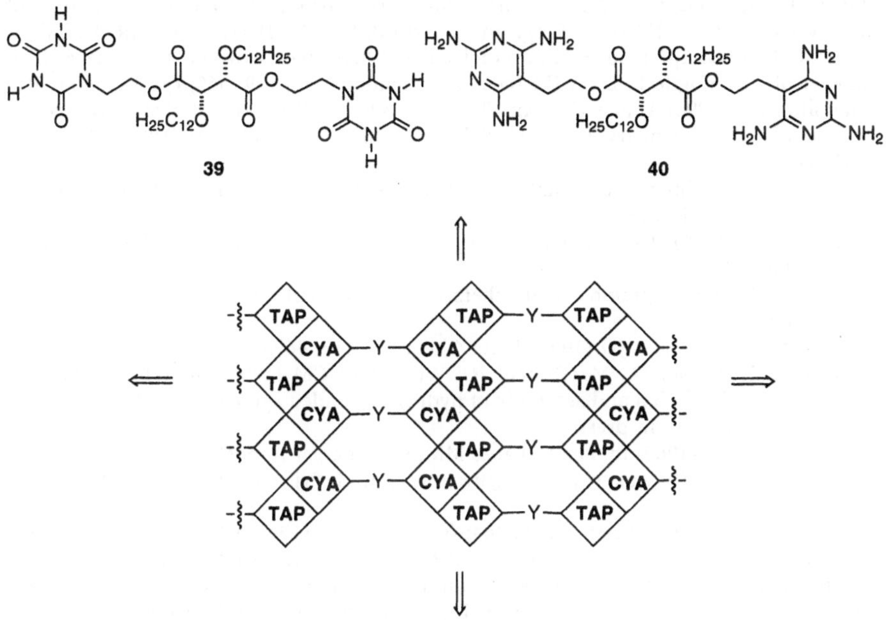

Fig. 27. Two-dimensional supramolecular network of H-bonded cross-linked main-chain polymers incorporating Janus recognition groups **5** and **6**

significance is the fact that recognition-directed self-assembly processes give access to functional materials and devices in a spontaneous but controlled fashion and thus bypass the need to resort to highly demanding fabrication procedures [7, 93]. They thus represent a major line of development of nanoscience and nanotechnology. Furthermore, in addition to the solid state, a most important aspect lies in the ability to direct organization and generate supramolecular architectures in a fluid medium, i.e. in solution. The achievements at hand and the perspectives they open up for progress in basic science, and for future applications, challenge the imagination of the chemist.

9
References

1. Latimer WM, Rodebush WH (1920) J Am Chem Soc 42: 1419
2. L(a) Wolf KL, Frahm H, Harms H (1937) Z Phys Chem Abt B 36: 237; (b) Wolf KL, Dunken H, Merkel KZ (1940) Phys Chem Abt B 46: 287; (c) Wolf KL, Wolff R (1949) Angew Chem 61: 191
3. (a) Lehn JM (1978) Pure Appl Chem 50: 871; (b) Lehn JM (1985) Science 227: 849; (c) Lehn JM (1988) Angew Chem Int Ed Engl 27: 89; (d) Lehn JM (1990) Angew Chem Int Ed Engl 29: 1304
4. For reviews see (a) Lawrence DS, Jiang T, Levett M (1995) Chem Rev 95: 2229; (b) Russell VA, Ward MD (1996) Chem Mater 8: 1654; (c) Bernstein J, Davis RE, Shimoni L, Chang NL (1995) Angew Chem Int Ed Engl 34: 1555; (d) Philp D, Stoddart JF (1996) Angew Chem Int Ed Engl 35: 1155; (e) Simanek EE, Li X, Choi IS, Whitesides GMA

(1996) Comprehensive Supramolecular Chemistry. Atwood JL, Davies JED, MacNicol DD, Vögtle F, Lehn JM (eds) Pergzamon, Oxford, 9: 595; (f) Fredericks JR, Hamilton ADA (1996) Comprehensive Supramolecular Chemistry. Atwood JL, Davies JED, MacNicol DD, Vögtle F, Lehn JM (eds) Pergzamon, Oxford, 9: 565; (g) Melendez RE, Hamilton AD (1998) Topics Current Chemistry 198: 97; (h) Toda F (1992) Advances in Supramolecular Chemistry 2: 141; (i) Subramanian S, Zaworotko MJ (1994) Chem Rev 137: 357

5. Whitesides GM, Simanek EE, Mathias JP, Seto CT, Chin DN, Mamman M, Gordon DM (1995) Acc Chem Res 28: 37
6. Desiraju G (1996) Crystal engineering: the design of organic solids. Elsevier, Amsterdam, 1989
7. Lehn JM (1995) Supramolecular chemistry – concepts and perspectives. VCH, Weinheim, chap 9
8. Simard M, Su D, Wuest JD (1991) J Am Chem Soc 113: 4696
9. Chin DN, Zerkowski JA, MacDonald JC, Whitesides GM (1998) In: Whitsell JT (ed) Strategies for the design and assembly of hydrogen-bonded aggregates in the solid state. John Wiley, London, in press
10. (a) McCrone WC (1965) In: Fox D, Labes MM, Weissberger A (eds) Polymorphism. John Wiley, New York; (b) Aakeröy CB, Nieuwenhuyzen M, Price SL (1998) J Am Chem Soc 120: 8986
11. (a) Desiraju GR (1997) Science 278: 404; (b) Perlstein J (1994) Chem Mater 6: 319; (c) Gavezotti A (1991) J Am Chem Soc 113: 4622; (d) Gavezotti A, Filippini G (1996) J Am Chem Soc 118: 7153; (e) Gavezotti A, Filippini G (1998) J Chem Soc Chem Commun 287; (f) Holden JR, Du Z, Ammon HL (1993) J Comput Chem 14: 422; (g) Chin DN, Palmore TR, Whitesides GM (1999) J Am Chem Soc 121: 2115; (h) Hofmann DWM, Lengauer T (1997) Acta Crystallogr A 53: 225
12. (a) Desiraju GR (1989) Crystal engineering: the design of organic solids. Elsevier, New York; (b) Desiraju GR (1995) Angew Chem Int Ed Engl 34: 2311; (c) Gavezotti A (1994) Acc Chem Res 27: 309; (d) Desiraju GR (1997) J Chem Soc Chem Commun 1495; (e) Braga D, Grepioni F, Desiraju GR (1998) Chem Rev 98: 1375
13. For reviews see (a) Corn RM, Higgins DA (1994) Chem Rev 94: 107; (b) Kanis DR, Ratner MA, Marks TJ (1994) Chem Rev 94: 195; (c) Etter MC, Huang KS, Frankenback GM, Adsmond DA (1991) In: Marder SR, Sohn JE, Studky GD (eds) Materials for non-linear optics: chemical perspectives, vol 455. American Chemical Society, Washington DC, p 446; (d) Prasad PN, Williams DJ (1991) Introduction to non-linear optical effects in molecules and polymers. John Wiley, New York
14. For reviews see (a) Torrance JB (1979) Acc Chem Res 12: 79; (b) Garito AF, Heeger AJ (1974) Acc Chem Res 7: 232
15. For reviews see (a) Wudl F (1984) Acc Chem Res 17: 227; (b) Williams JM, Beno MA, Wang HH, Leung PCW, Emge TJ, Geiser U, Carlson KD (1985) Acc Chem Res 18: 261; (c) Braga D, Grepioni F (1994) Acc Chem Res 27: 51; (d) Vögtle F (1991) Supramolecular chemistry. John Wiley, New York
16. For reviews on "organic zeolites" and "solid-state photochemistry" see (a) Gamelin JN, Jones R, Leibovitch M, Patrick B, Scheffer JR, Trotter J (1996) Acc Chem Res 29: 203; (b) Aoyama Y (1998) Top Curr Chem 198: 131 and references cited therein
17. (a) Jeffrey GA, Saenger W (1991) Hydrogen bonding in biological systems. Springer, Berlin Heidelberg New York; (b) Joesten MD, Schaad LJ (1974) Hydrogen bonding. M Dekker, New York; (c) Vinogradov SN, Linnel RH (1971) Hydrogen bonding. Van Nostrand Reihold, New York
18. (a) Speakman JC (1972) Struct Bond 12: 141; (b) McGregor DR, Speakman JC (1977) J Chem Soc Perkin Trans 2: 1740
19. Schiott B, Iverson BB, Madsen GKH, Bruice TC (1998) J Am Chem Soc 120: 12117 and references cited therein
20. (a) Taylor R, Kennard O (1984) Acc Chem Res 17: 320; (b) Scheiner S (1994) Acc Chem Res 27: 402

21. Perlmutter-Hayman B (1986) Acc Chem Res 19: 90
22. (a) Murray TJ, Zimmerman SC (1992) J Am Chem Soc 114: 4010; (b) Fenlon EE, Murray TJ, Baloga MH, Zimmerman SC (1993) J Org Chem 58: 6625; (c) Zimmerman SC, Murray TJ (1994) Tetrahedron Lett 35: 4077; (d) Bell DA, Anslyn EV (1995) Tetrahedron 51: 7161
23. (a) Jorgensen WL, Pranata J (1990) J Am Chem Soc 112: 2008; (b) Pranata J, Wierschke WL, Jorgensen WL (1991) J Am Chem Soc 113: 2810
24. Sartorius J, Schneider HJ (1996) Chem Eur J 2: 1446
25. Mammen M, Simanek EE, Whitesides GM (1996) J Am Chem Soc 118: 12614
26. (a) Beijer FH, Kooijman H, Spek AL, Sijbesma RP, Meijer EW (1998) Angew Chem Int Ed Engl 37: 75; (b) Corbin PS, Zimmerman SC (1998) J Am Chem Soc 120: 9710; (c) Lüning U, Kühl C (1998) Tetrahedron Lett 39: 5735
27. Zerkowski JA, Seto CT, Whitesides GM (1992) J Am Chem Soc 114: 5473
28. Drain CM, Russell KC, Lehn JM (1996) J Chem Soc Chem Commun 337
29. Russel KC, Leize E, Van Dorsselaer A, Lehn JM (1995) Angew Chem Int Ed Engl 34: 209
30. (a) Barr RG, Pinnavaia TJ (1986) J Chem Phys 90: 328; (b) Ciuchi F, Nicola GD, Franz H, Gottarelli G, Mariani P, Bossi MGP, Spada GP (1994) J Am Chem Soc 116: 7064
31. Seto CT, Whitesides GM (1990) J Am Chem Soc 112: 6409
32. (a) Seto CT, Whitesides GM (1993) J Am Chem Soc 115: 905; (b) Seto CT, Mathias JP, Whitesides GM (1993) J Am Chem Soc 115: 1321; (c) Seto CT, Whitesides GM (1991) J Am Chem Soc 113: 712; (d) Seto CT, Whitesides GM (1993) J Am Chem Soc 115: 1330; (e) Mathias JP, Seto CT, Simanek EE, Whitesides GM (1993) Angew Chem Int Ed Engl 32: 1766; (f) Mathias JP, Seto CT, Simanek EE, Whitesides GM (1994) J Am Chem Soc 116: 1725
33. (a) Vreekamp RH, van Duynhoven JPM, Hubert M, Verboom W, Reinhoudt DN (1996) Angew Chem Int Ed Engl 35: 1215; (b) Jolliffe KA, Timmerman P, Reinhoudt DN (1999) Angew Chem Int Ed Engl 38: 933
34. Yang J, Marendaz JL, Geib SJ, Hamilton AD (1994) Tetrahedron Lett 35: 3665
35. Zimmerman SC, Zeng F, Reichert DEC, Kolotuchin SV (1996) Science 271: 1025
36. Marsh A, Silvestri M, Lehn JM (1996) J Chem Soc Chem Commun 1527
37. Mascal M, Hext NM, Warmuth R, Moore MH, Turkenburg JP (1996) Angew Chem Int Ed Engl 35: 2204
38. Kolotuchin SV, Zimmerman SC (1998) J Am Chem Soc 120: 9092
39. Suárez M, Branda N, Lehn JM (1998) Helv Chim Acta 81: 1
40. (a) Brienne MJ, Gabard J, Leclercq M, Lehn JM, Cesario M, Pascard C, Chevé M, Dutruc-Rosset G (1994) Tetrahedron Lett 35: 8157; (b) Brienne MJ, Gabard J, Leclercq M, Lehn JM, Chevé M (1997) Helv Chim Acta 80: 856
41. (a) Wash PL, Maverick E, Chiefari J, Lightner DA (1997) J Am Chem Soc 119: 3802; (b) Ducharme Y, Wuest JD (1988) J Org Chem 53: 5787; (c) Gallent M, Viet MTP, Wuest JD (1992) J Org Chem 5: 2215; (d) Zimmerman SC, Duerr BF (1992) J Org Chem 57: 2215
42. Drain CM, Fischer R, Nolen EG, Lehn JM (1993) J Chem Soc Chem Commun 243
43. (a) Sessler JL, Wang R (1998) J Org Chem 63: 4079; (b) Sessler JL, Wang R (1998) Angew Chem Int Ed 37: 1726; (c) see also Kagechika H, Azumaya I, Tanatani A, Yamaguchi K, Shudo K (1999) Tetrahedron Lett 40: 3423
44. (a) Conn MM, Rebek J Jr (1997) Chem Rev 97: 1647; (b) Rebek J Jr (1996) Chem Soc Rev 255; (c) MacGillivray LR, Atwood JL (1999) Angew Chem Int Ed Engl 38: 1018
45. (a) Tokunaga Y, Rebek J Jr (1998) J Am Chem Soc 120: 66; (b) Castellano RK, Rebek J Jr (1998) J Am Chem Soc 120: 3657; (c) Ma S, Rudkevich DM, Rebek J Jr (1998) J Am Chem Soc 120: 4977; (d) Szabo T, Hilmersson G, Rebek J Jr. (1998) J Am Chem Soc 120: 6193; (e) Mecozzi S, Rebek J Jr (1998) Chem Eur J 4: 1016; (f) Hamann BC, Shimizu KD, Rebek J Jr (1996) Angew Chem Int Ed Engl 35: 1326; (g) Kang J, Rebek J Jr (1997) Nature 385: 50; (h) Mogck O, Paulus EF, Böhmer V, Thondorf I, Vogt W (1996) J Chem Soc Chem Commun 2533
46. Zerkowski JA, Seto CT, Wierda DA, Whitesides GM (1990) J Am Chem Soc 112: 9025

47. (a) Lehn JM, Mascal M, DeCian A, Fischer J (1990) J Chem Soc Chem Commun 479; (b) Russel KC, Lehn JM, Kyritsakas N, DeCian A, Fischer J (1998) New J Chem 123
48. Lehn JM, Mascal M, DeCian A, Fischer J (1992) J Chem Soc Perkin Trans 2 461
49. (a) Lehn JM (1995) Supramolecular chemistry – concepts and perspectives. VCH, Weinheim, p 121; (b) Mascal M, Fallon PS, Batsanov AS, Heywood BR, Champ S, Colclough M (1995) J Chem Soc Chem Commun 805; (c) Broughman DF, Caciuffo R, Horsewill AJ (1999) Nature 397: 241
50. MacDonald JC, Whitesides GM (1994) Chem Rev 94: 2383
51. (a) Watton HLL, Low JN, Tollin P, Howie AR (1988) Acta Crystallogr C 44: 1857; (b) Scheinbein J, Schempp E(1976) Acta Crystallogr B 32: 607
52. Alcala R, Martinez-Carrera S (1972) Acta Crystallogr B 28: 5
53. (a) Lewis FD, Yang JS, Stern CL (1996) J Am Chem Soc 118: 2772; (b) Leiserowitz L, Tuval M (1978) Acta Crysatallogr B 34: 1230
54. (a) Etter MC, Adsmond DA (1990) J Chem Soc Chem Commun 589; (b) Karle IL, Ranganathan D, Haridas V (1997) J Am Chem Soc 119: 2777; (c) Wash PL, Maverick E, Chiefari J, Lightner DA (1997) J AM Chem Soc 119: 3802; (d) Pedireddi VR, Chatterjee S, Ranganathan A, Rao CNR (1998) Tetrahedron 54: 9457
55. Schweibert KE, Chin DN, MacDonald JC, Whitesides GM (1996) J AM Chem Soc 118: 4018
56. (a) Zhao X, Chang YL, Fowler FW, Lauher JW (1990) J Am Chem Soc 112: 6627; (b) Lauher JW, Chang YL, Fowler FW (1992) Mol Cryst Liq Cryst 211: 99
57. Félix O, Hosseini MW, De Cian A, Fischer J (1997) Angew Chem Int Ed Engl 36: 102
58. Leiserowitz L, Schmidt GMJ (1969) J Chem Soc A 2372
59. (a) Hollingsworth MD, Brown ME, Hillier AC, Santarsiero BD, Chaney JD (1996) Science 273: 1355; (b) Pfeiffer P (1927) Organische Molekülverbindungen. Enke, Stuttgart
60. Palmore TR, McBride Mt (1998) J Chem Soc Chem Commun 145
61. Krische MJ, Lehn JM, Kyritsakas N, Fischer J, Wegelius EK, Nissinen MJ, Rissanen K (1998) Helv Chim Acta 81: 1921
62. Krische MJ, Lehn JM, Kyritsakas N, Fischer J (1998) Helv Chim Acta 81: 1909
63. Mascal M, Hansen J, Fallon PS, Blake AJ, Heywood BR, Moore MM, Turkenburg JP (1999) Chem Eur J 5: 381
64. (a) Chang YL, West MA, Fowler FW, Lauher JW (1993) J Am Chem Soc 11: 5991; (b) Coe S, Kane JJ, Nguyen TL, Toledo LM, Wininger E, Fowler FW, Lauher JW (1997) J Am Chem Soc 19: 86
65. Gong B, Zheng C, Skrzypczak-Jankun E, Yan Y, Zhang J (1998) J Am Chem Soc 120: 11194
66. (a) Russel VA, Etter MC, Ward MD (1994) J Am Chem Soc 116: 1941; (b) Russel VA, Etter MC, Ward MD (1994) Chem Mater 6: 1206; (C) Cai H, Hillier AC, Franklin K, Nunn CC, Ward MD (1994) Science 266: 1551
67. (a) Ermer O (1988) Angew Chem Int Engl 27: 829 b) Ermer O (1988) J Am Chem Soc 110: 2747; (c) Ermer O, Lindenberg L (1991) Helv Chim Acta 74: 825; (d) Ermer O, Eling A (1994) J Chem Soc Perkin Trans 2: 925
68. (a) Simard M, Su D, Wuest JD (1991) J Am Chem Soc 113: 4696; (b) Wang J, Simard M, Wuest JD (1994) J Am Chem Soc 116: 12119; (c) Brunet P, Simard M, Wuest JD (1997) J Am Chem Soc 119: 2737
69. (a) Swift JA, Reynolds AM, Ward MD (1998) Chem Mater 10: 4159; (b) Swift JA, Russel VA, Ward MD (1997) Adv Mater 9: 1183; (c) Swift JA, Pivovar AM, Reynolds AM, Ward MD (1998) J Am Chem Soc 120: 5887
70. Krische MJ, Krische AL, Lehn JM, Cheung E, Vaughn G (1999) CR Acad Sci in press
71. For reviews see (a) Lehn JM (1993) Makromol Chem Macromol Symp 69: 1; (b) Whitesides GM, Matthias JP, Secto CT (1991) Science 254: 1312; (c) Ozin GA (1992) Adv Mater 4: 612
72. For a review see Xia Y, Whitesides GM (1998) Angew Chem Int Ed Engl 37: 550
73. For reviews see (a) Paleos CM, Tsiourvas D (1995) Angew Chem Int Ed Engl 34: 1696; (b) Imrie CT (1995) TRIP 3: 22; (c) Ciferri A (1997) TRIP 5: 142

74. Demus D (1988) Mol Cryst Liq Cryst 165: 45
75. Brienne MJ, Gabard J, Lehn JM, Stibor I (1989) J Chem Soc Chem Commun 1868
76. (a) Bennet GM, Jones B (1939) J Chem Soc 420; (b) Hoffman S, Witkowski W, Borrman G, Schubert H (1978) Z Chem 18: 403
77. (a) Mallia VA, George M, Das S (1999) Chem Mater 11: 207; (b) Deschenaux R, Monnet F, Serrano E, Turpin F, Levelut AM (1998) Helv Chim Acta 81: 2072; (c) Fréchet JMJ, Kato T (1989) J Am Chem Soc 111: 8533; (d) Yu LJ (1993) Liq Cryst 14: 1303
78. Kleppinger R, Lillya CP, Yang C (1997) J Am Chem Soc 119: 4097
79. Suárez M, Lehn JM, Zimmerman SC, Skoulios A, Heinrich B (1998) J Am Chem Soc 120: 9526
80. (a) Palmans ARA, Vekemans JAJM, Fischer H, Hikemet RA, Meijer EW (1997) Chem Eur J 3: 300 (b) Palmans ARA, Vekemans JAJM, Hikemet RA, Fischer H, Meijer EW (1998) Adv Mater 10: 873
81. Yang W, Chai X, Chi L, Liu X, Cao Y, Lu R, Jiang Y, Tang X, Fuchs H, Li T (1999) Chem Eur J 5: 1144
82. (a) Fouquey C, Lehn JM (1990) Adv Mater 2: 254; (b) Gulik-krzywicki T, Fouquey C, Lehn JM (1993) Proc Natl Acad Sci USA 90: 163
83. Kotera M, Lehn JM, Vigneron JP (1994) J Chem Soc Chem Commun 197
84. Krische MJ, Pitsinos M, Petitjean A, Lehn JM to be published
85. (a) Sijbesma RP, Beijer FH, Brunsveld L, Folmer BJB, Hirschberg JHKK, Lange RFM, Lowe JKL, Meijer EW (1997) Science 278: 1601; (b) Folmer BJB, Cavini E, Sijbesma RP, Meijer EW (1998) J Chem Soc Chem Commun 1847
86. Castellano RK, Rudkevich DM, Rebek J Jr (1997) Proc Natl Acad Sci USA 94: 7132
87. Choi IS, Li X, Simanek EE, Akabe R, Whitesides GM (1999) Chem Mater 11: 684
88. (a) Alexander C, Jariwala CP, Lee CM, Griffin AC (1994) Macromol Symp 77: 283; (b) Lee CM, Jariwala CP, Griffin AC (1994) Polymer 35: 4550
89. Kihara H, Kato T, Uryu T, Fréchet JMJ (1996) Chem Mater 8: 961
90. (a) Edgecombe BD, Fréchet JMJ (1998) Chem Mater 10: 994; (b) Kato T, Kihara H, Kumar U, Uryu T, Fréchet JMJ (1994) Angew Chem Int Ed Engl 33: 1644; (c) Kato T, Kubota Y, Nakano M, Uryu T (1995) Chem Lett 1127
91. Marchi-Artzner V, Jullien L, Gullik-Krzywicki T, Lehn JM (1997) J Chem Soc Chem Commun 117
92. (a) Marguerettaz X, Redmond G, Rao SN, Fitzmaurice D (1996) Chem Eur J 2: 420; (b) Cusack L, Rao SN, Fitzmaurice D (1997) Chem Eur J 3: 202
93. Lehn JM (1999) In: Ungaro R, Dalcanale E (eds) Supramolecular science: where it is and where it is going. Kluwer, Dordrecht, p 287

Controlling Hydrogen Bonding:
From Molecular Recognition to Organogelation

Rosa E. Meléndez[1], Andrew J. Carr[2], Brian R. Linton[2], Andrew D. Hamilton[2]

Department of Chemistry, Yale University, New Haven, CT 06520, USA

[1] E-mail: *rosa.melendez@yale.edu*
[2] E-mail: *andrew.hamilton@yale.edu*

Hydrogen bonding has been studied extensively in solution, in the solid state and using theoretical methods. These studies complement each other and contribute to the progress made in understanding the behavior of this intermolecular interaction and its implications in chemistry, biology and physics. Extensive progress has been made in employing hydrogen bonding in fields such as self-assembly, host-guest recognition and crystal engineering. Recently, various hydrogen-bonding groups have been used in the design of organogelators, molecules that gel various solvents. In this chapter, we will outline the important work that has laid the foundations for the use of different hydrogen-bonding motifs in small host recognition studies and how this can be exploited in the design of self-assembling and gelating structures.

Keywords: Molecular recognition, Hydrogen bonding, Organogelation

Structure and Bonding, Vol. 96
© Springer Verlag Berlin Heidelberg 2000

1
Introduction

The field of molecular recognition involves the study of intermolecular interactions and their use in the design and synthesis of new molecules and supermolecules. The incorporation of functional groups into different molecular scaffolds that allow the recognition of other molecules has been demonstrated in many areas of organic chemistry. Some of the most studied noncovalent forces include van der Waals interactions between hydrophobic regions (π–π, alkyl–π, and alkyl–alkyl interactions), electrostatic interactions between cationic and anionic regions, and hydrogen bonding between donor and acceptor functional groups. These forces are weak compared to covalent bonds and are strongly influenced by the solvent and the complementarity of the interactions. However, they are extremely important due to their reversibility, and their ability to be switched on and off as a function of changes in their environment.

2
Hydrogen Bonding

Among the most studied of intermolecular interactions is hydrogen bonding [1, 2]. The hydrogen bond plays a critical role in a myriad of biological processes and has been a cornerstone in the rapid development of fields such as self-assembly [3] and crystal engineering [4, 5].

Although a precise definition of the hydrogen bond still remains the topic of some dispute [6], a simple description is that hydrogen bonds are formed when a donor (D) with an available acidic hydrogen is brought into close contact with an acceptor (A) possessing a lone pair of electrons (Fig. 1).

The geometry of the hydrogen bond has been extensively studied by statistical investigations [7, 8], X-ray diffraction analysis [9–11], and theoretical studies. Pioneering work, that set the stage for much of the current progress in supramolecular solid-state chemistry, concluded that hydrogen bonding is not a random event, but a highly ordered phenomenon [12–14]. The strength of the hydrogen bond can vary from 1 kcal/mol for C—H$\cdots$O hydrogen bonding [15, 16] to 40 kcal/mol for the HF_2^- ion [17, 18]. For hydrogen-bond acceptors, an increase in basicity is accompanied by an increase in the partial negative charge associated with the functional group. Furthermore, this increase in the net charge of the acceptor enhances the strength of the hydrogen bond, as demonstrated by Kelly's study of the binding of urea (1) (a hydrogen-bond donor) to a variety of anionic acceptors (see Table 1) [19].

$$A = F, Cl, O, S, N, \pi\text{-system}$$
$$D = F, Cl, O, S, N, C$$

Fig. 1. Hydrogen bond formed between an acidic hydrogen (D–H) and acceptor (A)

The hydrogen-bonding strength of a donor-acceptor (D/A) pair is greatly influenced by secondary electrostatic interactions [20]. Alternation of hydrogen-bond donors and acceptors in the same functional group lowers the association constant and thus the overall binding free energy of a pair of molecules. When a molecule is comprised of all donors and the corresponding partner of all acceptors, the secondary interactions are favorable, producing a stronger hydrogen-bonded complex (Fig. 2) [21]. Other studies on the role of secondary interactions have shown similar results [22–24]. The energetic contributions of individual secondary electrostatic interactions to these complexes in chloroform have been calculated by Jorgensen to be ±2.5 kcal/mol [25]. Recently, Schneider has studied a large number of examples from the literature and has measured the average contribution of a related series of secondary interactions to be ±0.7 kcal/mol [26].

The influence of solvent in hydrogen bonding has been studied by Lorenzi. In these studies it was shown that a small cyclic peptide does not dimerize in chloroform. In a less competitive solvent like carbon tetrachloride, however, a dimerization constant of 80 M^{-1} is observed [27]. The solvation of the binding site is a critical factor affecting the strength of binding. Wilcox has studied

Table 1. Association constants (M^{-1}) of anions binding to urea **1** in DMSO

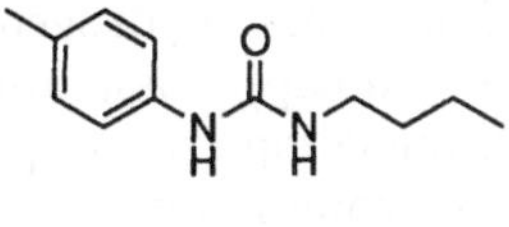

1

Guest	Ka	Guest	Ka
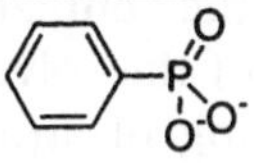	2500	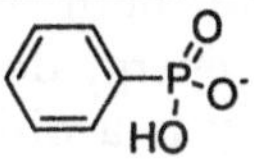	140
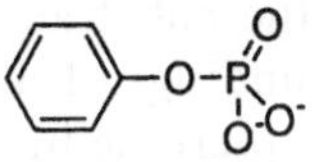	3600	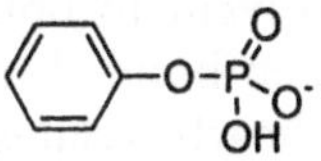	27
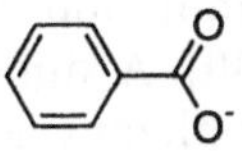	150	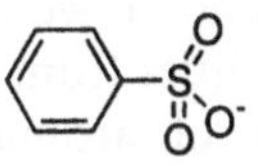	13

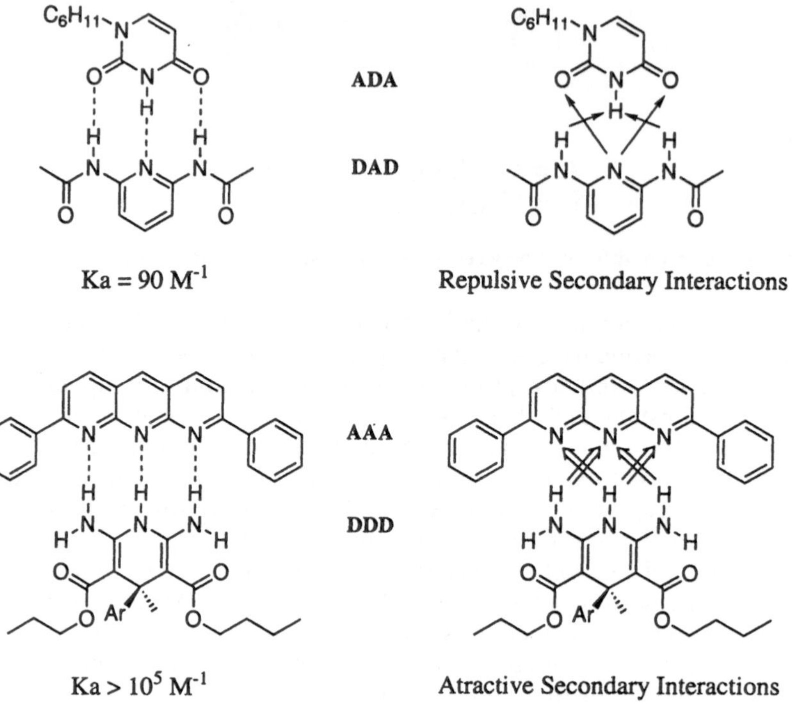

Fig. 2. Triple hydrogen-bonded donor-acceptor complexes and corresponding binding constants in chloroform

changes in the binding free energy as a function of the concentration of water in chloroform [28]. In a water-saturated chloroform environment, the hydrogen-bonding sites are solvated resulting in a decrease in the enthalpy component of the binding free energy. The release of ordered water molecules from the binding sites allows formation of more hydrogen bonds with the bulk solvent. This compensates for the enthalpy loss, and also leads to an entropically favorable component in the complexation process.

Enthalpy-entropy compensation is commonly observed in bimolecular association [29]. In the binding of guanidinium groups to carboxylates in competitive solvents (water/methanol mixtures), where both receptor and substrate are highly solvated, both binding affinity and the enthalpy component of the binding free energy decreases as the polarity of the solvent increases. This loss in enthalpy is a direct result of the breaking of strong hydrogen bonds from the solvent to both the receptor and the substrate in order to form the hydrogen-bonded complex. For the binding of bis-guanidinium derivatives to dicarboxylates, shown in Fig. 3, an increase in the percentage of water in methanol lowers the association constant and the driving force of complexation shifts from enthalpic to entropic [30]. A similar effect is seen in the binding of aromatic substrates to cyclophanes and cyclodextrins [31].

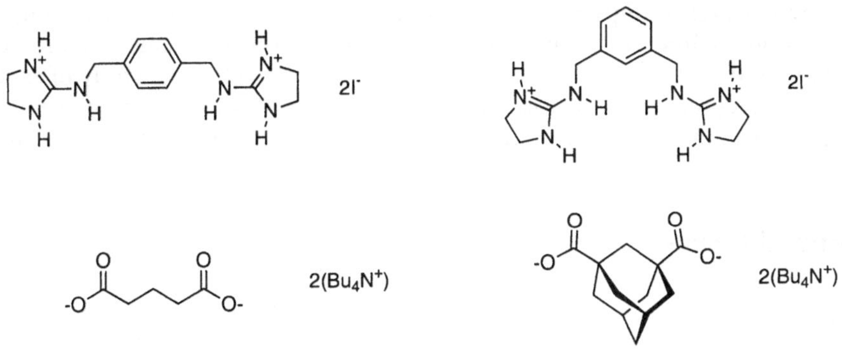

Fig. 3. Structures of bis-guanidiniums and bis-dicarboxylates used for solution studies in methanol/water mixtures

3
Organogelation

Gelation is a phenomenon that is familiar and easy to recognize. However, there is no precise definition for gelation and understanding the structure and properties of gels has become the focus of recent research [32]. An organogel is usually prepared by warming a gelator in an organic liquid until the solid dissolves, and then cooling the solution to below the gelation transition temperature [33]. The material that is formed has most likely a complex three-dimensional structure that involves encapsulation and immobilization of the solvent.

The structures of organogels have been studied by a variety of methods including neutron and X-ray scattering techniques, electron microscopy and infrared spectroscopy. Most of the imaging results indicate that the three-dimensional structures are fibrous [34].

Gels are usually classified as being either chemical or physical. In chemical gels both the fibers and the interactions between the fibers are covalent in nature. As a consequence the gels are not thermoreversible. Inorganic oxides, silica and cross-linked polymers belong to this class of gels. Physical gels usually have structures composed of small molecular subunits that are held together by non-covalent interactions. Low molecular weight organogelators are believed to assemble into filaments through hydrogen bonding, π–π stacking or other intermolecular forces [35].

Many of the small molecule organogelators reported in the recent literature contain hydrogen-bonding functional groups such as amides and ureas. Research in our laboratory has focused on the design, synthesis and investigation of molecular subunits that self-assemble into well-ordered structures in solution and the solid state [36, 37]. Some of the complementary hydrogen-bonding functional groups that we have investigated are depicted in Fig. 4.

Many research groups have been interested in using hydrogen bonding as the directional interaction between potentially gelling molecules in organic solvents. In this chapter we will show that lessons learned in the study of small molecule receptors and self-assembling systems can be directly applied in the design of structures capable of gelling a range of polar and non-polar solvents.

4
Functional Groups

4.1
Amides

4.1.1
Recognition Properties

Amides form linear arrays of hydrogen-bonded molecules in the solid state. One of the first crystal structural analyses of diamides, such as glutaramide, revealed that the geometry of these N—H $\cdots$ O hydrogen bonds is not linear [38]. The amide functional group forms one-dimensional patterns as well as eight-membered ring hydrogen-bond dimers (see Fig. 5). These cyclic dimers have been well documented for primary and secondary amides [39], as well as pyridones [40].

The closely positioned hydrogen-bond donor (amide) and hydrogen-bond acceptor groups (pyridine) in 2-amidopyridines form an eight-membered ring complex with carboxylic acids with complementary hydrogen-bonding sites. This interaction has been extensively used for the design of synthetic receptors in host-guest chemistry [41] and has also been used in the formation of hydrogen-bonded helices [42], as shown in Fig. 6. In the case of oligoanthranilamides, intramolecular amide hydrogen bonds determine the formation of helical structures [43]. Most recently, Crabtree has reported a family of aromatic diamides that act as anion receptors [44, 45].

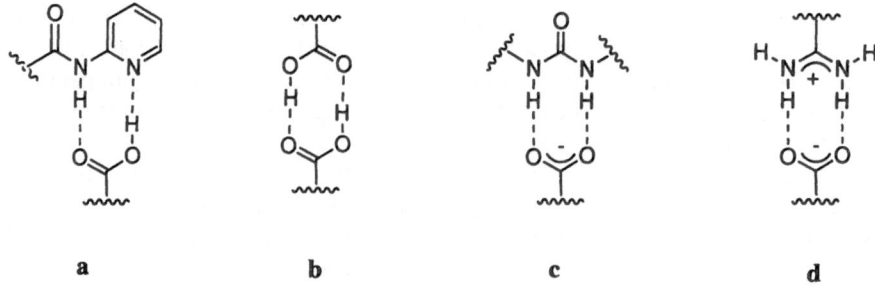

      a   **b**   **c**   **d**

Fig. 4a–d. Examples of complementary hydrogen-bonding functional groups. a 2-acylaminopyridine/carboxylic acid, b carboxylic acid dimer, c urea/carboxylate and d amidinium/carboxylate

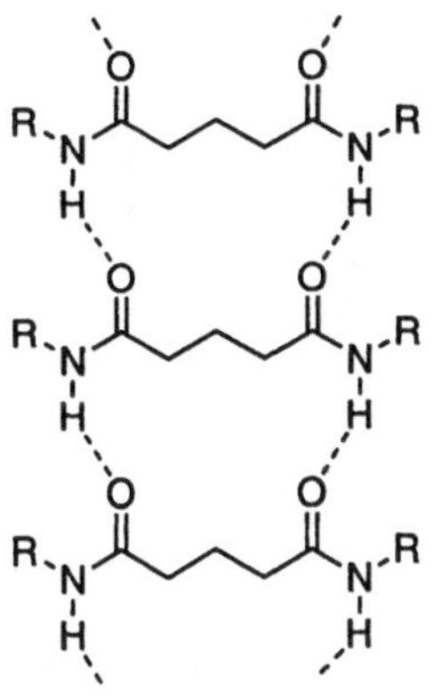

Fig. 5a,b. Hydrogen-bonding patterns formed by amides: **a** linear arrays and **b** dimers

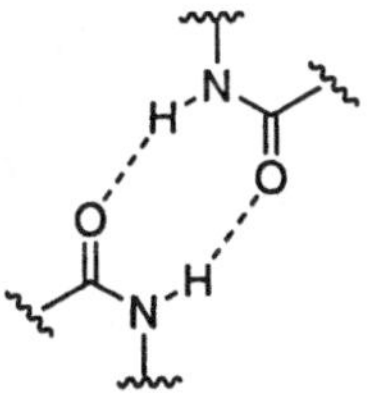

4.1.2
Gelation Properties

Hanabusa reported *trans*-cyclohexane-1,2-diamide **2** to be an effective organogelator for silicon oil and liquid paraffin at concentrations of 2 and 3 g/l, respectively [46]. Enantiomerically pure forms of the diamide gave stable gels. However, when a racemic mixture (**2** and **3**) was used, only a weak or unstable gel was observed. Interestingly the *cis*-1,2-diaminocyclohexane derivative **4** shows no gelation properties in any solvent (see Fig. 7). Circular dichroism (CD) studies indicated a chiral helical arrangement of the diamides. These results were supported by electron microscopy of a 2-acetonitrile gel, which showed the presence of helical superstructures. The authors reported that right- and left-handed helices are always observed for **2** and **3**, respectively.

The conclusion from these results was that helical superstructures could arise from the formation of stacked hydrogen-bonded rod-shaped aggregates of the type shown in Fig. 8. The orientation of the amide groups in an antiparallel fashion may be critical for this self-complementary interaction, since the *cis* derivative **4** cannot form these networks. This type of stacked amide hydrogen-bonding arrangement has been observed in a crystal structure of **5**, a cyclohexane-1,3,5-triamide derivative reported by Hamilton [47].

Shirota's group has studied molecules that are simple in structure and possess amide hydrogen-bonding functional groups around a rigid core (phenyl ring) [48]. Compound **6** shows gelation in various solvents, but when the hydrogen bonding is blocked, as in the *N*-methyl analog **7**, no gelation is observed. Increasing the relative distance between the hydrogen-bonding groups does not have an adverse effect on the gelling properties of these molecules (see **8** in Fig. 9) [49]. The thermoreversability of these gels was confirmed by differential scanning calorimetry.

Hanabusa has reported the design of long alkyl chain organogelators based around a more flexible core, *cis*-1,3,5-cyclohexanetricarboxamide [50]. The

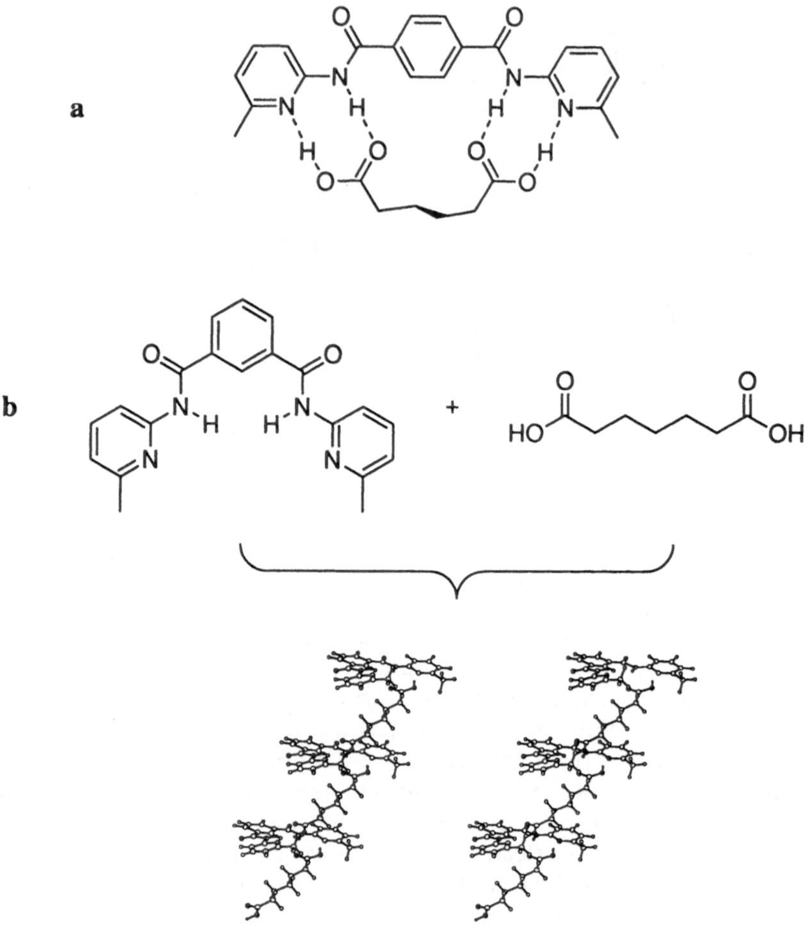

Fig. 6a,b. Complexes formed by 2-amidopyridines and dicarboxylic acids. **a** the complex assembles into a host-guest complex, whereas in **b** a helical structure is formed

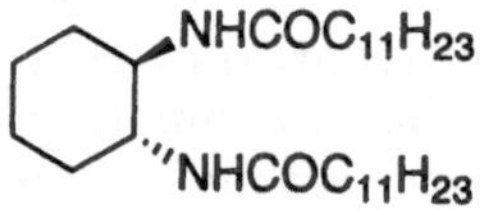

2 *trans (1R, 2R)*
3 *trans (1S, 2S)*

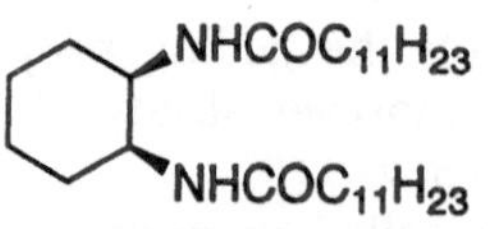

4 *cis* derivative

Fig. 7. Organogelators based on derivatives of *trans*-1,2-diaminocyclohexane

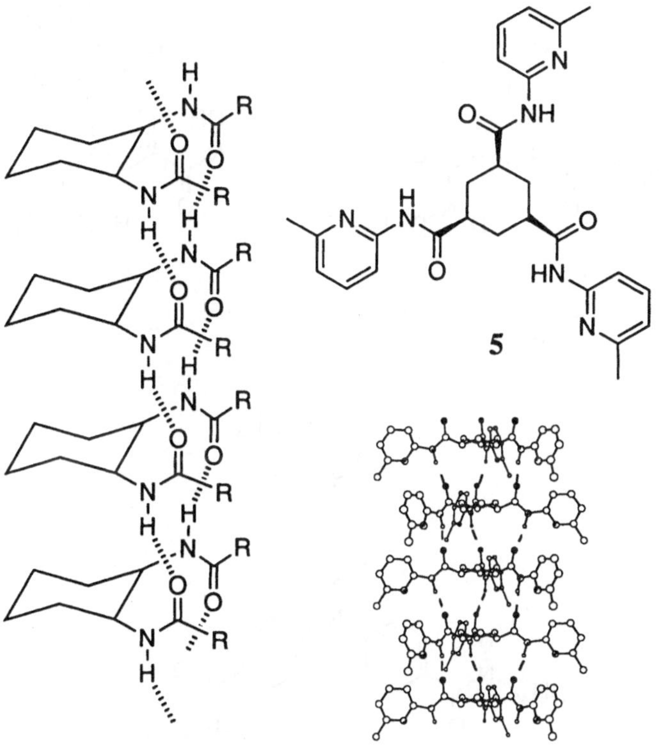

Fig. 8. a Hypothetical hydrogen-bonded strands formed by amides **2** and **3**. Crystal structure of **5** shown **b** in ball stick (hydrogen atoms removed for clarity)

shorter alkyl chain substituent **9** does not gel as well as **10**, **11** and **12** suggesting that intermolecular hydrophobic interaction among the alkyl chains plays a critical role in stabilizing the gel network (Fig. 10). At very low concentrations these molecules are able to increase the viscosity of different solvents. For example, 5 mg/ml of **10** in CCl_4 increases its viscosity to 2500 cP at 25 °C, from 0.908 cP in the absence of the gelling agent.

The recognition that organogels are novel soft solids formed by "trapping" a liquid within a matrix led Kato et al. to gel liquid crystals, **13** and **14** [51]. Gelating agent **2** was used at concentrations as low as 1 mol%. The gels were stable at room temperature and did not change to viscoelastic fluids over a period of several months. Measurements of the response to electric fields were made on gels of **2** and **13**. The threshold voltage of the gel at 5.0 V is larger than that of **13** alone (1.1 V). This is an important result since normally orientation of the solvent within an organogel cannot be achieved. The authors propose a structure for the gel that resembles the cartoon in Fig. 11. These types of soft structures have great potential in dynamic materials such as electrooptical devices.

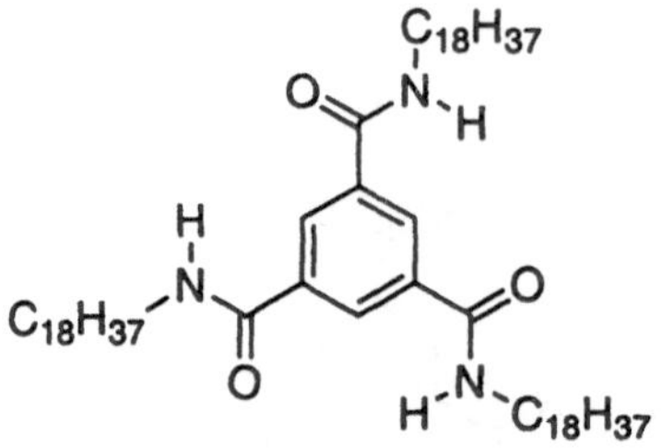

6 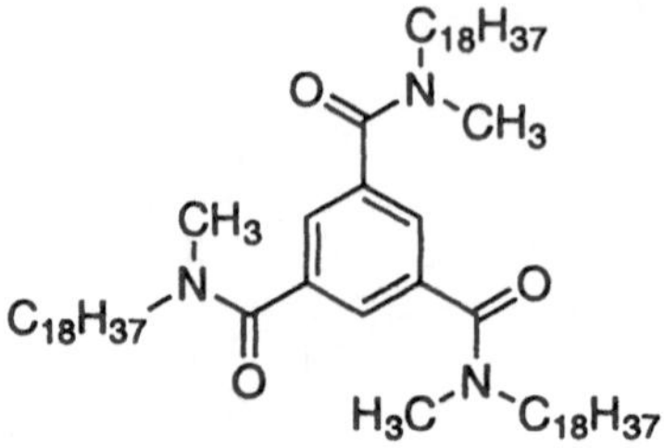   **7**

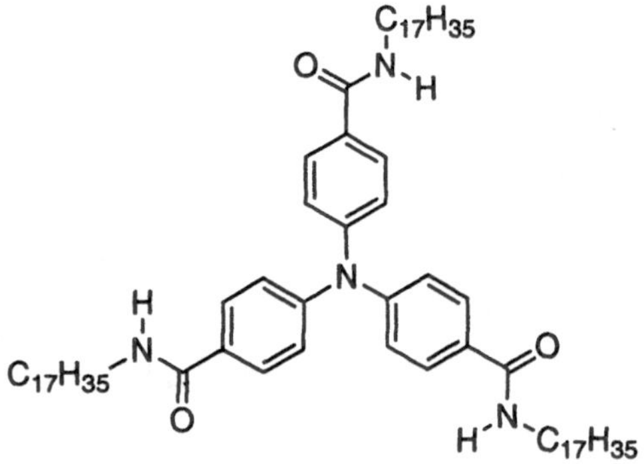

8

Fig. 9. Low molecular weight organogelators

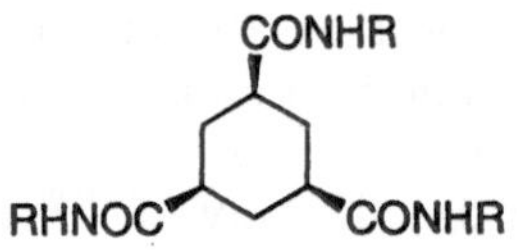

9; R = $CH_2(CH_2)_4CH_3$
10; R = $CH_2(CH_2)_{10}CH_3$
11; R = $CH_2(CH_2)_{16}CH_3$
12; R = $CH_2CH_2CH(CH_3)(CH_2)_3CH(CH_3)_2$

Fig. 10. Long alkyl chain organogelators 9, 10, 11 and 12 are amide derivatives around a cyclic core

Masuda made use of the amide hydrogen bond as a stabilizing force in the polymerization of gels formed by 1-aldosamide 15 [52]. The results obtained from energy-filtering transmission electron microscopy show formation of robust nanofibers. The formation of hydrogen-bonding contacts between the aminosaccharides was consistent with evidence from the infrared stretching bands. Moreover, the 1-galactosamide homolog 16 formed amorphous solids and not fibers, presumably due to steric hindrance between adjacent pyranose rings (see Fig. 12). Polymerization of 15 to form the linked polydiacetylene structure was confirmed by UV absorption and gel permeation chromatography.

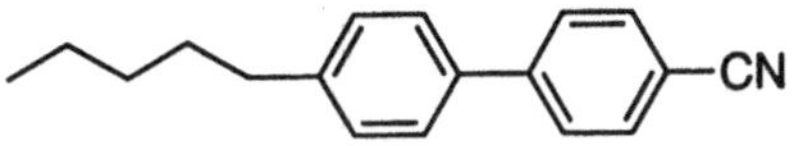

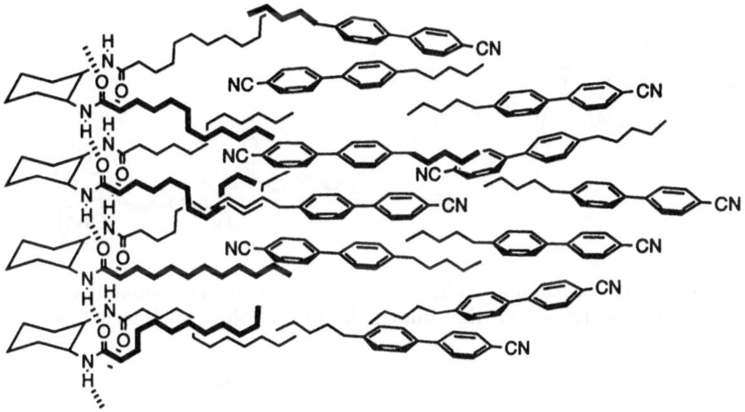

Fig. 11. Liquid crystals **13** and **14** were gelled by **2**. The structure of the gels is believed to be as depicted in the illustration above

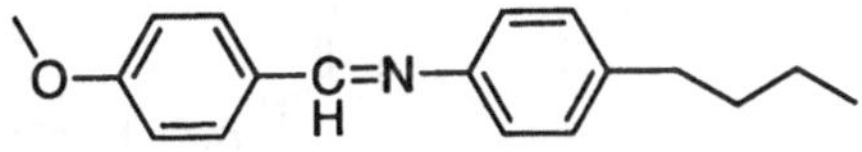

Shinkai used **17** as an organogelator that can be polymerized in situ (see Fig. 13) [53]. The absorption spectra of the gels before and after photoirradiation show a distinct change that is consistent with cross-linking to generate a more stable covalently linked matrix.

4.2
Ureas

4.2.1
Recognition Properties

The association of carboxylates with various simple receptors capable of bidentate hydrogen bonding has been widely investigated [54]. Urea, thiourea and guanidinium groups have been incorporated into synthetic receptors and shown to bind with tetramethylammonium (TMA) acetate in dimethyl sulfoxide (DMSO), as detailed in Table 2. The binding affinity of this bidentate interaction increases uniformly with the increasing acidity of the hydrogen-bond donor protons.

Other groups have also exploited these simple motifs for the recognition of carboxylates (Fig. 14). Wilcox used ureas such as **18** to complex various anions including carboxylates [55]. Similar receptors have been utilized by Curran to alter the stereochemistry of the alkylation of α-sulfinyl radicals [56] as well as accelerate a Claisen rearrangement [57]. Moran also created a urea-based

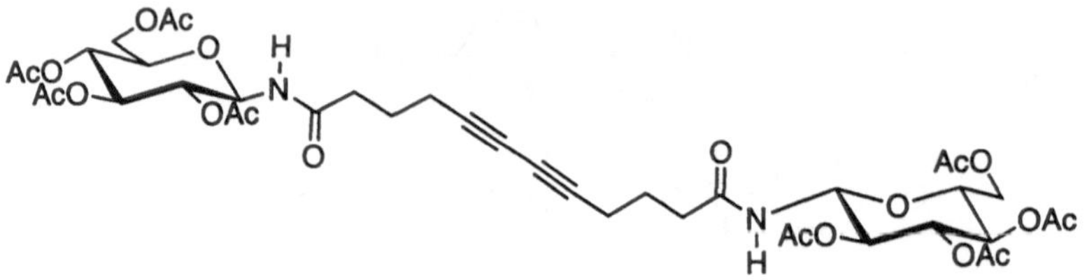

Fig. 12. Polymerizable aldosamide 15 is stabilized by one-dimensional hydrogen bonds, whereas 16 has axial acetyl groups that hinder the formation of such structures

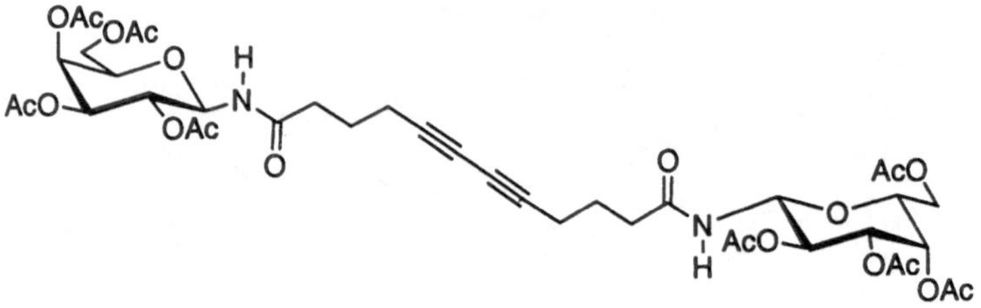

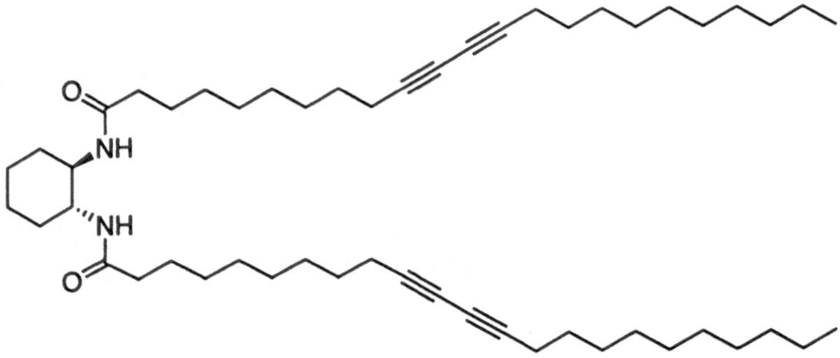

Fig. 13. Bis-diacetylene gelator 17 can be cross-linked in situ

receptor **19** which adds additional hydrogen-bond donors to promote stronger association with carboxylates. Receptors lacking the peripheral amide groups demonstrate weaker association [58]. A bis-urea **20** developed by Rebek has been shown to complex carboxylates with modest enantioselectivity when R is a chiral group [59].

Kelly has also developed the ditopic receptor **21** which can bind guests through the formation of four hydrogen bonds. Bis-anionic guests form strong complexes (Table 3), but again neutral groups show no association. Determination of association in less polar solvents was impossible due to the insolubility of this receptor [19].

A systematic study by Etter of the host-guest behavior of diarylureas shows that the formation of bidentate hydrogen bonds predominates in the solid state [60]. Graph sets and hydrogen-bonding rules for diarylureas were derived from this data and used to predict hydrogen-bonding patterns in related structures. Lauher successfully predicted the formation of a bidentate chain

Table 2. Binding of TMA acetate in DMSO

Receptor	Ka(M⁻¹)	pKa
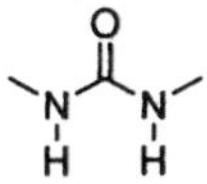	45	27
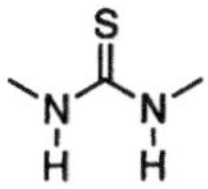	350	21
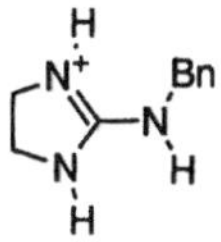	12000	13

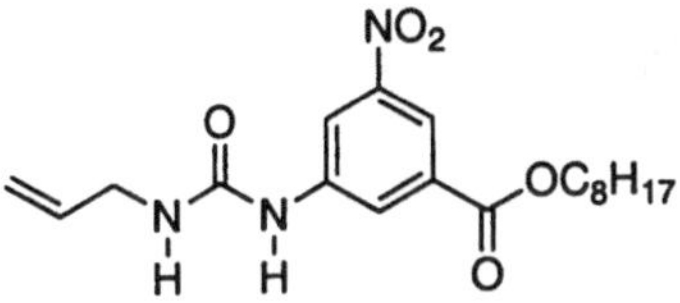

18 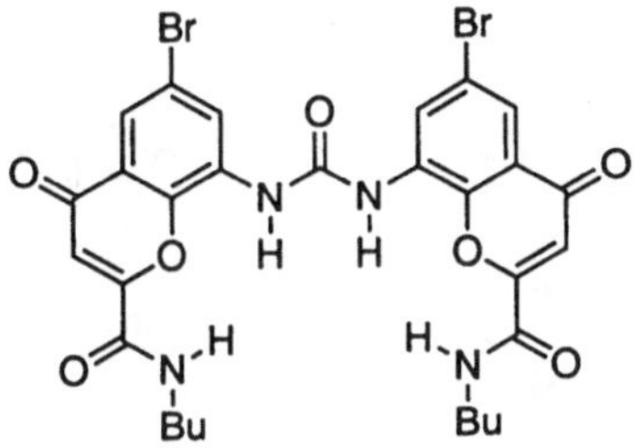   **19**

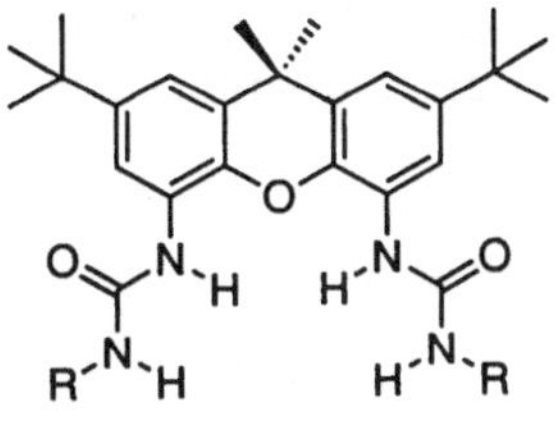

20

Fig. 14. Urea molecules used for binding of carboxylates. Molecule **19** has additional hydrogen-bonding sites and **20** has chiral R groups for enantioselective recognition

pattern formed by ureylenedicarboxylic acids. The analysis of six different structures led to the conclusion that this interaction persists even in the presence of other strong hydrogen-bonding groups (e.g., carboxylic acid) [61, 62]. Compound **22** shows the formation of catenated bidentate urea interactions, whereas in ureylenedicarboxylic acids of type **23** both urea–urea and carboxylic acid dimer motifs are present, creating a stable sheet

Table 3. Association constants for **19** with guest molecules in DMSO

19

Guest	Ka	Guest	Ka
	63000		87
	34000		86
	745		30
	104		No binding

arrangement (see Fig. 15). Although the bidentate interaction is polar, the adjacent strands in **23** adopt an antiparallel orientation, therefore canceling polarity in the sheets.

4.2.2
Gelation Properties

Hanabusa studied bis-urea derivatives as organogelators (Fig. 16) [63]. Molecules **24** and **25** show gelation only in toluene and tetrachloromethane, respectively. However, **26**, **27** and **28** show remarkable physical gelation in a range of different solvents. Systematic variation of the structure suggests that the long alkyl substituents play an important role in gelation. Similar to their results with 1,2-bisamidocyclohexanes, the *cis*-1,2-bis(dodecyluriedo)cyclo-hexane **29** shows no gelation towards any liquid.

Kellogg and Feringa reported the design of a related series of bis-urea molecules as organogelators, as shown in Fig. 17. These gels are thermore-versible and some are stable at temperatures of 100 °C. The gels were shown by electron microscopy to take the form of very thin rectangular sheets [64].

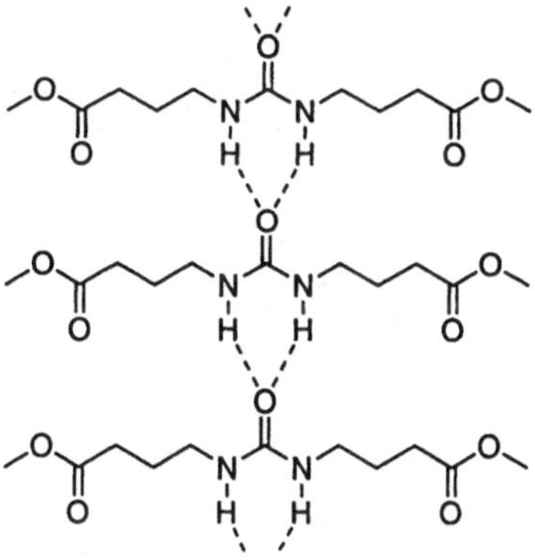

22

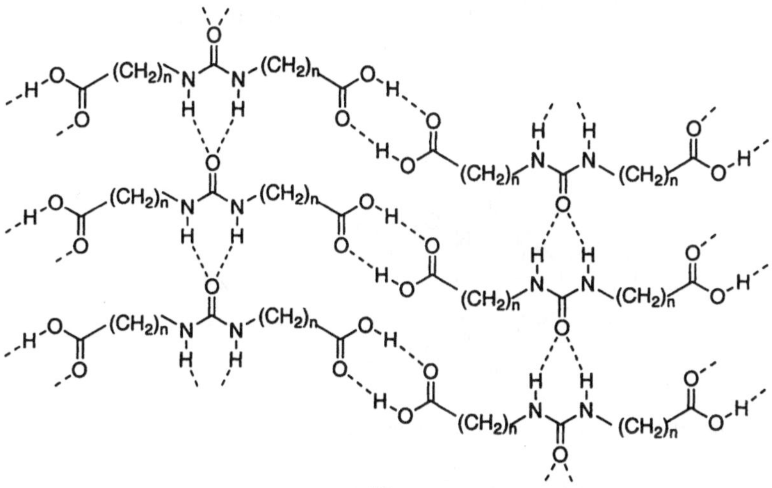

23

n = 1, 2, 3

Fig. 15. Formation of bidentate hydrogen-bonded chains in **22** prevails in the presence of other hydrogen-bonding groups **23**

Hamilton has published organogelation studies on the family of bis-ureas shown in Fig. 18 [65]. These compounds gel solvents and mixtures of solvents at 5 °C. The crystal structure of compound **30** confirms the extensive formation of one-dimensional hydrogen-bonding networks by both urea groups. The urea hydrogen-bonding distances are within the expected range NH $\cdots$ CO, 2.18, 2.23 Å.

Kellogg and Feringa have conducted extensive work on cyclic bis-urea molecules derived from *trans*-1,2-diaminocyclohexane and 1,2-diaminobenzene. In particular they studied aliphatic, aromatic and ester derivatives and demonstrated that some of these molecules gel organic solvents below 1 w/v%.

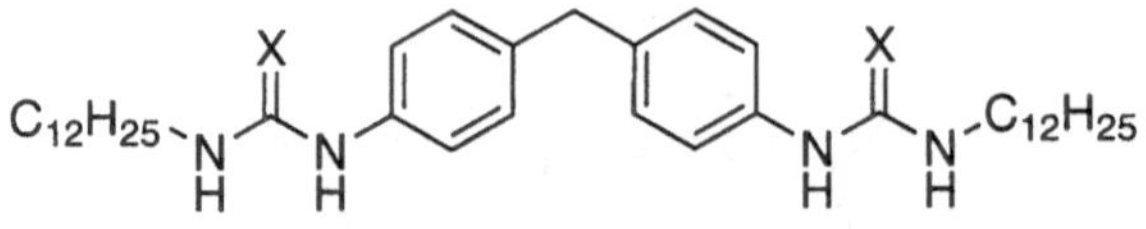

24; X = O
25; X = S

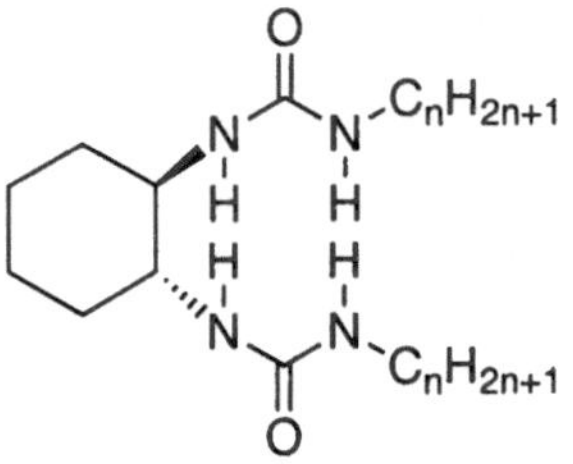

trans(1R,2R)
26; n = 4
27; n = 12
28; n = 18

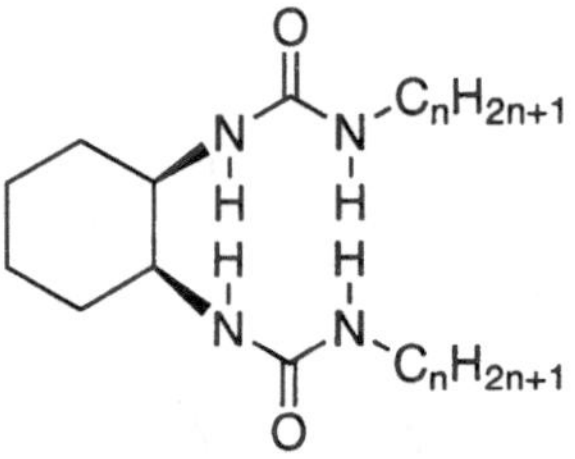

29

Fig. 16. Organogelators based on bis-urea derivatives

They prepared **31** and **32** as candidates for polymerizable organogels, using a methacrylate substituent [66]. Compound **31** only forms gels in tetralin, whereas **32** forms optically transparent gels in a variety of organic solvents. Polymerization was achieved by photoirradiation and resulted in an increase in the turbidity and stability of the gels. After polymerization and removal of the solvent from the gel by freeze-drying, a highly porous material was obtained. Another design for functionalized gels involved incorporating thiophene and bisthiophene in the spacer between the urea groups, as in compounds **33** and **34** (see Fig. 19). Thiophene groups are promising organic semiconductors and in this case show efficient charge transport within the organogels [67].

Molecular modeling studies reveal that a trans antiparallel orientation of the substituents on the ureas is energetically preferred over the parallel arrangement (Fig. 20). The one-dimensional aggregates formed by an antiparallel orientation lead to extensive hydrogen bonds with translation symmetry whereas the parallel orientation requires a screw axis to form one-dimensional aggregates [68]. The crystal structure of **35** shows a parallel orientation of the urea substituents around the benzene ring. The authors have conducted

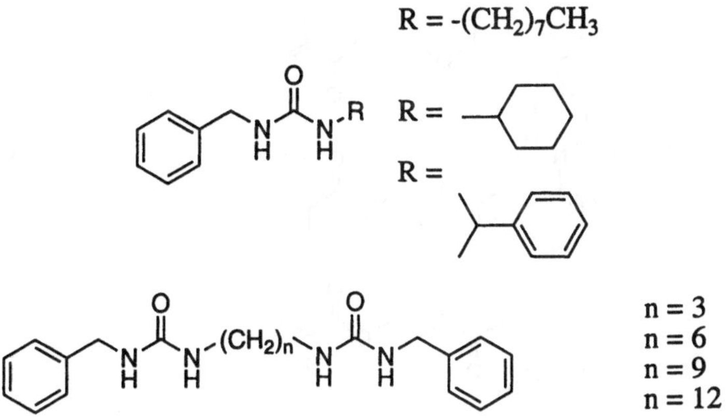

$$R = -(CH_2)_7CH_3$$

$$R =$$

$$R =$$

$$n = 3$$
$$n = 6$$
$$n = 9$$
$$n = 12$$

Fig. 17. Simple alkyl bis-urea organogelators

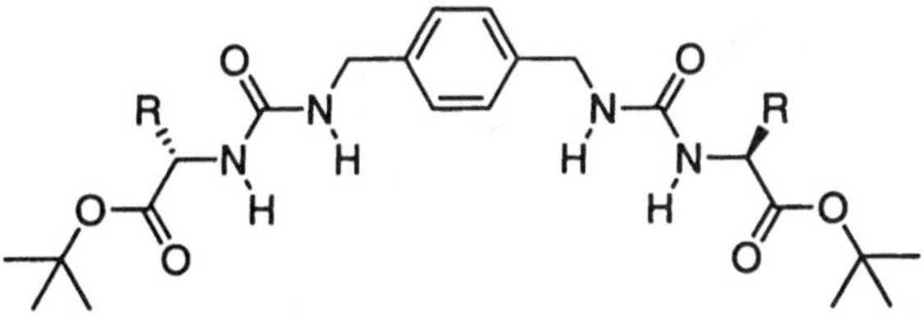

$$R = -CH_3, -CH_2Ph, -CH(CH_3)_2 \ (\mathbf{30})$$

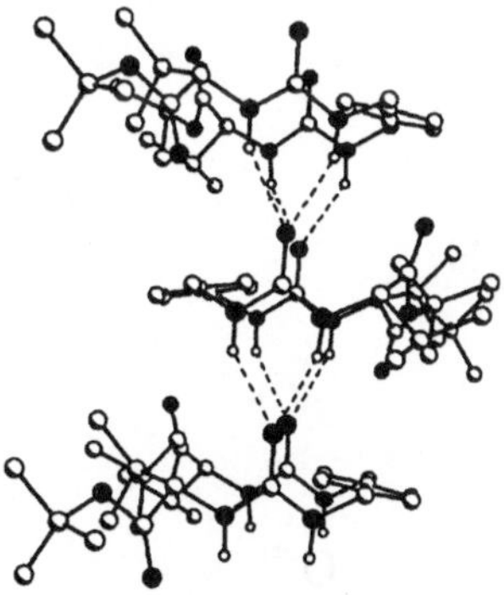

30

Fig. 18. Crystal structure of **30**, an amino acid bis-urea derivative. The urea groups are in a parallel orientation and form the expected bidentate hydrogen bonds

gelation studies in a variety of organic solvents and have used several different techniques to elucidate the structure of their gels, including infrared spectroscopy, differential scanning calorimetry, small-angle X-ray scattering, single-crystal X-ray diffraction and electron microscopy.

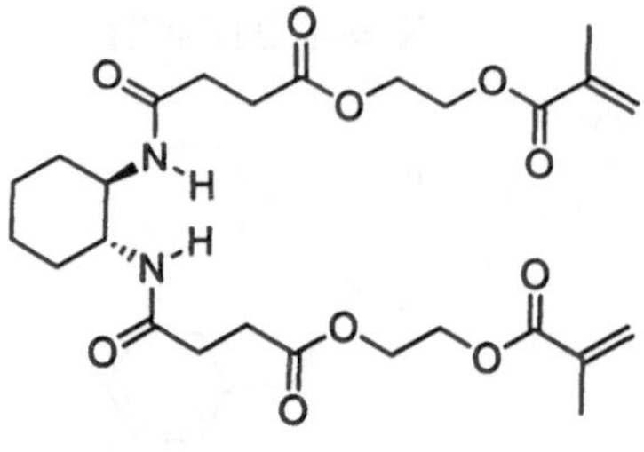

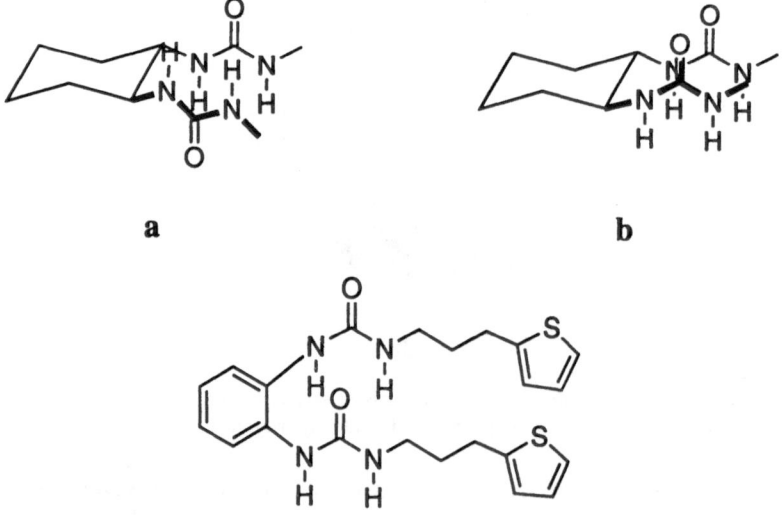

31

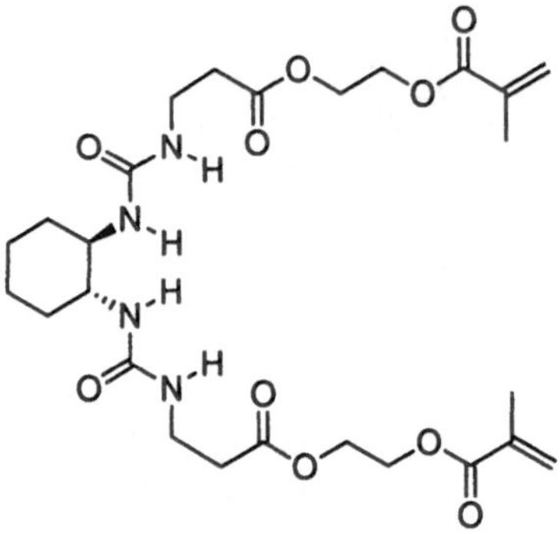

32

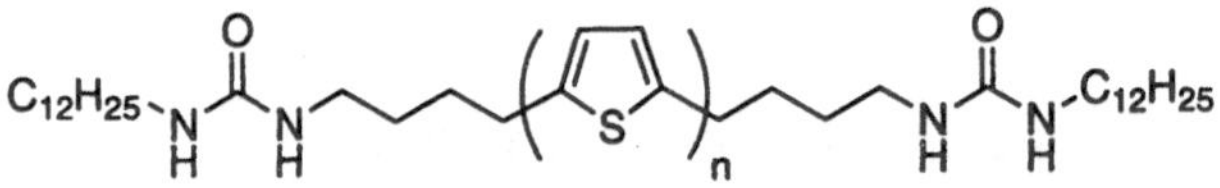

33 (n=1)
34 (n=2)

Fig. 19. Organogels can be functionlized for polymerization (31 and 32) or for charge transport (33 and 34)

a   **b**

Fig. 20. Molecular modeling reveals that *a* antiparallel orientation of bis ureas is favorable over *b* parallel orientation. However, the crystal structure of 35 shows parallel orientation

4.3
Barbiturate-Melamine

4.3.1
Recognition Properties

A classical two-component hydrogen-bonded system in self-assembly is represented by the melamine-barbituric/cyanuric acid complex. Whitesides and Lehn have exploited the complementarity of this interaction in the preferential formation of cyclic aggregates over the corresponding linear forms [69–71]. Cocrystals of **36** with **37**, **38** and **39** indicate that the size of the substituent on the melamine component is essential in controlling the final supramolecular architecture (see Fig. 21). When relatively small substituents are present on the melamine subunit, one-dimensional tapes are formed. However, bulkier substituents enhance the formation of cyclic aggregates.

4.3.2
Gelation Properties

In a related study, Hanabusa's group has incorporated long alkyl chain substituents onto similar melamine and barbiturate subunits [72]. They have demonstrated that a 1:1 mixture of **40** and **41** was capable of gelling *N,N*-dimethylformamide, chloroform, tetrachloromethane and cyclohexane at concentrations as low as 0.04 mol/ml (Fig. 22). The resulting stoichiometry of the system indicated that each **40** and **41** pair was able to gel approximately 460 molecules of cyclohexane.

Another related example is the cholesterol-substituted melamine/cyanuric acid pair **42** and **43** (Fig. 23), which has been used to investigate the mechanism of gelation [73]. Findings from these studies indicate that the hydrogen-bonding aggregates in the gels are neither linear nor cyclic but mismatched. This arrangement leaves hydrogen-bond donor or acceptor sites free in each component and results in extensive cross-linking within the matrix. It should be noted that steroid derivatives themselves are excellent organogelators [33].

4.4
2,6-Diamidopyridines

4.4.1
Recognition Properties

Hamilton has extensively studied 2,6-diamidopyridines as receptors for barbituric acid derivatives. The incorporation of this hydrogen-bonding unit into a macrocyclic structure positions all hydrogen-bonding groups inward, allowing for recognition of the appropriate sized guest [74]. Studies in solution and in the solid state have confirmed the formation of the host-guest complexes shown in Fig. 24 [75]. These types of synthetic receptors coupled to

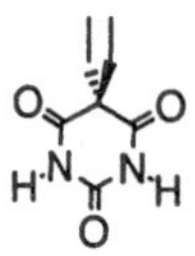

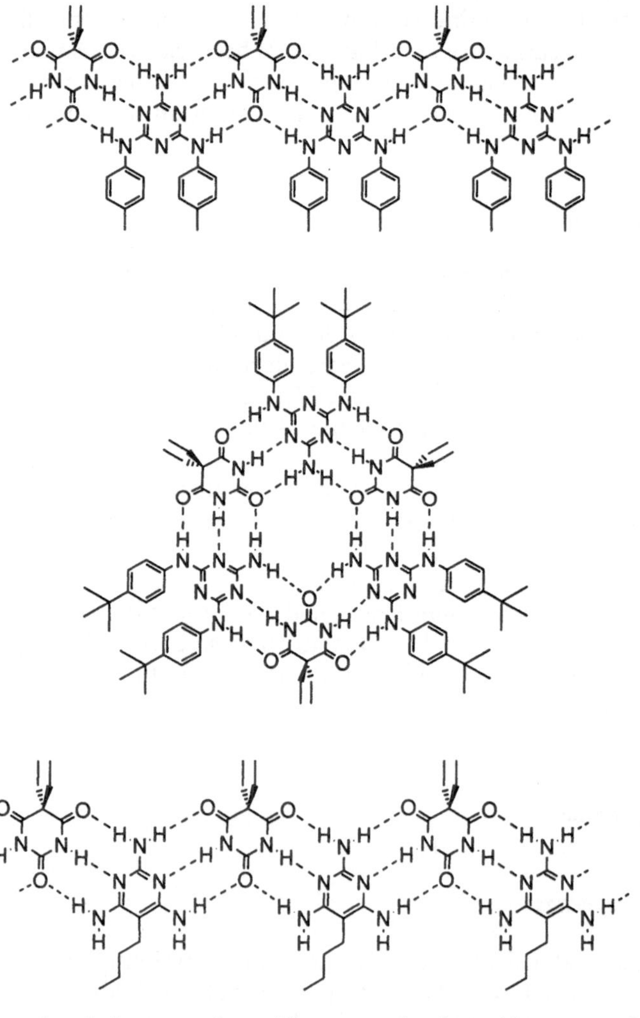

Fig. 21. Hydrogen-bonded patterns formed in cocrystals of **36** with **37**, **38** and **39**

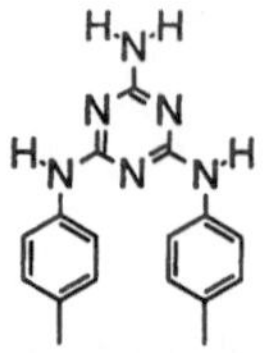

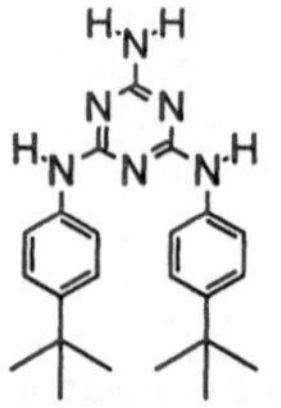

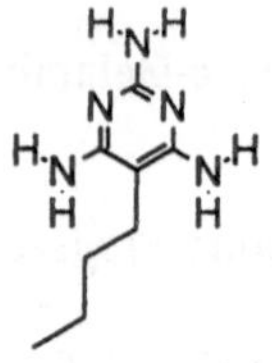

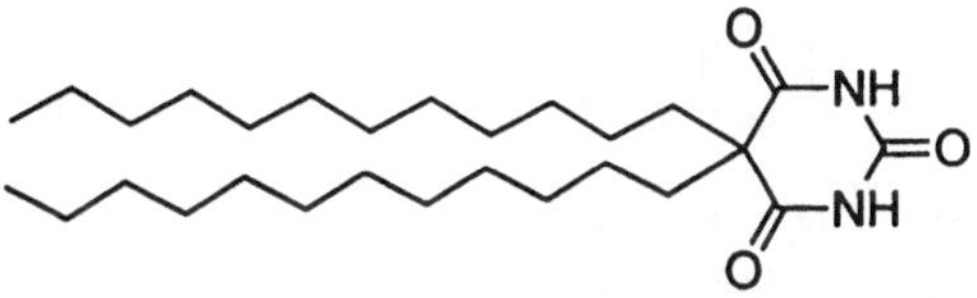

Fig. 22. Two-component complementary organogelators

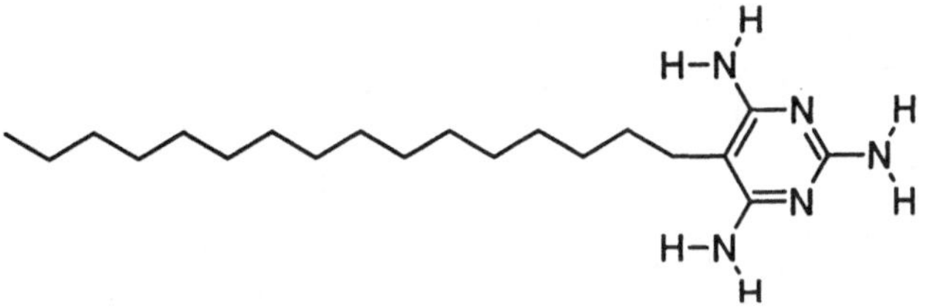

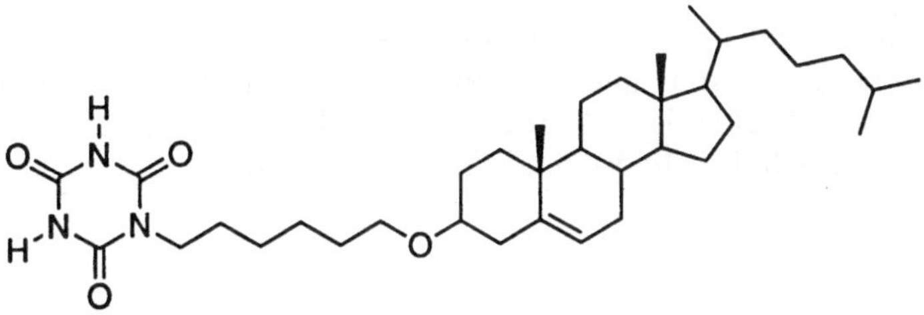

Fig. 23. Cholesterol derivatives of isocyanuric acid **42** and 2,4,6-triaminopyrimidine **43**

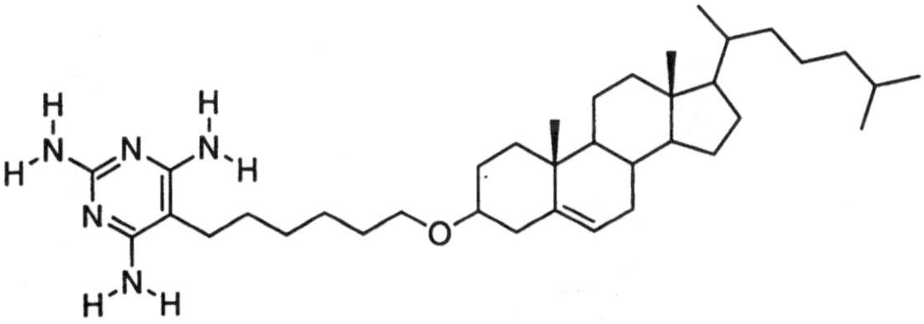

an appended thiol nucleophile also show acceleration in the thiolysis reactions of barbiturate active ester derivatives [76].

4.4.2
Gelation Properties

The same type of barbital derivatives has been used by Shinkai to develop host-guest organogelators [77]. The pairs formed by **44:45** and **46:45** were found to gel organic solvents by different mechanisms (see Fig. 25). As expected, the **46:45** pair forms the complementary complex through six

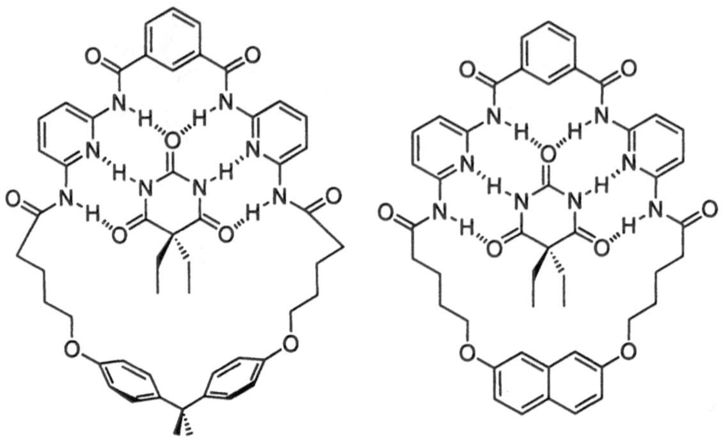

Fig. 24. Synthetic macrocyclic receptors for barbiturates

hydrogen bonds and leads to gels by intermolecular stacking, whereas in **44:45** the alignment of barbiturate guest and diamidopyridine host is not as optimal. This leads to free N—H and C=O groups that are believed to form gels through a network of cross-linked hydrogen-bonding interactions.

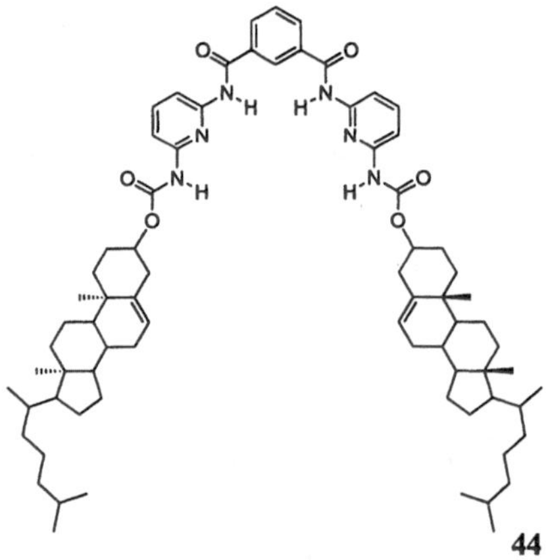

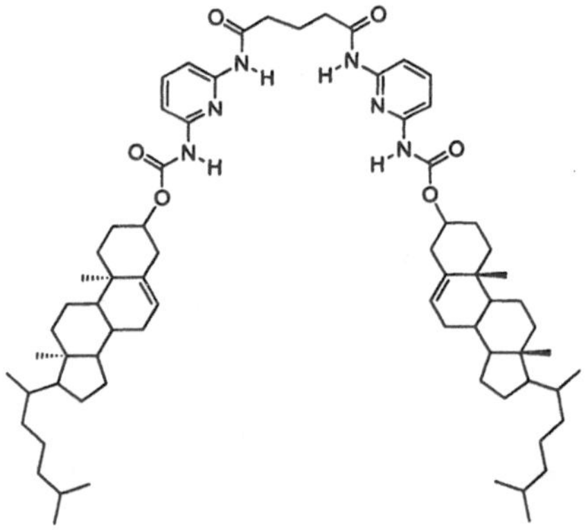

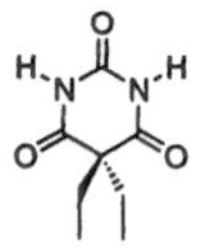

Fig. 25. Cholesterol derivatives of hydrogen-bonding host **44** and **46** form hydrogen-bonded gels with barbital **45**

4.5
Other Hydrogen-Bonded Organogelators

Surfactants such as sodium bis(2-ethylhexyl)sulfosuccinate (AOT) **47** were found to transform into organogelators by addition of suitable phenols in non-polar solvents [78]. The authors suggest that hydrogen bonding may occur between the head group of AOT and the hydroxyl group of several phenol molecules. Furthermore, NMR investigations indicate that the phenol-AOT aggregates form stacked strands within the gel matrix (see Fig. 26).

Hanabusa has studied small peptides with long alkyl substituents, for example **48**, and found them to gel organic solvents at very low concentrations [79, 80]. The aggregates are mainly assembled by $\pi-\pi$, dipole–dipole and hydrophobic interactions, as well as hydrogen bonding. The long chain alkyl amide **49** forms gels that exhibit fibrous morphology in transmission electron micrographs [81]. The Hanabusa group has also investigated cyclic peptides of general structure **50**, where gelation was observed in a wide variety of solvents, including edible oils, gylceryl esters, alcohols and aromatic molecules (see Fig. 27) [82].

Shinkai has used monosaccharide derivatives of **51, 52** and **53** as organogelators. The attraction of this approach is that a large selection of chiral sugar segments are readily available from natural or synthetic sources [83]. These structures were further modified by incorporating a *p*-nitrophenol group as a chromophore. Interestingly, **56** was too soluble in organic solvents to act as a gelator while compound **55** was sparingly soluble. D-Glucose derivative **54** was found to be an excellent gelator and has been studied by many techniques (see Fig. 28). Scanning electron microscope images of dry samples of **54** show formation of fibrils from benzene, carbon tetrachloride and cyclohexane. From methylcyclohexane, the structure of the fibrils adopts a helical arrangement and the chirality of these helical aggregates was confirmed by circular dichroism [84].

Fig. 26. Schematic representation of the proposed gel structure of **47** and phenol

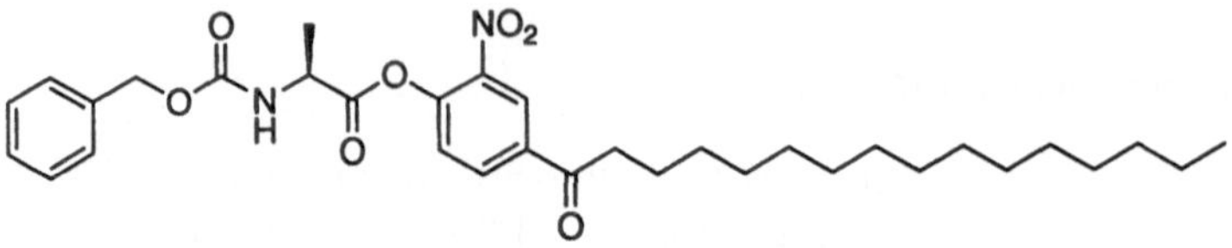

Fig. 27. Linear and cyclic peptides as gelling agents of organic solvents

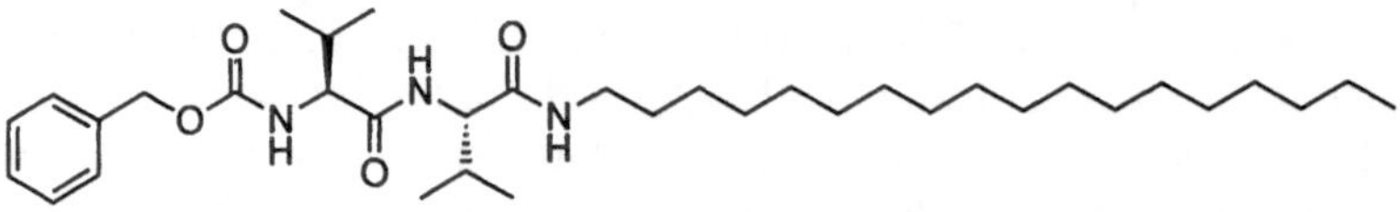

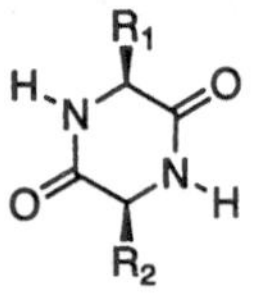

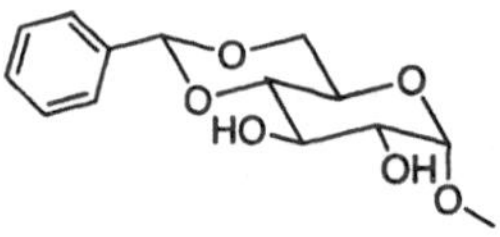

Fig. 28. Two families of sugar-based gelators. Substitution and conformation have a remarkable effect on the gelation abilities of these compounds

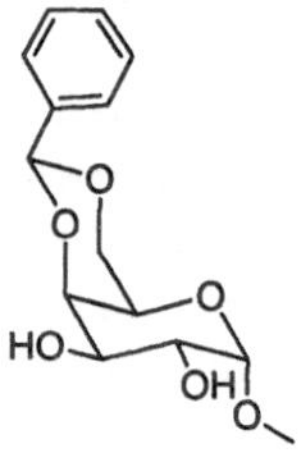

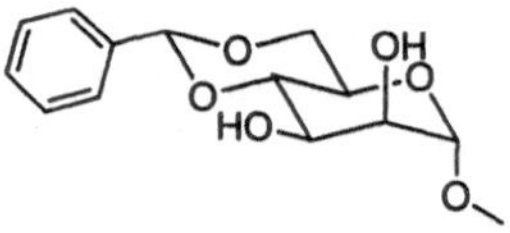

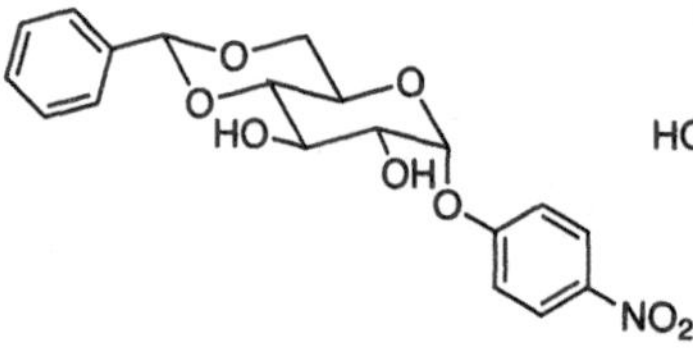

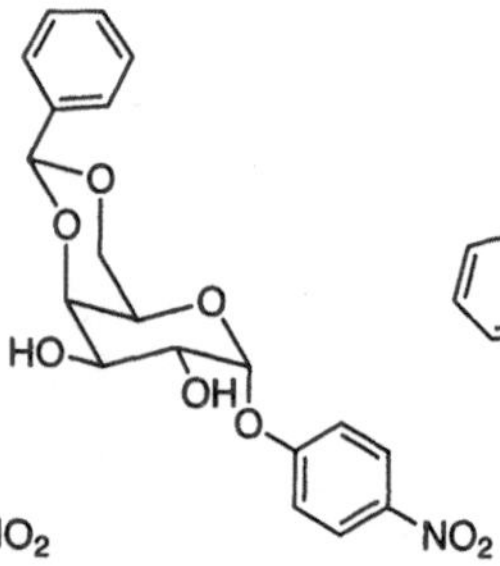

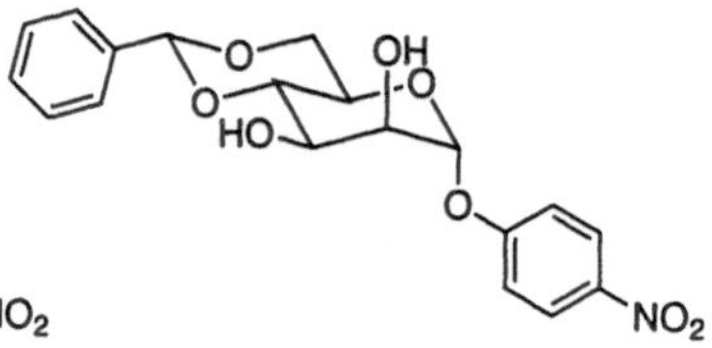

5
Non-Hydrogen-Bonded Organogelators

The main focus of this chapter has been organogelators in which the key self-assembling interactions are well-defined hydrogen-bonding patterns. However, there are many other organogelators that do not use hydrogen bonding to organize the gel matrix. These include fatty acids, surfactants, anthracene and steroid derivatives [33]. Some of the first organogelators reported were fatty acids such as **57** [85]. Surfactants have also been found to gel a variety of organic solvents, as well as water. Compound **58** behaves both as a gelator and as a surfactant [86] and compound **59** forms gels in some chlorinated solvents that remain unchanged after several months (see Fig. 29) [87].

Anthracene **60** has been classified as a very good organogelator, since the stiochiometry of the gel is one molecule of **60** to 40,000 molecules of methanol [88]. These flat structures are believed to form gelling aggregates through π–π stacking interactions between adjacent aromatic rings. In an attempt to study the structure of these aggregates, partial hydrogenation of the anthracene ring was effected. Surprisingly, compounds **61** and **62** gel a large number of organic solvents even though only moderate spectral shifts from π–π interactions are observed [89]. Anthraquinone **63** also forms thermoreversible gels in aliphatic alcohols and amines at low concentrations. The interesting photochromatic properties of these compounds led to a study of the different substitution patterns on the anthraquinone core [90], and the dialkoxy-2,3-anthraquinone derivatives (such as **63**) were found to be the most efficient systems (see Fig. 30).

Weiss has reported detailed studies of a family of anthracene organogelators [91] composed of 2-substituted anthracenyl groups coupled through a spacer

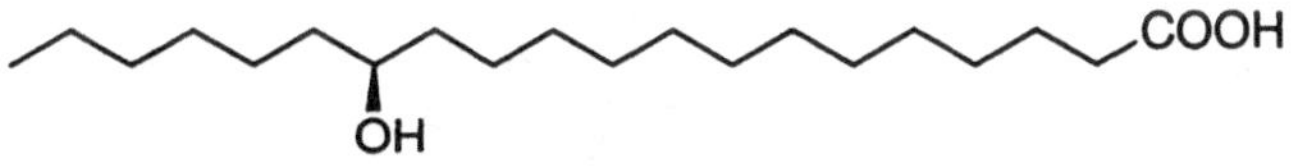

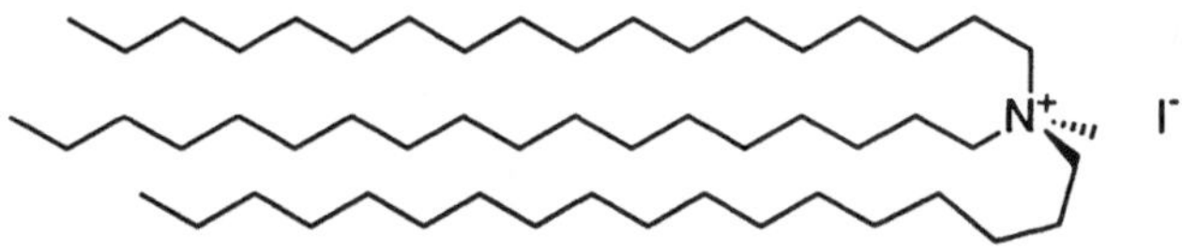

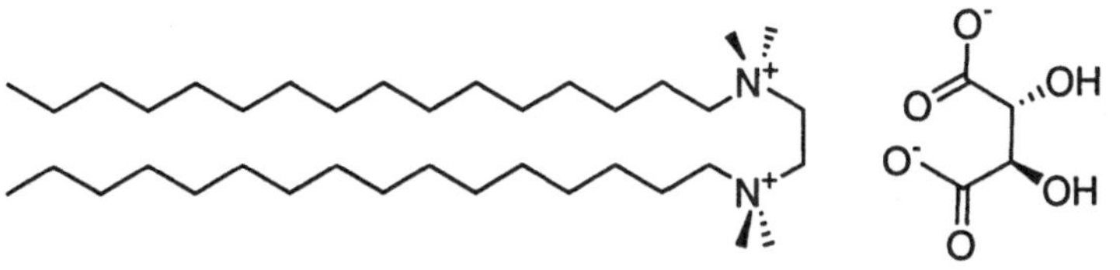

Fig. 29. Examples of fatty acid and surfactant organogelators

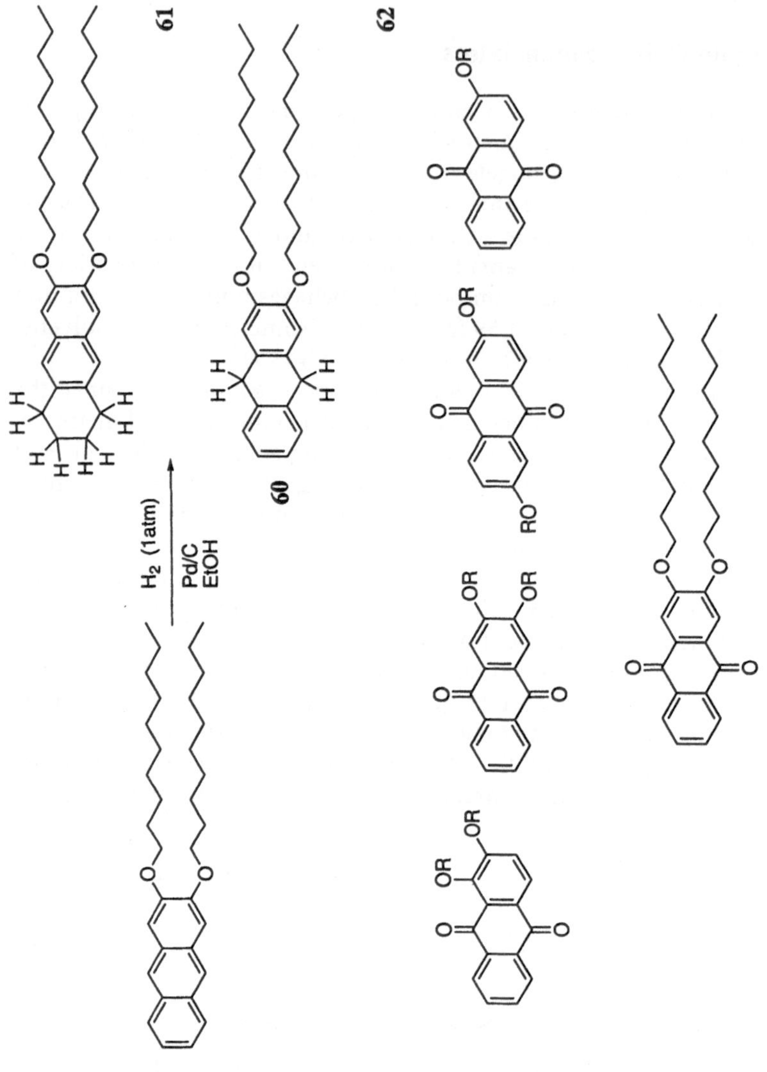

Fig. 30. Anthracene and anthraquinone derivatives are non-hydrogen-bonding gelators. In anthraquinone systems 2,3-dialkoxy derivatives such as **63** are found to be the most efficient

to C3 of a steroid group. These compounds (including **64** and **65**, Fig. 31) show potent gelation of a range of solvents and the nature of the matrix has been studied by fluorescence, circular dichroism, X-ray diffraction, NMR, microscopy and photochemistry [92].

Of the other cholesterol derivatives that have recently been reported, compound **66** (Fig. 32) bears a porphyrin moiety with potential photochemical

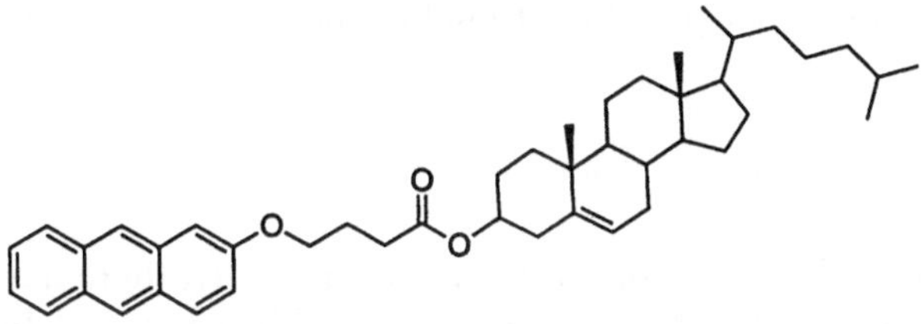

64

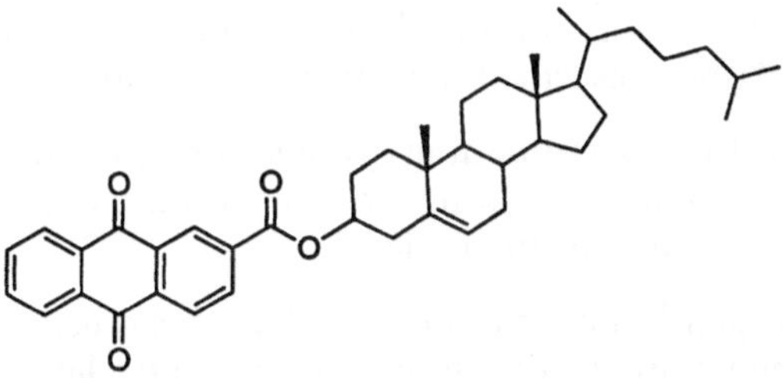

65

Fig. 31. Examples of cholestryl-anthracyl **64** and cholestryl-anthraquinyl **65** organogelators

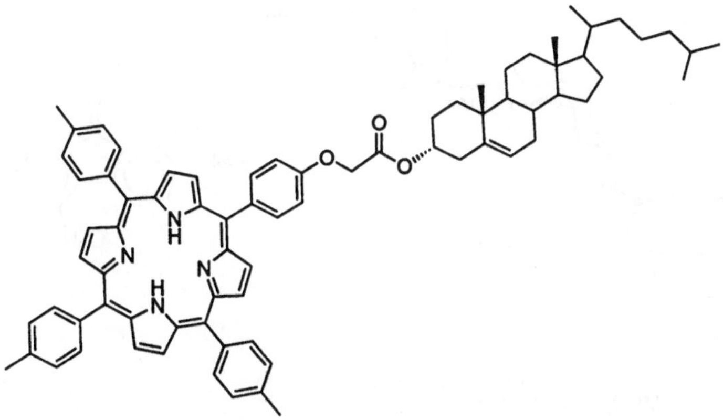

66

Fig. 32. Porphyrin **66** gels organic solvents and has potential photochemical properties

and catalytic properties [93]. Interestingly, the naturally occurring steroid with S configuration at C3 gels certain solvents, whereas the R isomer shows no gelation properties.

Bile acid derivatives as organogelators were first reported by Hanabusa [94]. However, recently, Maitra has reported bile acid derivatives functionalized at C3 with aromatic groups that gel organic solvents in the presence of trinitrofluorenone 67. The intensity in the color of the gels formed by 68 and 69 increases substantially during gelation in the presence of 67 due to donor-acceptor interactions between the aromatic rings (see Fig. 33) [95].

6
Conclusions

Although organogelation is still in its infancy, recently, there have been many efforts to understand physical and thermoreversible gelation. A broad range of small organic molecules have been designed and studied to throw light on the forces involved in generating the primary structures of these networks. Because the morphology of gels is in between that of a solid and a liquid, the methods employed to study them have often been borrowed from solution and solid-state chemistry. An important observation from this survey is that many of the efficient gelators described have three common features:

1. the presence of hydrogen-bonding groups in one or more components,
2. long alkyl substituents emanating from a central core, and
3. stereogenic centers within the molecule.

Many of the hydrogen-bonding motifs that have been used in the field of self-assembly have been employed to design new organogelators. The interactions that in earlier studies were directed towards the formation of small, well-defined aggregates can by careful design be diverted for the formation of extensive networks within a gel matrix. In some cases interesting function-

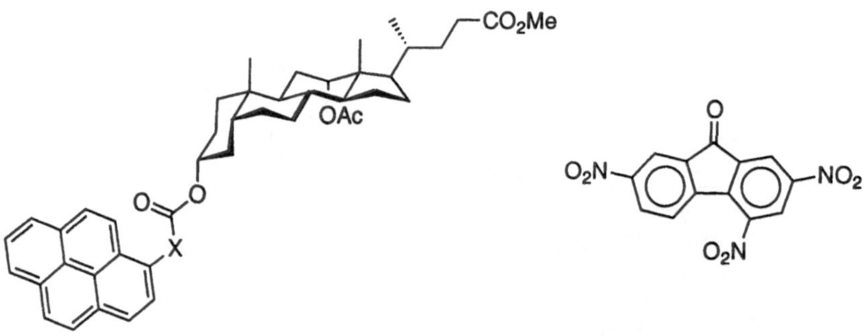

68 X = NH (carbamate)
69 X = --- (ester)   67

Fig. 33. Donor-acceptor gels are formed by 68 and 69 in the presence of 67

alities have been incorporated to create "functional gels". The potential for structural control and functional variety suggests that these types of compounds will find many novel applications in the near future.

Acknowledgements. The work described here from Yale University and the University of Pittsburgh was supported by the National Science Foundation.

7
References

1. Jeffrey GA, Saenger W (ed) (1991) Hydrogen bonding in biological structures. Springer, Berlin Heidelberg New York
2. Jeffrey GA (1997) An introduction to hydrogen bonding. Oxford University, New York
3. Philp D, Stoddart JF (1996) Angew Chem Int Ed Engl 35: 1154
4. Desiraju GR (1989) Crystal engineering, the design of organic solids. Elsevier, Amsterdam
5. Desiraju GR (1995) The crystal as a supramolecular entity, vol 2. John Wiley, Chichester
6. Martin TW, Derewenda ZS (1999) Nat Struct Biol 6: 403
7. Taylor R, Kennard O (1983) Acta Crystallogr Sect B 39: 133
8. Taylor R, Kennard O (1984) Acc Chem Res 17: 320
9. Etter MC (1990) Acc Chem Res 23: 120
10. Leiserowitz L (1976) Acta Crystallogr B 32: 775
11. MacDonald JC, Whitesides GM (1994) Chem Rev 94: 2383
12. Etter MC (1991) J Phys Chem 95: 4601
13. Zaworotko MJ (1994) Chem Soc Rev 283
14. Aakeröy CB, Seddon KR (1993) Chem Soc Rev 397
15. Desiraju GR (1991) Acc Chem Res 24: 290
16. Desiraju GR (1996) Acc Chem Res 29: 441
17. Emsley J (1980) Chem Soc Rev 9: 91
18. McMahon TB, Larson JW (1982) J Am Chem Soc 104: 5848
19. Kelly TR, Kim MH (1994) J Am Chem Soc 116: 7072
20. Pranata J, Wierschke SG, Jorgensen WL (1991) J Am Chem Soc 113: 2810
21. Murry TJ, Zimmerman SC (1992) J Am Chem Soc 114: 4010
22. Hamilton AD, Engen DJV (1987) J Am Chem Soc 109: 5035
23. Kyugoku Y, Lord RC, Rich A (1967) Proc Natl Acad Sci USA 57: 250
24. Park TK, Schroeder J, Rebek J Jr (1991) J Am Chem Soc 113: 5125
25. Jorgensen WL, Pranata J (1990) J Am Chem Soc 112: 2008
26. Sartorius J, Schneider H-J (1996) Chem Eur J 2: 1446
27. Sun X, Lorenzi GP (1994) Helv Chim Acta 77: 1520
28. Adrian JC, Wilcox CS (1991) J Am Chem Soc 113: 678
29. Rekharsky MV, Inoue Y (1998) Chem Rev 5: 1875
30. Berger M, Schmidtchen FP (1998) Angew Chem Int Ed Engl 37: 2694
31. Inoue Y, Liu Y, Tong L-H, Jin D-S (1993) J Am Chem Soc 115: 475
32. Almdal K, Dyre J, Hvidt S, Kramer O (1993) Polymer Gels Networks 1: 5
33. Terech P, Weiss RG (1997) Chem Rev 97: 3133
34. Geiger C, Stanescu M, Chen L, Whitten DG (1999) Langmuir 15: 2241
35. Esch JV, Schoonbeek F, de Loos M, Veen EM, Kellogg RM, Feringa BL (1999) In: Ungaro R, Dalcanale E (eds) Supramolecular science: where it is and where it is going. Kluwer Academic Publishers, Netherlands
36. Garcia-Tellado F, Geib SJ, Goswami S, Hamilton AD (1991) J Am Chem Soc 113: 9265
37. Yang J, Marendaz J-L, Geib SJ, Hamilton AD (1994) Tetrahedron Lett 35: 3665
38. Leiserowitz L, Tuval M (1978) Acta Crystallogr B 34: 1230

39. Leiserowitz L, Schmidt GMJ (1969) J Chem Soc 2372
40. Ducharme Y, Wuest JD (1988) J Org Chem 53: 5787
41. Garcia-Tellado F, Goswami S, Chang SK, Geib SJ, Hamilton AD (1990) J Am Chem Soc 112: 7393
42. Geib SJ, Vicent C, Fan E, Hamilton AD (1993) Angew Chem Int Ed Engl 32: 119
43. Hamuro Y, Geib SJ, Hamilton AD (1994) Angew Chem Int Ed Engl 33: 446
44. Kavallieratos K, de Gala SR, Austin DJ, Crabtree RH (1997) J Am Chem Soc 119: 2325
45. Kavallieratos K, Bertao CM, Crabtree RH (1999) J Org Chem 64: 1675
46. Hanabusa K, Yamada M, Kimura M, Shirai H (1996) Angew Chem Int Ed Engl 35: 1949
47. Fan E, Yang J, Geib SJ, Stoner TC, Hopkins MD, Hamilton AD (1995) J Chem Soc Chem Commun 1251
48. Yasuda Y, Iishi E, Inada H, Shirota Y (1996) Chem Lett 575
49. Yasuda Y, Takebe Y, Fukumoto M, Inada H, Shirota Y (1996) Adv Mater 8: 740
50. Hanabusa K, Kawakami A, Kimura M, Shirai H (1997) Chem Lett 191
51. Kato T, Kutsuna T, Hanabusa K, Ukon M (1998) Adv Mater 10: 606
52. Masuda M, Hanada T, Yase K, Shimizu T (1998) Macromolecules 31: 9403
53. Inoue K, Ono Y, Kanekiyo Y, Hanabusa K, Shinkai S (1999) Chem Lett 429
54. Fan E, Arman SAV, Kincaid S, Hamilton AD (1993) J Am Chem Soc 115: 369
55. Smith PJ, Reddington MV, Wilcox CS (1992) Tetrahedron Lett 33: 6085
56. Curran DP, Kuo LH (1994) J Org Chem 59: 3259
57. Curran DP, Kuo LH (1995) Tetrahedron Lett 36: 6647
58. Raposo C, Crego M, Mussons ML, Caballero MC, Moran JR (1994) Tetrahedron Lett 36: 3409
59. Hamann BC, Branda NR, Rebek J Jr (1993) Tetrahedron Lett 34: 6837
60. Etter MC, Urbańczyk-Lipkowska Z, Zia-Ebrahimi M, Panunto TW (1990) J Am Chem Soc 112: 8415
61. Zhao X, Chang Y-L, Fowler FW, Lauher JW (1990) J Am Chem Soc 112: 6627
62. Chang Y-L, West M-A, Fowler FW, Lauher JW (1993) J Am Chem Soc 115: 5991
63. Hanabusa K, Shimura K, Hirose K, Kimura M, Shirai H (1996) Chem Lett 885
64. van Esch J, Kellogg RM, Feringa BL (1997) Tetrahedron Lett 38: 281
65. Carr AJ, Melendez R, Geib SJ, Hamilton AD (1998) Tetrahedron Lett 39: 7447
66. deLoos M, van Esch J, Stokroos I, Kellogg RM, Feringa BL (1997) J Am Chem Soc 119: 12675
67. Schoonbeek FS, van Esch JH, Wegewijs B, Rep DBA, de Haas MP, Klapwijk TM, Kellogg RM, Feringa BL (1999) Angew Chem Int Ed Engl 38: 1393
68. van Esch J, Schoonbeek F, de Loos M, Kooijman H, Spek AL, Kellogg RM, Feringa BL (1999) Chem Eur J 5: 937
69. Zerkowski JA, MacDonald JC, Seto CT, Wierda DA, Whitesides GM (1994) J Am Chem Soc 116: 2382
70. Zerkowski JA, Seto CT, Whitesides GM (1992) J Am Chem Soc 114: 5473
71. Lehn J-M, Mascal M, De Cian A, Fischer J (1990) J Chem Soc Chem Commun 479
72. Hanabusa K, Miki T, Taguchi Y, Koyama T, Shirai H (1993) J Chem Soc Chem Commun 1382
73. Won S, Shinkai S (1997) Nanotechnology 8: 179
74. Chang S-K, Hamilton AD (1988) J Am Chem Soc 110: 1318
75. Chang S-K, Van Engen D, Fan E, Hamilton AD (1991) J Am Chem Soc 113: 7640
76. Tecilla P, Jubian V, Hamilton AD (1995) Tetrahedron 51: 435
77. Inoue K, Ono Y, Kanekiyo Y, Ishi-i T, Yoshihara K, Shinkai S (1999) J Org Chem 64: 2933
78. Tata M, John VT, Waguespack YY, McPherson GL (1994) J Am Chem Soc 116: 9464
79. Hanabusa K, Okui K, Karaki K, Koyama T, Shirai H (1992) J Chem Soc Chem Commun 137
80. Hanabusa K, Okui K, Karaki K, Kimura M, Shirai H (1997) J Colloid Interface Sci 195: 86
81. Hanabusa K, Tange J, Taguchi Y, Koyama T, Shirai H (1993) J Chem Soc Chem Commun 390

82. Hanabusa K, Matsumoto Y, Miki T, Koyama T, Shirai H (1994) J Chem Soc Chem Commun 1401
83. Yoza K, Ono Y, Yoshihara K, Akao T, Shinmori H, Takeuchi M, Shinkai S, Reinhoudt DN (1998) Chem Commun 907
84. Amanokura N, Yoza K, Shinmori H, Shinkai S, Reinhoudt DN (1998) J Chem Soc Perkin Trans 2 2585
85. Tachibana T, Mori T, Hori K (1980) Bull Chem Soc Jpn 53: 1714
86. Lu L, Weiss RG (1996) Chem Comm 2029
87. Oda R, Huc I, Candau SJ (1998) Angew Chem Int Ed Engl 37: 2689
88. Brotin T, Utermöhlen R, Fages F, Bouas-Laurent H, Desvergne J-P (1991) J Chem Soc Chem Commun 416
89. Placin F, Colomès M, Desvergne J-P (1997) Tetrahedron Lett 38: 2665
90. Clavier GM, Brugger J-F, Bouas-Laurent H, Pozzo J-L (1998) J Chem Soc Perkin Trans 2: 2527
91. Lin Y-C, Kachar B, Weiss RG (1989) J Am Chem Soc 111: 5542
92. Terech P, Furman I, Weiss RG (1995) J Phys Chem 99: 9558
93. Tian HJ, Inoue K, Yoza K, Ishi-i T, Shinkai S (1998) Chem Lett 871
94. Hishikawa Y, Sada K, Watanabe R, Miyata M, Hanabusa K (1998) Chem Lett 795
95. Maitra U, Kumar PV, Chandra N, D'Souza LJ, Prasanna MD, Raju AR (1999) Chem Commun 595

Heteroaromatic Modules for Self-Assembly Using Multiple Hydrogen Bonds

Steven C. Zimmerman, Perry S. Corbin

Department of Chemistry, University of Illinois, Urbana, IL 61801, USA
E-mail: sczimmer@uiuc.edu

Hydrogen bonding is a directional and moderately strong intermolecular force. Compounds that present multiple hydrogen-bond donor and acceptor groups have proven to be extremely important in creating new self-assembled structures. A review of several classes of organic compounds capable of multiple hydrogen-bond recognition is presented with a focus on the factors that contribute to complex stability.

Keywords: Hydrogen bonding, Molecular recognition, Self-assembly, Complexation, Heterocyclic compounds

1
Introduction

Efforts to create chemical models of biochemical systems or processes have been termed biomimetic chemistry by Breslow [1]. Self-assembly, an essential process for creating and maintaining the organization of complex biological systems, has provided a special challenge for biomimetic chemists. This challenge is essentially one of chemical information storage and processing [2, 3]. For example, one can ask the question how can an organic molecule be constructed such that it will recognize itself or other molecules and assemble into a specific structure? One of the simplest approaches uses arrays of hydrogen-bond donor and acceptor groups and, thus, borrows directly from the genetic storage mechanism of DNA base-pairing.

This chapter, which serves as an updated and expanded version of a previous review [4], will focus on organic modules that recognize themselves or complementary modules by formation of multiple hydrogen bonds. We arbitrarily define multiple as more than two. This review is not intended to be comprehensive, but illustrative of basic concepts. As a result, many important studies are not covered, and a detailed description of the synthesis of these modules is beyond the scope of the discussion. However, the usefulness of these modules will clearly depend on the overall yield and number of steps needed to produce them, so synthetic accessibility is considered in the evaluation of several of the modules. Particular focus will be placed on the strength and specificity of complexation, with a detailed consideration of the factors that contribute to both. Several of the modules described herein were designed for use in host-guest systems or to answer fundamental questions concerning hydrogen-bond complexation. Many others were created explicitly for use in self-assembly. A few selected examples of self-assembling systems are described to illustrate the application of these heteroaromatic modules.

2
Heteroaromatic Compounds
with Complementary Hydrogen-Bonding Arrays

As noted in the introduction, the use of heteroaromatic modules as recognition units for self-assembly is inspired by Nature's use of purines and pyrimidines as the storage units of genetic information. Numerous model studies of DNA and RNA base-pairing provide a considerable body of quantitative binding data (*vide infra*), as do the many host-guest studies that have targeted the nucleobases with heteroaromatic hosts [5–10]. Heteroaromatic compounds have several additional advantages, foremost of which is the geometrically well-defined, often linear array of hydrogen-bond donor (D) and acceptor (A) groups presented on the edges of the heteroaromatic system. Furthermore, there are many general synthetic approaches to heteroaromatic compounds that can be used to prepare new modules. The main disadvantages of using heteroaromatic modules in self-assembly are that they are often poorly soluble and it is difficult to control and even establish the protomeric (tautomeric) form of many heteroaromatic compounds (see Sect. 3.5). In the sections that follow, examples of heteroaromatic modules and their corresponding complexes are presented.

2.1
DAD and ADA Heteroaromatic Modules

The DAD · ADA array is the most well-studied and most commonly used hydrogen-bonding motif. This particular array is not found in the common DNA base-pairs, but is present in base-pairs between 2-aminoadenine (A') and thymine (T) (e.g., **3 · 5**, Fig. 1). Such A'T base-pairs are found naturally in the cyanophage S-2L [11] and, in synthetic oligonucleotides, have been found to increase duplex stability and specificity relative to adenine · thymine (AT) base-pairs [12, 13]. This property makes the A'T base-pair of interest in developing more effective antisense and antigene therapeutic agents. The increased stability of the A'T base-pair in oligonucleotides is reflected in a 0–2 and 2.5–6.5 °C increase in melting temperature (T_m) per A $\rightarrow$ A' substitution for duplex DNA and PNA · NA duplexes, respectively [12, 13]. This increase in T_m is consistent with model studies of base-pairing in organic solvents, in which the A · T base-pair has an association constant (K_{assoc}) of 90 M^{-1} in chloroform and the A'T base-pair (i.e., **3 · 5**) has a K_{assoc} of 210 M^{-1} [5].

A series of DAD · ADA complexes and their corresponding K_{assoc} values in chloroform is displayed in Fig. 1. As can be seen, the values range from 65 to 900 M^{-1}. Although this is a considerable spread for complexes that contain the same number and arrangement of hydrogen bonds, the free energy ($\Delta G°$) difference is only 1.5 kcal mol^{-1}. Furthermore, many of the association constants in Fig. 1 were measured in different laboratories. Thus, it is likely that some of the differences may be attributed to variability in the conditions for the binding studies. For example, complexes **6 · 7** and **20 · 21** are structurally very similar, yet their complexation free energies, measured in different laboratories, differ by ca. 1.0 kcal mol^{-1}. One variable that is usually not controlled in binding assays is the water content in the chloroform, which is likely to have a small effect on the K_{assoc} values, and a larger effect on the enthalpy of binding [20].

The most commonly used ADA modules contain the pyrimidine-2,4-dione nucleus. Not surprisingly, *N*-alkylation of thymine or uracil with an alkyl halide provides a simple, one-step method of functionalizing this module. Other ADA units include simple imides (e.g., **10**), heterocycle **13**, which was used by Kelly in a bisubstrate reaction template [17], and the anthyridinone or anthyridan units in **15** and **18**, respectively. The latter modules are synthesized by double Friedländer condensation of 2,6-diaminopyridine-3,5-dicarboxal-

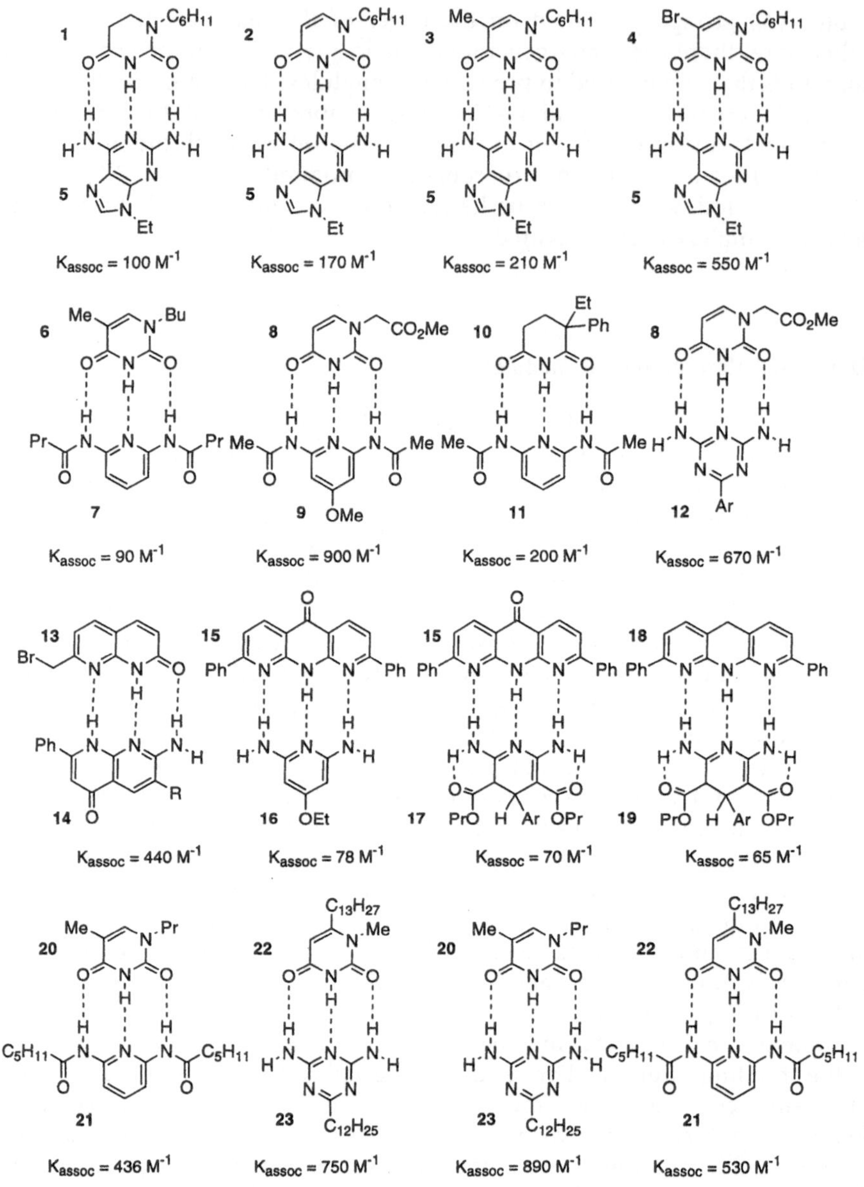

Fig. 1. Triply hydrogen-bonded complexes containing the ADA · DAD arrangement. Association constants are measured in chloroform-*d*. Complexes 5 · 1–4 [5]; 6 · 7 [14]; 8 · 9, 8 · 12 [15]; 10 · 11 [16]; 13 · 14 [17]; 15 · 16, 15 · 17, 18 · 19 [4, 18]; 20 · 21, 22 · 23, 20 · 23, 22 · 21 [19]

dehyde and acetylaromatics or heteroaromatics, followed by either oxidative (15) or reductive workup (18) [21].

As noted above, 2-aminoadenine is a commonly used DAD module. In addition, the use of 2,6-diamidopyridines as the DAD complement to ADA

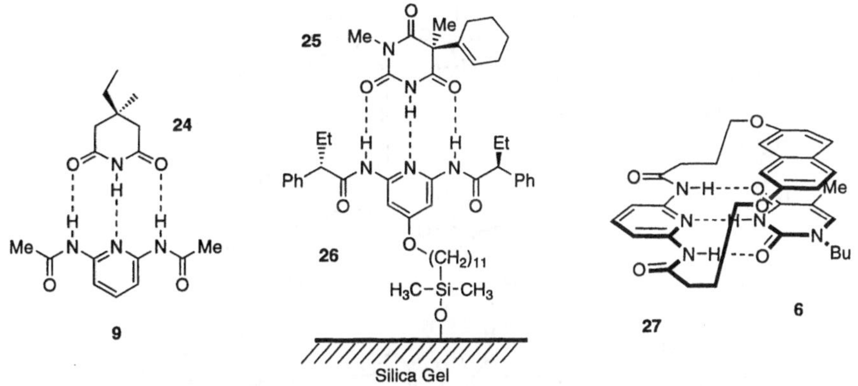

Fig. 2. Three ADA · DAD complexes featuring a 2,6-diamidopyridine unit as the DAD module

units has been described by Feibush, Karger et al. in the development of chemically bonded stationary phases for barbiturates, glutarimides, succinimides, and hydantoins [22]. The DAD · ADA pairing was clearly seen in an X-ray analysis of crystalline complex 9 · 24 (Fig. 2). Phase 26 was also effectively used for chiral separations, including the near base-line resolution of the enantiomers of hexobarbital (25). Hamilton has studied [23] thymine receptor 27, which was shown to complex 6 in chloroform with $K_{assoc} = 290$ M^{-1} (vs 90 M^{-1} for 6 · 7, Fig. 1). The added stability of 6 · 27 reflects π-stacking interactions in the complex. The naphthalene · thymine interaction was also seen in the solid-state structure. In these examples, the amide groups provide a convenient point of attachment to other compounds; however, unsymmetrically acylated diaminopyridines are not trivial to obtain. The 4-alkoxy group in compounds such as 9 provides another point of attachment, but its synthesis requires several steps. Related modules are the di- and triaminopyrimidines and di- and triaminotriazines, which are discussed in the next section.

2.1.1
Dimeric or Ditopic ADA and DAD Modules

There are two types of ditopic ADA and DAD modules, those synthesized by covalently linking modules, each containing one such hydrogen-bonding array, and those in which the hydrogen-bonding arrays are contained within a single heteroaromatic system. Along these lines, dimeric systems containing ADA subunits are readily available by alkylating suitable thymine or uracil analogs with an alkyl dibromide. Leonard's synthesis [24] of bis-thymine 28 (Fig. 3) illustrates this simple approach to dimeric modules. Interestingly, photolysis of 28 at 300 nm produces the *cis*-syn dimer 29 which projects a very rigid and somewhat splayed arrangement of the two ADA hydrogen-bonding arrays. In addition, Sessler has synthesized a rigid unit 30 as one in a series of "artificial dinucleotides", some of which form DNA-like duplexes [25]. These compounds are readily prepared by Sonagashira reactions with the appropri-

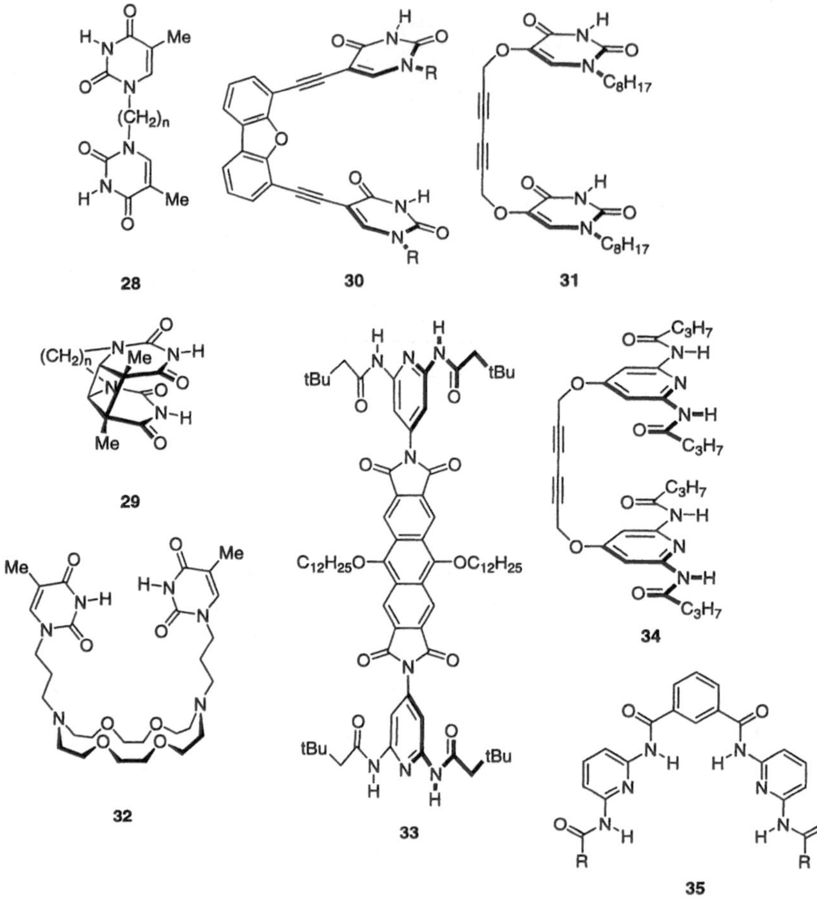

Fig. 3. Dimeric (ditopic) DAD and ADA modules constructed by covalently linking two identical modules

ate halogenated chromophores. Another example is Hamilton's linking of two 1-octyl-5-hydroxythymine units together with a diyne spacer to give **31** [26]. The connection of the subunits through the 5-alkoxy groups presumably allowed N-1 to carry the solubilizing octyl group. A final example is found in Gokel's extensive studies [27] of self-assembling ionophores (e.g., **32**), wherein base-pairing interactions create hosts capable of complexing complementary guest molecules.

Dimers of the DAD module are also illustrated in Fig. 3. Compound **34**, which holds two diamidopyridine units semi-rigidly with an interchromophore distance of about 10 Å, is analogous and complementary to **31** [26]. Likewise, Lehn developed the rigid, rod-like bis-diamidopyridine subunit **33** in which the hydrogen-bonding sites are directed along the long axis of the molecule [28]. As will be discussed in Sect. 5, such compounds are useful for creating polymeric

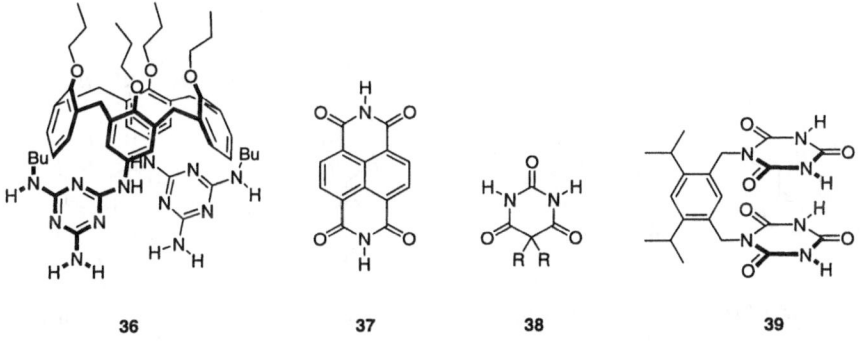

Fig. 4. Intrinsic ditopic modules containing two DAD or two ADA modules. Compounds **36** and **39** have two such modules linked by a semi-rigid scaffold

assemblies, whereas dimers such as **29–32**, **34**, and **35** are inclined to form closed, discrete aggregates with appropriate complementary partners.

In addition, the calix[4]arene-containing dimer of the triaminotriazine subunit, **36**, (Fig. 4) contains four recognition sites with two edges of each heteroaromatic group presenting the DAD hydrogen-bonding array [29]. Heterocycles such as the triaminotriazine unit, where the two hydrogen-bonding arrays are contained within a single heteroaromatic chromophore, may be termed *intrinsic* ditopic systems (Fig. 4). A ditopic ADA unit is found in commercially available naphthalene diimide (**37**) and in substituted barbituric acid (**38**), both of which have been used extensively as modules in self-assembly (*vide infra*). Compound **39** is analogous to **37** in presenting four recognition sites, but in this case, the ADA array is found in *N*-alkylated cyanuric acid derivatives [30]. The aminotriazene, barbituric acid, and cyanuric acid modules are readily available and easily derivatized, so the main effort is in synthesizing the dimers.

2.2
DDA and AAD Heteroaromatic Modules

The CG base-pair (i.e., **40** · **41**, R = ribose) is the prototypical DDA · AAD complex (Fig. 5). The AAD (cytosine) array is also found in 2-amidonaphthyridines (e.g., **42** and **44**). 2-Amino-5,7-dimethyl-1,8-naphthyridine is readily synthesized by the acid-catalyzed reaction of 2,6-diaminopyridine and acetylacetone [31] and is commercially available. Subsequent reaction with acetic anhydride readily affords **44**. In contrast, the DDA array is less common and less accessible synthetically. This motif is found primarily in the 6-amino-2(1*H*)-pyridone (**43**) [17], guanine (**41**, **47**), and 7-deazaguanine (**45**) [18] chromophores.

Guanine and guanosine are, of course, commercially available and can be derivatized in numerous ways. Along these lines, Sessler has used [32] a palladium-catalyzed Stille coupling to connect a porphyrin system to 8-bromoguanosine for use in a self-assembling, photosynthetic model system

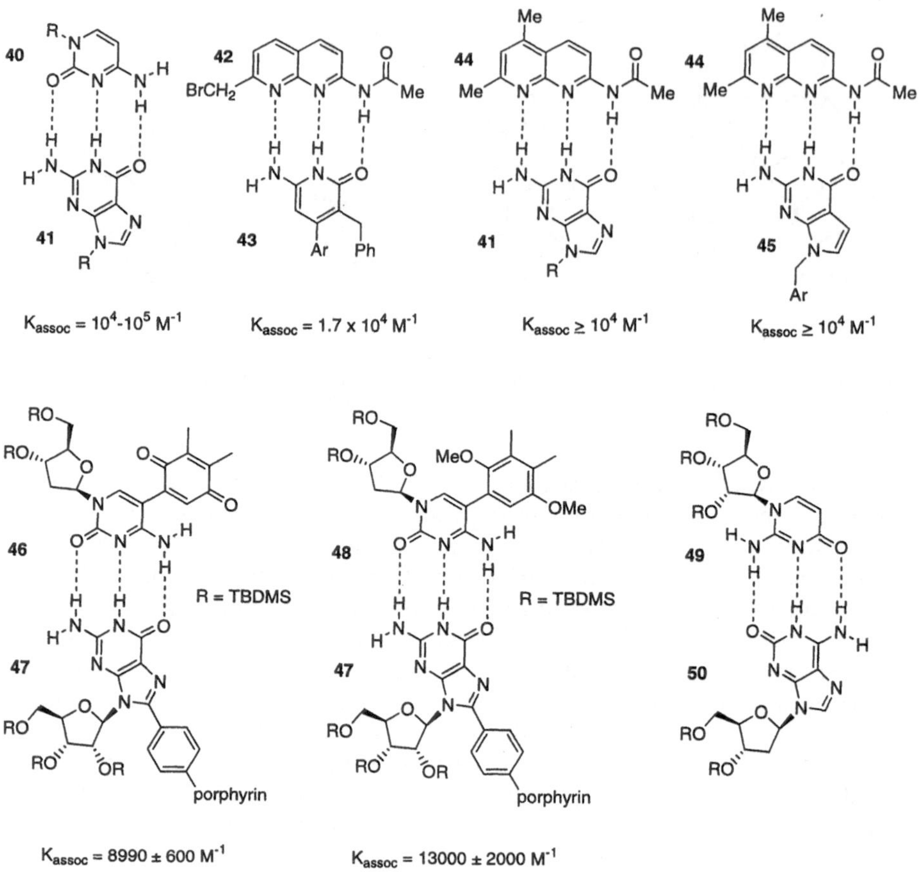

$K_{assoc} = 10^4\text{-}10^5\ M^{-1}$ $K_{assoc} = 1.7 \times 10^4\ M^{-1}$ $K_{assoc} \geq 10^4\ M^{-1}$ $K_{assoc} \geq 10^4\ M^{-1}$

$K_{assoc} = 8990 \pm 600\ M^{-1}$ $K_{assoc} = 13000 \pm 2000\ M^{-1}$

Fig. 5. Complexes containing the DDA · AAD motif. Association constants are measured in chloroform-*d*

(see **46** · **47** and **47** · **48**). Likewise, isoguanosine can base-pair with isocytidine (see **49** · **50**). An RNA · DNA 12-mer containing this non-natural base-pair exhibited the same melting temperature as the same sequence with a GC pair [33]. To our knowledge, the stability of isoG · isoC, isoG · C, or G · isoC complexes have not been measured in organic solvents.

The association constants listed in Fig. 5 appear to lie approximately in the $10^4\text{-}10^5\ M^{-1}$ range, which is two to three orders of magnitude higher than observed for the ADA · DAD systems. However, the difference may be even greater. Kyogoku and co-workers noted that their measured K_{assoc} values were only approximate due to the experimental method used [5], and Murray and Zimmerman found that **41** and **45** dimerized strongly in chloroform [4]. By extension, **43** and **47** are likely to dimerize, but this self-association was not taken into account when the K_{assoc} values in Fig. 5 were measured. Thus, the values reported are likely lower limits.

2.2.1
Dimeric or Ditopic AAD and DDA Modules

Several dimeric AAD and DDA modules have been reported that use porphyrin or a porphyrin analog as the scaffold (Fig. 6). For example, Sessler has covalently linked two guanine units to a sapphyrin (i.e., **51**) as a synthetic receptor for nucleotide phosphates [34] and has synthesized a bis-guanine analog of **47** [25]. Similarly, Hisatome has prepared [35] base-pair analogs by linking complementary nucleobases to aryl porphyrins (e.g., **52**). In the synthesis of **52** the guanine units are attached by an alkylation reaction as the last step. Evidence was presented for intramolecular base-pairing. In a related development, Sessler and Wang synthesized [25] and studied cytosine-guanine "heterodimer" **53** as part of their effort to create artificial duplexes (see **30**, *vide supra*). In this case, the bases are rigidly held apart so that base-pairing is forced to occur intermolecularly. Matile et al. have reported the synthesis of bis-guanine **54** as a first step toward a membrane spanning pore created by the tetramerization of polyphenylene-based guanine oligomers [36]. The authors noted solubility problems, which, in fact, are significant for most guanine-based compounds.

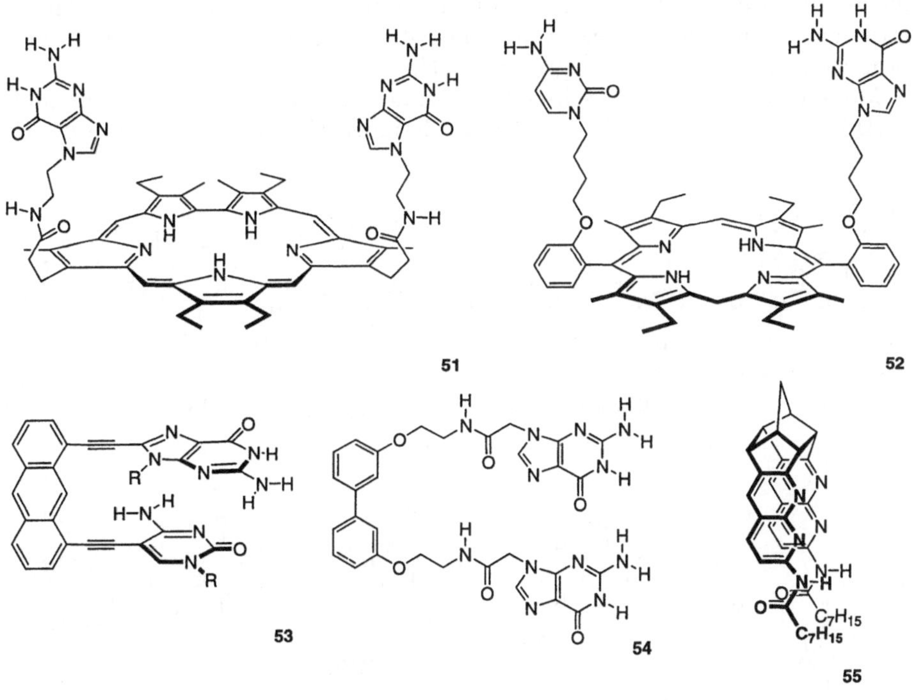

Fig. 6. Dimeric (ditopic) DDA and AAD modules constructed by covalently linking two identical modules

In contrast to the DDA units described above, there are comparatively few dimeric AAD modules. Zimmerman and co-workers have developed a convenient synthesis of 2,6-diaminopyridine-3-carboxaldehyde (DPC), which undergoes a Friedländer reaction with ketones to produce 2-amino-1,8-naphthyridines [37]. For example, the reaction of DPC with tetracyclo[6.3.0.0^{4,11}.0^{5,9}]undecane-2,7-dione followed by acylation with octanoic anhydride produced 55. A number of diketones reacted similarly with DPC to produce bis-2-amino-1,8-naphthyridines with varying chromophore orientations.

In addition to the dimeric AAD/DDA units mentioned above, several intrinsic ditopic AAD/DDA modules have been reported (Fig. 7). In parallel, independent studies Lehn and Zimmerman developed tricyclic heteroaromatic compounds containing two DDA sites or two AAD sites [38–42]. Lehn has referred to these modules as Janus molecules [38]. Compounds 56 and 57 contain two cytosine nuclei (AAD array) fused to a central pyridine or benzene unit, respectively. Compound 56 is available in three steps from appropriate N-alkylureas [41], and 57 is available in approximately nine steps from resorcinol and 4-dodecylaniline [38]. The ditopic guanine analogs 58 and 59 contain two 5-amino-2(1H)-pyridone rings fused to a central pyrrole or benzene unit, respectively, and were designed as complements to 56 and 57. Compound 58 is synthesized in five steps from methyl acetoacetate [41], and 59 is available in three steps from pyromellitic anhydride [38]. Whereas the 60° angle between the hydrogen-bonding arrays in the other heteroaromatic compounds in Fig. 7 is designed to produce closed-cyclic aggregates, the angle in 58 was intended to produce helical, polymeric aggregates [41].

Modules 60 and 61 are self-complementary, containing both a DDA and AAD hydrogen-bonding site. Both were designed to self-assemble into cyclic

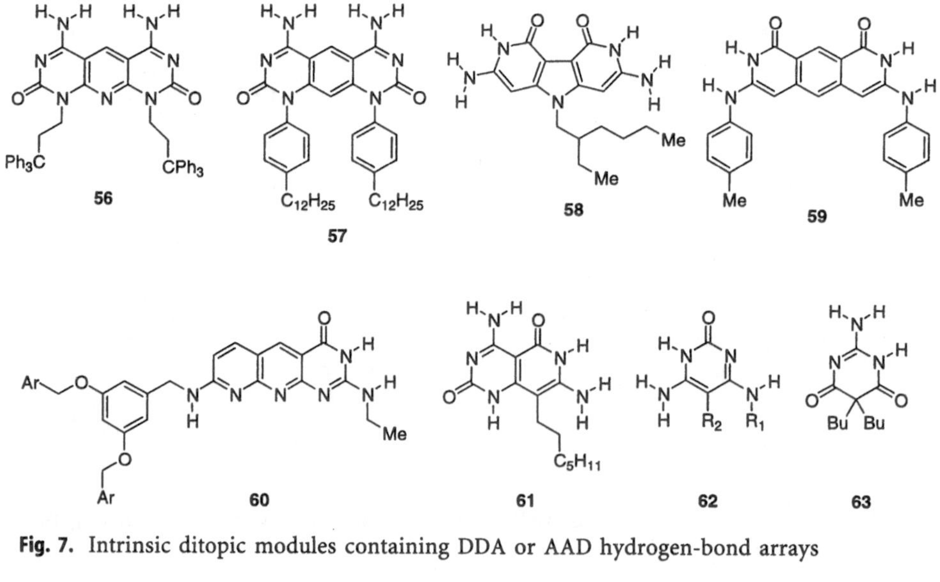

Fig. 7. Intrinsic ditopic modules containing DDA or AAD hydrogen-bond arrays

hexamers. Compound **60** is available from a multi-step synthesis [42]. Module **61** was prepared by Mascal in five steps starting from commercially available malononitrile dimer (2-amino-1-propene-1,1,3-tricarbonitrile) [43]. Janus modules **62** and **63**, both readily synthesized, contain a DDA and AAD hydrogen-bonding site [40]. Thus, they are self-complementary and complementary to one another. The self-assembly of most of these compounds has been studied and is discussed in Sect. 4.

2.3
DDD and AAA Heteroaromatic Modules

There are very few heteroaromatic modules that present the DDD or AAA hydrogen-bonding motif (Fig. 8). The latter is found in the 1,9,10-anthyridine (1,8,9-triaza-anthracene) nucleus (e.g, **64**), which can be conveniently synthesized in a single step by a double Friedländer reaction of 2,6-diamino-3,5-pyridine dicarboxaldehyde and ketones [21]. One of the only DDD modules, dihydropyridine **65**, is also readily available via a single step Hantsch synthesis using methyl carbamimidoyl acetate and 3-nitrobenzaldehyde [21]. The Hantsch synthesis using 2-nitrobenzaldehyde affords the analogous dihydropyridine, which serves as a DAD module because it exists exclusively in the 3,4-dihydro form (i.e., **17**, Ar = 2-nitrophenyl) in both DMF and CDCl$_3$ [44]. Unfortunately, the dihydropyridines are prone to oxidation. In addition to this neutral DDD · AAA complex, Anslyn has demonstrated [45] that ethyl 2,6-diaminonicotinate in its protonated form (**66**), with a lipophilic tetraarylborate counterion, will dissolve in organic solvents and tightly complex **64** (Ar = phenyl). The authors note that although the measured pK$_a$ values are consistent with complex **64 · 66**, the microenvironment of the complex could lead to a degree of proton transfer, which would lead to a complex with the DAD · ADA hydrogen-bonding arrangement. The cationic hydrogen bonding

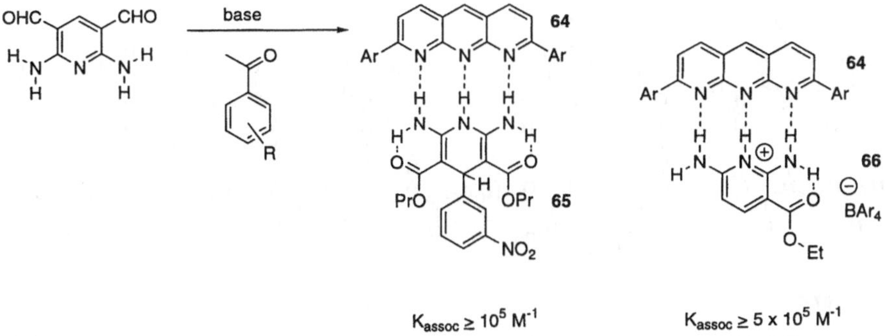

Fig. 8. Synthesis of the diaryl-1,9,10-anthyridine (AAA) module **64** and its complex with DDD modules, dihydropyridine **65** and pyridinium **66**. Association constants are measured in chloroform-*d*

is expected to provide additional stability, and the oxidative instability of **65** is avoided by use of **66**.

2.4
ADAD Heteroaromatic Modules

One advantage of using modules with multiple hydrogen-bond donor and acceptor groups is the potential for creating self-assembled structures from multiple, complementary pairs of modules. Most self-assembly processes studied to date use a single recognition motif, but it is clear that more complicated assemblies will require a higher information content contained within two or more pairs of complementary recognition units. With two hydrogen-bonding sites, there are three modules possible that form two complexes: AA · DD and (AD)$_2$. As seen in the previous sections, a linear array of three sites can be arranged in six ways that form three hydrogen-bond motifs: ADA · DAD, AAD · DDA, and AAA · DDD. Four sites increase the number of arrays to ten, with six complexes possible, due to the fact that two of the modules are self-complementary: (ADAD)$_2$, (AADD)$_2$, ADDA · DAAD, ADDD · DAAA, DADD · ADAA, and AAAA · DDDD. The first three of these motifs have been studied and are discussed in this and the following section.

Meijer and co-workers have shown [46] that acylation of both diamino-triazines and diaminopyrimidines affords heteroaromatic systems containing the self-complementary DADA motif (see **67–70**, Fig. 9). Thus, well-established DAD modules can be readily converted into DADA units. The dimerization constants for **67–70**, measured in CDCl$_3$, vary considerably with the corresponding complexation free energies differing by over 4 kcal mol^{-1} from strongest to weakest. This considerable variation for complexes with the same number of hydrogen bonds, and nominally the same hydrogen-bonding motif and geometry, will be discussed in more detail in Sect. 3. Two other DADA units have been reported as protomers of DDAA systems and, thus, will be discussed in the next section.

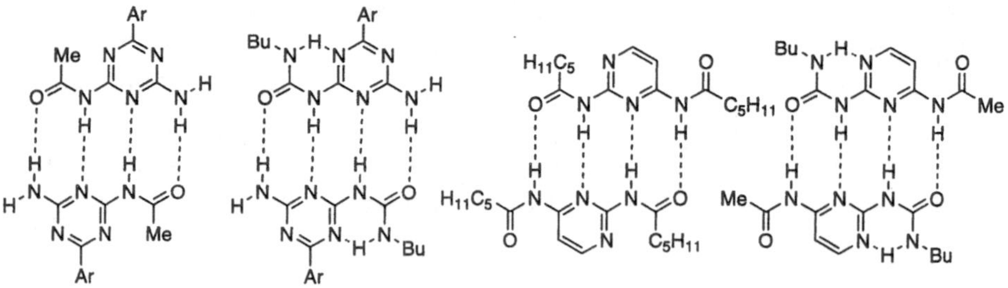

67·67

$K_{dimer} = 530 \ M^{-1}$

68·68

$K_{dimer} = 2 \times 10^4 \ M^{-1}$

69·69

$K_{dimer} = 170 \ M^{-1}$

70·70

$K_{dimer} = 2 \times 10^5 \ M^{-1}$

Fig. 9. Dimers of acylated diaminotriazine and -pyrimidine DADA modules. Association constants are measured in chloroform-*d*

2.5
DDAA Heteroaromatic Modules

Corbin and Zimmerman [47] and Meijer and co-workers [48, 49] have recently developed related DDAA modules (Figs. 10 and 11). The Corbin-Zimmerman module was created by efforts to meet several design criteria: (1) irrelevance of

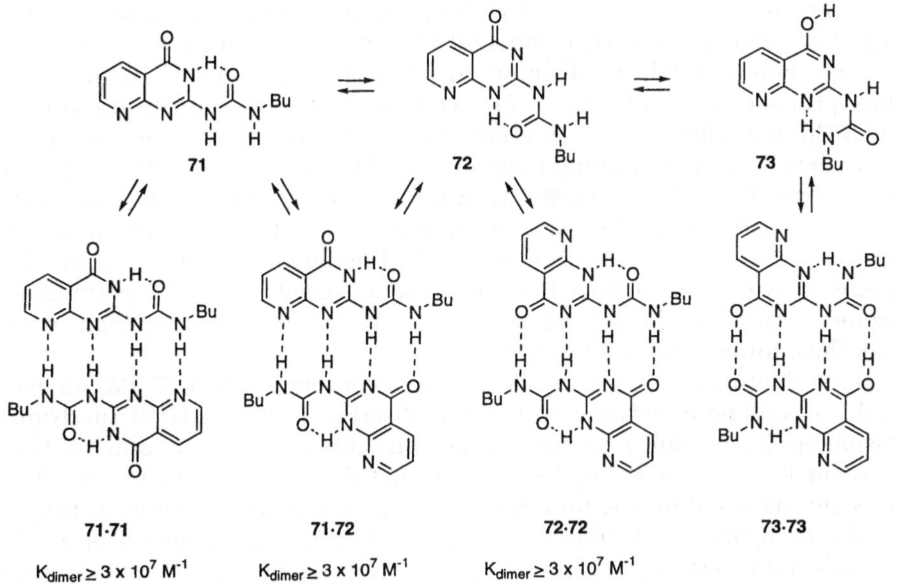

Fig. 10. Conformational, protomeric, and self-association equilibria of **71**, a DDAA (ADAD) module

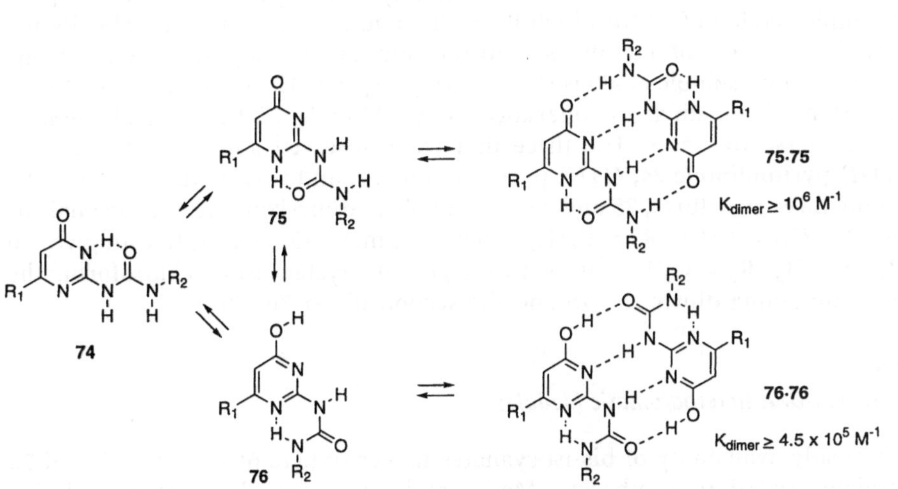

Fig. 11. Conformational, protomeric, and self-association equilibria of **74**, a DDAA (ADAD) module

protomeric form, (2) very strong self-association, (3) inexpensive starting materials, (4) a scalable, short synthesis.

One of the problems with using heteroaromatic modules in self-assembly is controlling the protomeric equilibrium. The presence of undesired protomers may carry the wrong assembly instructions and will decrease the affinity with which the correct pairing occurs. This issue is taken up in greater detail in Sect. 3.

Compound 71 meets essentially all of the aforementioned design criteria (Fig. 10). Along these lines, compound 71 was synthesized in two steps from commercially available 2,6-diamino-3H-pyrimidin-4-one and 1,1,3,3-tetrame-thoxypropane [47, 50]. Protomer 71 was shown to exist in equilibrium with 72 in CDCl$_3$ and toluene-d_8, as ascertained by ^{1}H NMR studies. Moreover, both protomers associated forming homo dimers 71 · 71 (13%) and 72 · 72 (39%) and hetero dimer 71 · 72 (46%), with the relative percentages in toluene-d_8 shown in parentheses. The spatial arrangement of the alkyl substituent is maintained in each of these structures, which is important for its use in self-assembly. A minor protomer was tentatively assigned as 73. This pyrimidino protomer contains the self-complementary DADA hydrogen-bonding motif and, thus, formed dimer 73 · 73 (2%).

Upon diluting a CDCl$_3$ solution of 71 to a concentration of 12 μM, no new peaks or shifting of existing peaks were observed in the ^{1}H NMR spectrum. Assuming ≥95% dimer formation at this concentration, a dimerization constant $K_{dimer} \geq 3 \times 10^7$ M^{-1} can be estimated. This is the highest stability constant measured to date for a neutral complex with four or fewer hydrogen bonds. As tightly as 71 and its protomers are associated, addition of a 2,7-diamido-1,8-naphthyridine led to complete dissociation of the dimers (see next section). The utility of 71 in creating discrete self-assembled structures in solution has been demonstrated [51].

Meijer and co-workers have extensively investigated module 74 [48]. A simple version of 74 (in which R$_1$ = CH$_3$ and R$_2$ = C$_4$H$_9$) is available in one step by reaction of methyl isocytosine and butyl isocyanate, respectively, whereas more soluble analogs (R$_1$ = C$_{13}$H$_{27}$; R$_2$ = C$_4$H$_9$ or Ar) are available in two steps from ethyl 3-oxohexadecanoate. Module 74 has several possible conformers/protomers. The three that were observed and characterized are 6(1H)-pyrimidinone 74, 4(1H)-pyrimidinone 75, and pyrimidinol 76 (Fig. 11). Of the three structures, 75 and 76 are capable of dimerizing. The X-ray analysis of 75 (R$_1$ = CH$_3$; R$_2$ = C$_4$H$_9$) shows dimer 75 · 75. Interestingly, 76 (R$_1$ = C$_6$H$_5$; R$_2$ = C$_4$H$_9$) forms two types of crystals from chloroform, the first containing dimer 75 · 75 and the second dimer 76 · 76.

2.5.1
Dimeric DDAA Heteroaromatic Modules

The ready availability of bis-isocyanates makes dimers of modules 71 and 75 straightforward to synthesize. Meijer and co-workers have pioneered this approach [49] to create hydrogen-bonded, supramolecular polymers of considerable size and molecular weight (see Sect. 4). A different dimeric

DDAA module developed by Sessler uses a single oxygen atom to accept two hydrogen bonds [52]. This rigid module (77, R = tris-O-TBDMS ribose), which is related to dimer 53, is shown in Fig. 12. Its self-assembled structure is shown here rather than in Sect. 4 to illustrate the non-standard pairing proposed to hold 77 · 77 together.

2.6
DAAD and ADDA Heteroaromatic Modules

Compound 71 can exist in additional conformations (e.g., 78, 79), as shown in Fig. 13. These conformations contain a non-self-complementary ADDA hydrogen-bond array, which is complementary to diamidonaphthyridine 80. In fact, two equivalents of 80 are sufficient to entirely convert the homodimers and heterodimer of 71 into complex 78 · 80 or 79 · 80 [47]. These complexes are, therefore, more stable than the already highly stable dimers. Preliminary computational work suggests that this is an intrinsic property of the complexes rather than an energetic preference for conformer 78 or 79. Also shown in Fig. 13 is the ADDA · DAAD complex formed between 80 and bis-2-pyridylurea 81 reported by Lüning [53]. Despite the structural similarity of

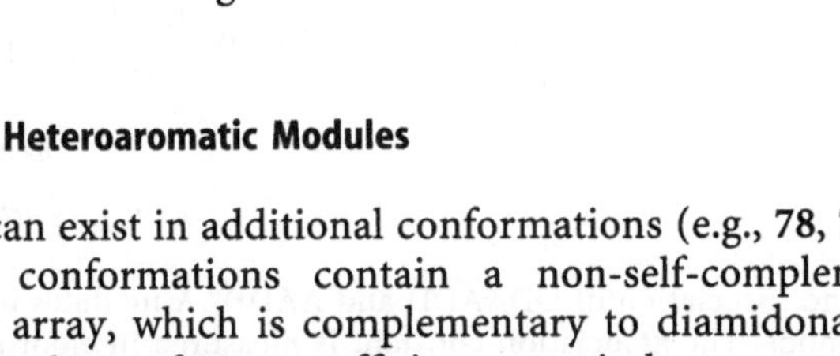

77·77

Fig. 12. Dimer formed from AADD module 77. Guanosine carbonyl acts as an AA unit by accepting two hydrogen bonds

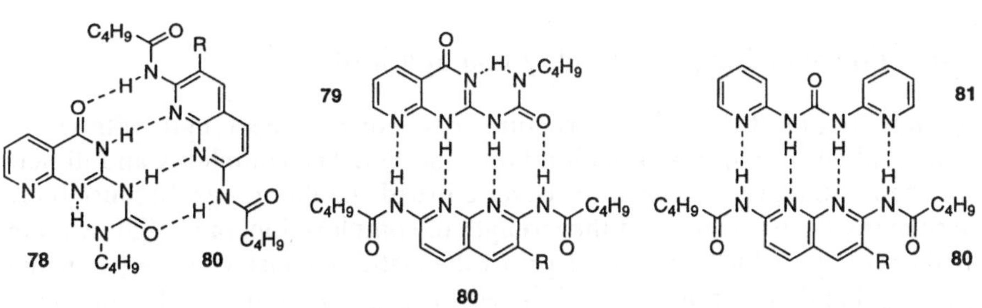

$K_{assoc} \geq 3 \times 10^7$ M^{-1} $K_{assoc} = 2000$ M^{-1}

Fig. 13. Three complexes containing a DAAD · ADDA motif. In complexes with 78 (79) (R = H in 80) and in complex with 81 (R = CN in 80). Association constants are measured in chloroform-d

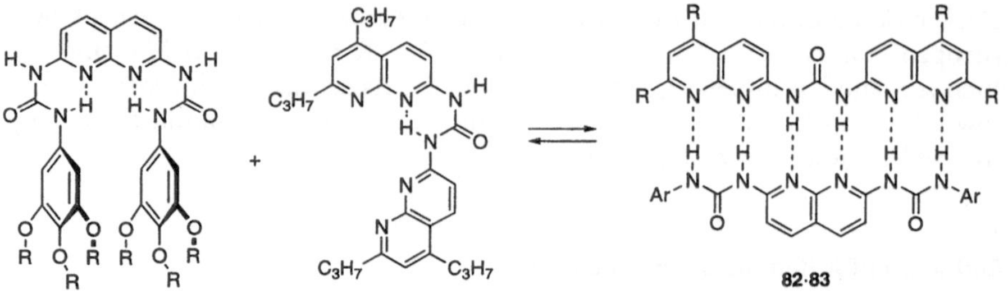

Fig. 14. Unfolding and association of DDAADD and AADDAA modules to form a complex with six hydrogen bonds. The association constant is measured in chloroform-*d*

$78 \cdot 80$ and $80 \cdot 81$, the K_{assoc} of $78 \cdot 80$ is only 2000 M^{-1}. This lower stability is due to a lower degree of preorganization, in particular an intramolecular hydrogen bond within **81**, which must be broken before complexation. Such issues are further addressed in Sect. 3.

2.7
AADDAA and DDAADD Heteroaromatic Modules

Corbin and Zimmerman have recently shown that a DDAADD module, bis-ureido-naphthyridine **82**, pairs with an AADDAA module, bis-naphthyrdinyl-urea **83**, by forming complex $82 \cdot 83$, which contains six hydrogen bonds (Fig. 14) [54]. The K_{assoc} in $CDCl_3$ is estimated to be between 10^5 and 10^6 M^{-1}. Compound **82** exists in a folded state with two intramolecular hydrogen bonds both in solution and in the solid state, and **83** contains one intramolecular hydrogen bond in solution. Thus, the formation of six intermolecular hydrogen bonds in $82 \cdot 83$ comes at the price of three intramolecular hydrogen bonds in the components. Complex $82 \cdot 83$ can be regarded as a model of an oligomeric or even polymeric naphthyridinylurea that may form a folded structure or self-associate through formation of multiple hydrogen bonds.

3
Factors Affecting Complex Stability and Specificity

The stability of self-assembled structures depends on the strength with which the individual recognition units bind one another. The complexes and dimers described in the previous sections cover a broad stability range. Because there are so many factors that affect the strength of complexation, one might wonder whether it is possible to predict a priori the stability constant for a particular hydrogen-bonded complex. Certainly, advances in computational approaches are bringing closer the day when one will be able to calculate absolute free energies of complexation before even setting foot in the laboratory [55]. As will be discussed below, there are some general trends based on considerable experimental data, and possibly even empirical relationships, that allow at least a qualitative prediction of stability for several classes of hydrogen-bonded complexes.

The specificity with which individual components assemble is a critical factor in determining whether the correct self-assembled structure is formed. Often very strong complexes are formed selectively, but this need not be the case. One advantage of using a single recognition unit is the lack of competing complex equilibria in the self-association process. With two complementary recognition units one must be concerned with self-association of the components competing with complex formation.

3.1
Primary Hydrogen-Bond Strengths

A central determinant of stability of any hydrogen-bonded complex will be the strength of the individual hydrogen bonds. The primary hydrogen-bond strengths depend, in turn, on the acidity of the donor site and the basicity of the acceptor site. For example, the log K_{assoc} values for complexes between phenol and nineteen substituted pyridines linearly correlated with the gas-phase proton affinity of the same pyridines [56]. Recognizing the importance of primary hydrogen-bond strengths is one thing, but it is quite another to actually determine these strengths in the more complicated multiply hydro-gen-bonded complexes described herein. Indeed, it is not generally possible to factor out the relative contribution of individual contacts in complicated complexes such as base-pairs (see Fig. 1).

However, it is possible to order the stability of structurally similar complexes based on measured pK_a values or expected relative acidities or basicities. For example, the increase in stability observed in the complexes between aminoadenine 5 and uracil derivatives 1–4 (Fig. 1) was attributed to the increase in acidity for the uracil imino proton across the series from 1 to 4 [5]. Likewise, the observation that the K_{assoc} of 8 · 9 is ten times larger than that of 6 · 7 can be attributed to the increased basicity of the methoxy-substituted pyridine. However, as noted previously [4], the K_{assoc} values of 6 · 7 and 8 · 9 were measured in different laboratories, so a direct comparison must be made with caution.

3.2
Number of Hydrogen Bonds

In an early analysis, a linear correlation was made between the complexation free energy ($\Delta G°$) in chloroform and the number of hydrogen bonds in eight hydrogen-bonded complexes formed from amide or imide and amino- or amidopyridine components [57]. From the linear correlation it was concluded that each hydrogen bond contributed approximately 1.2 kcal mol^{-1} to the free energy of complexation.

Even if similar primary hydrogen-bond strengths are found in these complexes, such a correlation is surprising because it seemingly requires a linear correlation between the complexation entropy and the number of

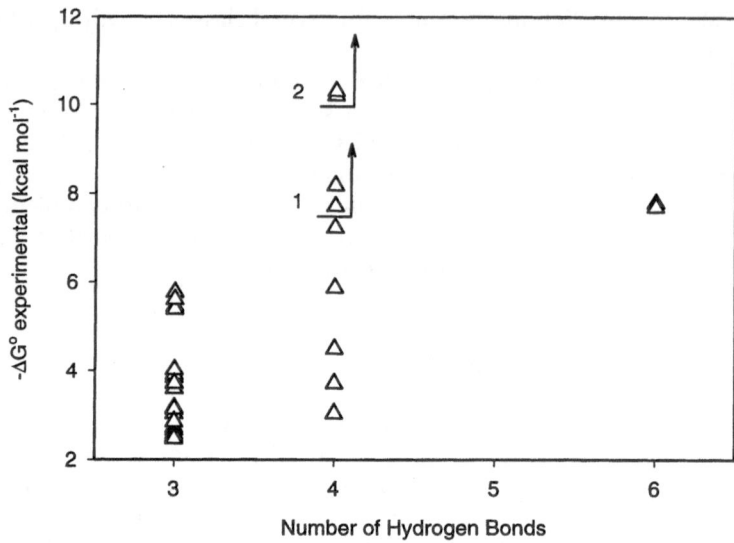

Fig. 15. Plot of free energies of complexation in chloroform-*d* vs number of hydrogen bonds for all complexes in this review. Data point 1 (75 · 75 and 76 · 76) and 2 (71 · 71 and 78 · 80) are plotted with $\Delta G°$ values that are lower limits. Actual values are likely to be considerably higher

hydrogen bonds. No attempts to generalize this type of correlation have been reported, and the analysis described above remains limited to carefully chosen complexes with similar types of hydrogen-bonding arrays. Indeed, there are now many examples where the more stable complex of a pair contains fewer hydrogen bonds. This situation can be seen graphically in Fig. 15, where the $\Delta G°$ values for each of the complexes discussed in this review are plotted against the number of hydrogen bonds.

3.3
Arrangement of Hydrogen-Bond Donor and Acceptor Groups

Jorgensen observed that base-pairs 2 · 5 and 6 · 7 have similar K_{assoc} values (ca. 100 M^{-1}), but that two other triply hydrogen-bonded complexes (40 · 41 and 42 · 43) displayed K_{assoc} values that were at least two orders of magnitude higher [58]. The differences, which were confirmed computationally, were attributed to secondary hydrogen bonding or secondary electrostatic interactions arising from the different arrangements of the hydrogen-bond donor and acceptor groups within the complexes (see Fig. 16). In short, Jorgensen proposed that the adjacent donor and acceptor groups are sufficiently close to make a significant contribution to complex stability. In the DAD · ADA motif of complexes 2 · 5 and 6 · 7, each of the secondary contacts are repulsive and destabilize the complex. In the DDA · AAD motif of complexes 40 · 41 and 42 · 43 two attractive secondary contacts offset two repulsive ones, which explains the higher stability of the DDA · AAD complexes. Jorgensen further

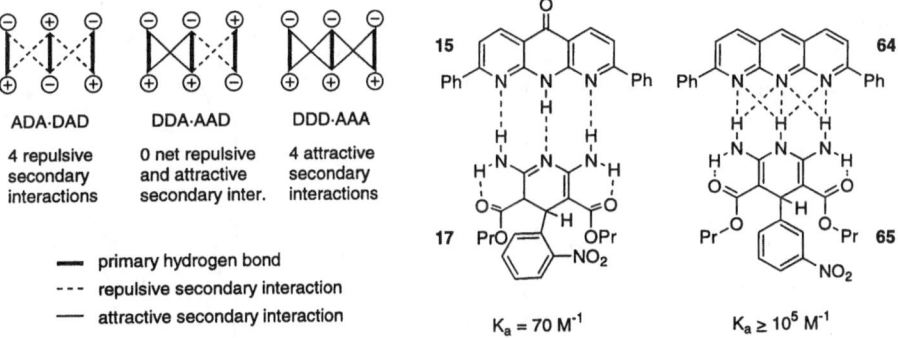

Fig. 16. Jorgensen secondary electrostatic model and its illustration with dramatically different K_{assoc} seen for DAD · ADA and DDD · AAA complexes

proposed that a complex with the AAA · DDD motif would be expected to be particularly robust as a result of four attractive secondary hydrogen bonds.

As described above, Murray and Zimmerman reported [18] the first example of an AAA · DDD complex (**64 · 65**). The ability to switch the protomeric form of **17** to **65** by a remote substituent effect, and construct ADA and AAA modules **15**, **18** and **64** with the 2,7-diphenylanthyridine nucleus, allowed for a careful comparison of structurally similar complexes. The >10³-fold difference in K_{assoc} for **15 · 17** (**18 · 19**) and **64 · 65**, and the high stability of the latter, was consistent with the Jorgensen proposal. The four attractive interactions in **64 · 65** are shown as hydrogen bonds in Fig. 16. The K_{assoc} trends in Figs. 1, 5, 8, 10 and 11 for closely related complexes are generally in agreement with the proposal that the arrangement of the hydrogen-bond donor and acceptor groups in a given module is a critical factor in determining complex stability. For example, the DDAA dimer **75 · 75** has a K_{assoc} value ca. 4 orders of magnitude higher that the related DADA dimer **76 · 76** in DMSO/chloroform mixtures, although their K_{assoc} are indistinguishably strong in pure chloroform [48]. Furthermore, systems with two hydrogen bonds and "overhanging" secondary interactions, which are not covered by this review, are also consistent with the proposal [58, 59]. On the other hand, there are some clear exceptions. For example, DAAD · ADDA complex **78 · 80** is strong enough to dissociate the AADD dimers of **71** even though the former contains two net repulsive secondary interactions, while the latter have two net attractive secondary contacts. Furthermore, the wide range of K_{assoc} values in Fig. 9 indicates that other factors are significant in determining complex stability because each of these complexes has the same number of hydrogen bonds and the same ADAD array.

The extension of the secondary hydrogen-bonding concept to peptides was investigated by Gellman [60], who tested the hypothesis that glycine dimer **84** would be less stable than complex **85 · 86** when constrained to be planar (Fig. 17). Foldamers **87** and **89**, cleverly designed to test this hypothesis, provided a scaffold wherein a single intramolecular hydrogen bond preorganizes the system for the second internal hydrogen bond, which is of the glycine

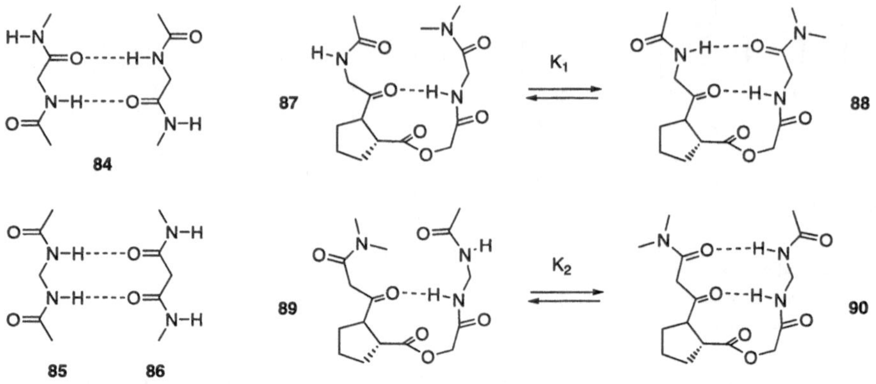

Fig. 17. A test of Jorgensen's secondary electrostatic hypothesis in flexible peptide models

dimer-type (88) or analogous to that in complex 85 · 86 (90). The experimentally determined K_1 and K_2 values (Fig. 17) were nearly identical, and the authors concluded that dipole–dipole repulsions within the arms of 90 favor alternative conformations. In relation to the heteroaromatic modules described herein, in the chemical synthesis of compounds such as 64 and 65, the energy price is already paid to align the dipoles.

The tremendous scatter seen in the plot of $\Delta G°$ versus number of hydrogen bonds (Fig. 15) is clearly due in part to the differences in secondary hydrogen bonding. This raises the question of whether consideration of both factors could lead to the ability to predict complexation free energies. Along these lines, Schneider has proposed [61], based on multiple linear regression analysis, that there is a linear relationship between the observed and calculated free energies of complexation if each primary hydrogen bond contributes 1.9 kcal mol^{-1} to the total energy and each secondary contact ±0.69 kcal - mol^{-1}, with the sign of the latter depending on whether it is attractive or repulsive. The author's plot of $\Delta G°_{calcd}$ versus $\Delta G°_{expt}$ for 58 different complexes gave a slope of 0.84, with an $R^2 = 0.834$, and an average difference between $\Delta G°$ values of just ±0.4 kcal mol^{-1}. This remarkable agreement suggests that the stability of multiply hydrogen-bonded complexes can be predicted a priori using these empirical increments. However, complexes have been reported where the $\Delta G°_{calcd}$ and the $\Delta G°_{expt}$ values are not in good agreement [46, 53].

A plot of $\Delta G°_{calcd}$ versus $\Delta G°_{expt}$ is shown in Fig. 18 for the complexes described in this review, along with the best-fit line (solid) and the line expected for an ideal one-to-one correlation (dotted). Although a few points fall on the best-fit line, the slope is 0.68 and the $R^2 = 0.535$. Furthermore, the four DDA · AAD complexes that have $\Delta G°$ values that fall on the solid line (data points 1 in Fig. 18) are likely, as described in Sect. 2.2, to have complexation energies in excess of the reported values. This is represented by the up-arrow in Fig. 18. Data point 2 for complex 78 · 80 is plotted with $K_{assoc} = 3 \times 10^7$ M^{-1}. Because this is a lower limit, its true K_{assoc} exceeds its calculated K_{assoc} of 3.6×10^4 M^{-1} by several orders of magnitude. The data in Fig. 18 suggest that

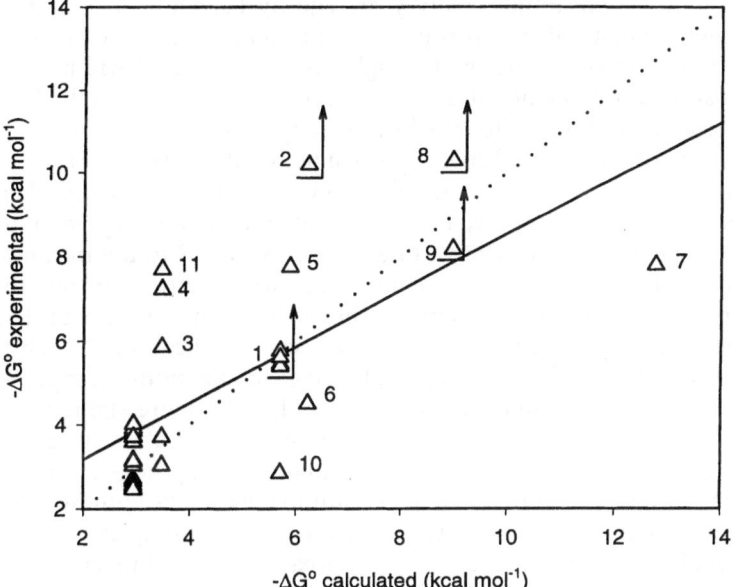

Fig. 18. Plot of free energies of complexation for all complexes in this review vs calculated $\Delta G°$ values using [61]. The *solid line* represents the linear regression fit of the data and the *dotted line* is expected for a 1:1 correlation of experimental and calculated values. Data points: 1 (DDA · ADD complexes), 2 (78 · 80), 3 (68 · 68), 4 (70 · 70), 5 (35 · 38), 6 (80 · 81), 7 (82 · 83), 8 (71 · 71), 9 (75 · 75), 10 (41 · 91), 11 (76 · 76). The *up-arrows* indicate points whose $\Delta G°$ values are lower limits

the previous analysis is not general and that other factors beyond the number of primary and secondary hydrogen bonds must be considered to predict complex strength. In the case of **78 · 80**, preliminary computational work suggests that the primary hydrogen bonding is especially strong [47]. Other factors that must be considered are discussed in the sections below.

3.4
Preorganization

As alluded to above, one of the central tenets of host-guest chemistry is the idea that the most effective hosts are rigidly held in the conformation needed for binding and are poorly solvated. Cram has stated this idea most explicitly in his principle of preorganization [62]. This principle states that complex stability will be highest when the host and guest are organized for binding and low solvation. It has been known for many years that crown ethers complex metal ions more strongly than their acyclic analogs (podands) [63]. Zimmerman and co-workers have quantified the free energy benefit of freezing single bond rotations in a host-guest (molecular tweezers) system [64]. Each free rotation cost ca. 1 kcal mol^{-1} in complexation free energy. This value is comparable to the energy cost of freezing single bond rotations in intramolecular reactions determined many years ago by Page and Jencks [65] and a more recent estimate of the entropic cost for complexation by Williams [66]. Clearly, systems that require many single bonds to be fixed will pay a serious energy cost. It should be noted that a very recent analysis of acyclic host-guest systems reached the opposite conclusion, suggesting a much lower complexation energy cost of 0.3 kcal mol^{-1} per single bond rotation [67].

Several related complexes in this review differ in their degree of preorganization. For example, Meijer has attributed the higher stability of dimers 68 · 68 and 70 · 70 relative to 67 · 67 and 69 · 69 to the intramolecular hydrogen bond in the former complexes, which preorganizes the four hydrogen-bond donor and acceptor groups. Each intramolecular hydrogen bond increases the complex stability by ca. 1–2 kcal mol^{-1}. This value contains more than just the entropic advantage, as the non-hydrogen-bonded amides in the monomer likely adopt the enthalpically more stable trans form.

Complexes 80 · 81 and 82 · 83 are significantly weaker than expected; particularly the latter complex, which contains six hydrogen bonds and a DDAADD · AADDAA motif. An explanation for the weaker binding is simple. These components are poorly organized for complexation as a result of an intramolecular hydrogen bond in pyridyl urea 81 and naphthyridinyl urea 83. In fact, the solid-state and solution structure of 82 reveal two additional intramolecular hydrogen bonds that must be broken prior to formation of 82 · 83. Thus, three hydrogen bonds must be broken to form the six hydrogen bonds in this complex.

In addition to intramolecular hydrogen bonding, other conformational issues may affect the level of preorganization. Along these lines, Hamilton found [68] that ethoxynaphthyridine 91 bound triacetyl guanosine 41 with K_{assoc} = 126 M^{-1} (Fig. 19), which is at least two orders of magnitude lower than the K_{assoc} of a very similar complex, 41 · 44. Murray and Zimmerman proposed [69] that when the ethoxy group is in its preferred orientation there are steric interactions with the guanine amino group (see 41 · 91'). Thus, there is an energy cost for producing the less stable conformation of the ethoxy group in

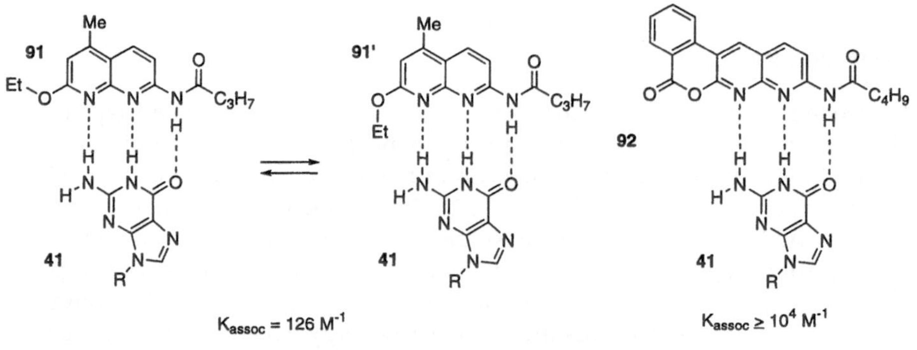

K_{assoc} = 126 M^{-1} K_{assoc} ≥ 10^4 M^{-1}

Fig. 19. Conformational equilibrium of ethoxynaphthyridine 91 and its complex with 41. Compound 92 contains an oxy substituent but is constrained in a ring. Association constants are measured in chloroform-d

41 · 91. Evidence for this hypothesis came from studies of **92**, which contains an alkoxy group in the 8-position that is "tied back" in a lactone ring. In contrast to **41 · 91**, the K_{assoc} of **41 · 92** was $\geq 10^4$ M^{-1}, which was expected of the DDA · AAD-type complex.

3.5
Protomeric Form: Effect on Complex Strength and Specificity

Watson and Crick suggested in 1953 that base substitution errors in DNA might occur as a result of the mis-pairing of minor protomeric forms of the four natural bases [70]. For example, the enol form of thymine (T_{enol}) is complementary to guanosine (G). Although such protomeric forms have been observed directly in the gas phase using model compounds [71], and in the X-ray structure of non-natural base-pairs [72], direct evidence for their importance in naturally occurring DNA is lacking. Nonetheless, many agree that the mis-insertion rate due to protomerism is likely to be in the range of 10^{-4}–10^{-6} [72, 73]. How is this frequency of mis-insertion consistent with the high fidelity with which DNA is replicated (i.e., as few mistakes as one in 10^9 bases)? The seemingly inconsistent result is easily explained: DNA polymerases have a built-in proofreading function, and a second level of proofreading takes place post-replication.

Biomimetic chemists do not have the luxury of adding enzymes to check for mistakes in self-assembly. When modules contain the incorrect protomeric form, they carry misinformation that may translate into the wrong structure being assembled. Alternatively, a price may be paid in complex or assembly stability to shift the protomeric equilibrium toward the desired form. For example, Meyer has studied [44] the substituent effects on the protomeric equilibrium of **65** by ^{1}H NMR spectroscopy and has found that in dimethylformamide-d_7 (DMF-d_7) the 1,4-dihydro protomeric form exists exclusively. In chloroform-d (CDCl$_3$), however, a ca. 70:30 mixture of 1,4-dihydro (**65**) and 3,4-dihydro (**17**) protomers (slow exchange) were present. Thus, although complex **64 · 65** forms quantitatively from a 1:1 mixture of its components, an energy price must be paid to shift the protomeric equilibrium. Clearly, in this case, the energy price is very small (<1 kcal mol^{-1}). In cases where the desired protomer is disfavored, the energy price will be higher.

Remarkably, few of the heteroaromatic compounds used as modules for self-assembly have had their protomeric equilibria examined. Many investigators simply assume the form they desire is the favored one. In part this is because determining the protomeric form with certainty is difficult and usually requires the synthesis of N- and O-methyl derivatives as bond-fixed analogs of the possible protomers. Even then spectroscopic analysis may be insufficient. Beak has shown [74] that many of the protomeric forms assigned to heteroaromatic compounds are incorrect because they were determined under conditions where the compounds were fully self-associated, which affects the protomeric equilibria.

Meijer and co-workers have studied [48] the protomeric equilibria of dimers **75 · 75** and **76 · 76**, and found the equilibrium to be affected by the electronic nature of the substituent (Fig. 11). For example, **76** was the exclusive form when $R_1 = CF_3$; $R_2 = C_4H_9$. In studies carried out in THF-d_8, **76** was the preferred protomer in the monomeric state when $R_2 = C_4H_9$. In mixtures of CDCl$_3$ and DMSO-d_6, the 6(1H)-pyrimidinone form (i.e., **74**) is preferred in the monomeric state. The dimerization constants for **75 · 75** and **76 · 76** were estimated to be very strong in CDCl$_3$, at least 4.5×10^5 and 10^6 M^{-1}, respectively. In CDCl$_3$/DMSO-d_6 mixtures, the dimerization constant for the DADA motif of **76 · 76** is ca. 10^4-fold weaker than that for the DDAA motif in **75 · 75**. All these data suggest that the very strong dimerization of **75** is weakened by protomerism. However, the effect has not been quantified and is likely irrelevant for most applications given the very high stability of the dimer.

As described above, heteroaromatic module **71** was specifically designed [47] so that each major protomeric form would be capable of strong dimerization. To our knowledge, it is the first module designed to meet this criterion. As shown in Fig. 10, both the 1H- and 3H-protomers are indeed present in apolar organic solvents in homodimers **71 · 71** and **72 · 72** and heterodimer **71 · 72**. The K_{dimer} values in chloroform-d are immeasurably high, at least 3×10^7 M^{-1}. Although not by design, another heteroaromatic compound is known wherein two protomeric forms are capable of dimerizing. Specifically, 2(1H)-pyridone and 2-hydroxypyridine can each dimerize or form a complex together. However, the 2-pyridone dimerization constant is only ca. 100 M^{-1}, and the heterodimer forms in a head-to-head arrangement whereas the homodimers are head-to-tail [47]. As described above, the dimers of **71** are expected to maintain a very similar spatial arrangement of the N-alkyl groups.

3.6
Other Factors

There are many other factors that will determine the strength with which heteroaromatic modules pair. Electrostatic effects, those beyond the primary and secondary hydrogen-bonding contacts discussed above, are likely to play an important role. Ion-pair hydrogen bonds are found to be stronger than neutral ones both in biological systems [75] and in abiotic complexes [76]. Jorgensen has noted that the CG base-pair may enjoy additional stabilization from the large dipole moment of the component bases, which interact favorably in the complex [58]. There is a possibility that complexes with favorable secondary hydrogen bonding will also have significant molecular dipolar attractions. For example, complexes **64 · 65** and **64 · 66** contain components whose molecular structures suggest large dipole moments that align in a favorable way in the complexes.

Another obvious factor that contributes to the pairing strength is the extent of solvation of the components and the complex itself. Chloroform is used in many complexation and self-assembly studies because it is not highly competitive for hydrogen-bond donor and acceptor sites, yet it is often one of the best solvents for dissolving heteroaromatic compounds. It is widely known that pairing strengths increase in even less polar solvents such as the hydrocarbons and are greatly diminished in protic and other highly polar solvents such as dimethyl sulfoxide. Most designers of hydrogen-bonding modules rarely consider the issue of solvation, yet it may play a decisive role in determining the stability of an assembly. Hard data is hard to come by, but one might imagine that the modules described above containing the DDD and AAA hydrogen-bonding arrays may be difficult to solvate fully.

4
Selected Examples of Self-Assembling Systems

Many of the modules described above were designed to form closed or discrete aggregates. In some cases the modules simply dimerize or complex a complementary partner (see Figs. 1, 2, 5, and 8–14). The linking of modules to increase complex or dimer stability is a common theme in the area of self-assembly. For example, Hamilton has shown [26] that the K_{assoc} for 31 · 34, 4,500 M^{-1}, is about 10-fold higher than the analogous single base-pair (Fig. 20). A larger increase is predicted but strain in the macrocyclic complex was proposed to diminish the stability of 31 · 34. In a related study, Sessler reported that dimer 53 · 53 (Fig. 20) did not show appreciable dissociation upon dilution (^{1}H NMR monitoring) in chloroform-d, whereas complexes such as 40 · 41 do fully dissociate [25]. Most remarkable is dimer 77 · 77 (Fig. 12), which does not dissociate, even in the highly competitive solvent DMSO [52].

Whereas the hydrogen-bonding sites in the modules described above are roughly in parallel planes or slightly diverging, the ADA sites in ditopic receptor 35 converge and are designed to complex the intrinsic ditopic guest 38. Thus, macrocyclic barbituric acid receptor 35 complexes diethyl barbituric acid (38) with a K_{assoc} of approximately 10^5–10^6 M^{-1} (Fig. 20) [77], a value that is 2–4 orders of magnitude higher than the DAD · ADA hydrogen-bonded complexes in Fig. 1. A 2 + 2 self-assembled structure, $(93)_2(94)_2$ (Fig. 21), reported by Lehn [78] takes advantage of the intrinsic ditopic DAD array in triaminotriazine module 93 and its complementarity to uracil.

Another common strategy for creating closed assemblies is the use of intrinsic ditopic modules in which two complementary hydrogen-bonding

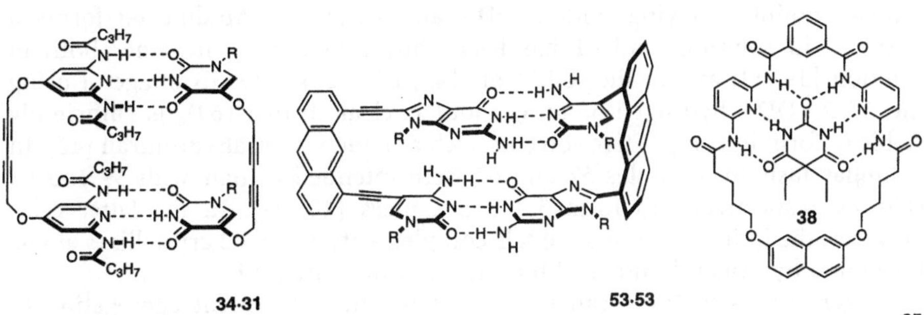

34·31   53·53   35

Fig. 20. Self-assembled structures involving dimeric modules

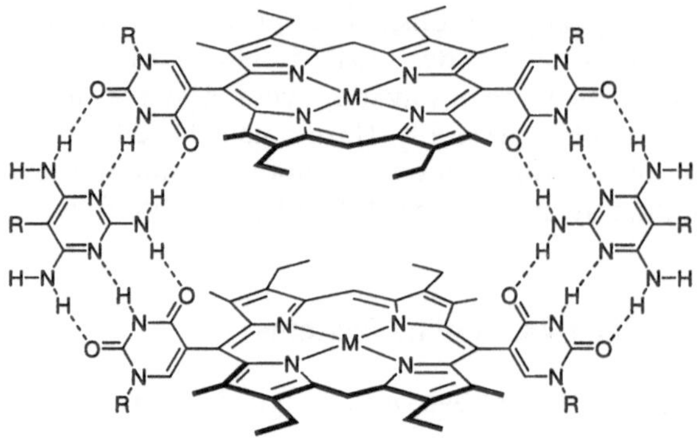

Fig. 21. A dimeric porphyrin via 2 + 2 self-assembly. R = decyl

sites are fixed at a 60° angle. As seen in Fig. 22, such modules are capable of forming cyclic hexamers. One of the most studied motifs uses analogs of two complementary, ditopic modules: cyanuric or barbituric acid and melamine. In this case, the hexamer **38 · 95** is formed from three molecules of **38** and three of **95**. This assembly (Ar = 4-*tert*-butylphenyl) is formed in the solid state, but the association in solution is weak [30, 79]. The structure is greatly stabilized, however, by covalently linking the melamine units, as in **36**, or the cyanuric acid modules, as in **39** (Fig. 4). An example of such an assembly is $(36)_3(38)_6$ (Fig. 22). This general strategy has been extensively investigated by both Whitesides [30, 79, 80] and Reinhoudt [29, 81].

The advantage of the melamine · cyanuric acid approach is the comparative ease with which these modules can be derivatized. Indeed, a multitude of different monomeric, dimeric, and trimeric modules have been synthesized and studied. The disadvantage of the approach is two-fold. The DAD · ADA motif is the weakest of the triply hydrogen-bonded arrays (*vide supra*), and its symmetry means that the components lack sufficient information to form the cyclic aggregate. Indeed, the same basic modules have been used to form polymeric aggregates (see below). To counter this problem, hexamers $(60)_6$ [42] and $(61)_6$ (R = C_7H_{15}) [43] were designed to form from a single, intrinsic ditopic module carrying both a DDA and AAD site. Module **60** forms a hexamer in solution, and **61** has been shown to form a hexamer both in solution [39, 42] and in the solid state [43] (Fig. 22). The advantage of using the AAD · DDA hydrogen-bonding motif is clear in that $(60)_6$ is sufficiently stable to form in competitive solvents such as aqueous tetrahydrofuran [42]. In a similar fashion, modules **57** and **62** were intended to pair with **59** and **63** (Fig. 7), respectively, to form 3 + 3 hexamers [40, 41]. In the latter case, however, the individual units are self-complementary and **62** crystallizes alone, forming a hydrogen-bonded ribbon in the solid state [40].

Larger superstructures can be formed by the subsequent aggregation of discrete assemblies or directly from self-assembling subunits that are designed to form polymeric aggregates. Along these lines, Kimizuka and Kunitake have

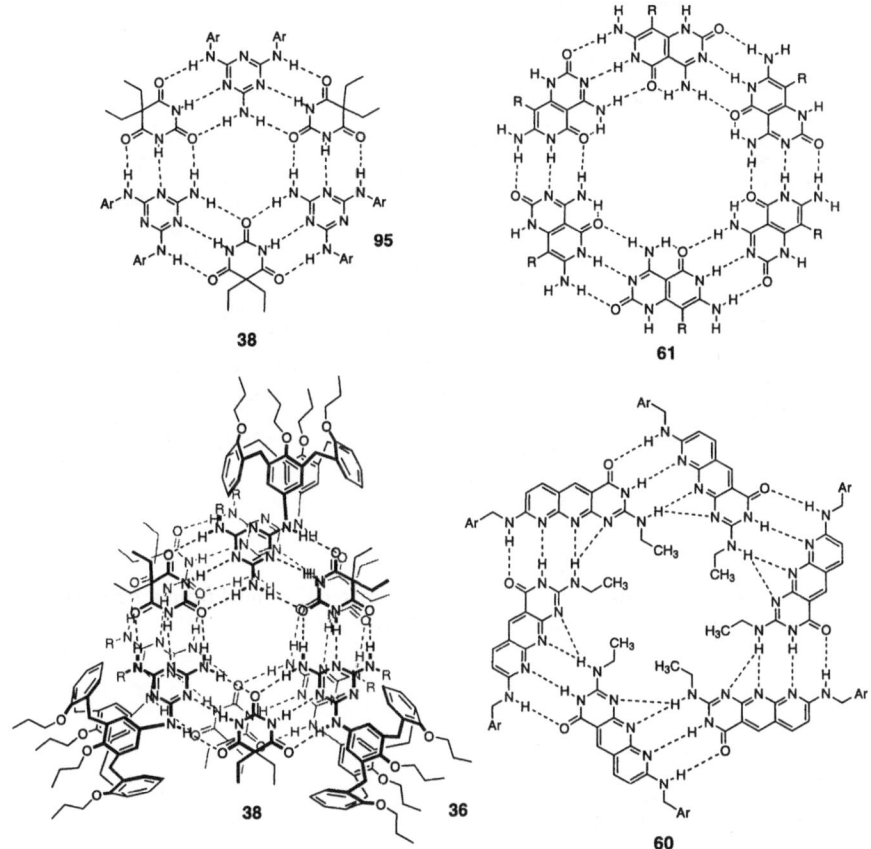

Fig. 22. Cyclic "hexameric" aggregates $(38)_3(95)_3$, $(60)_6$, $(61)_6$, and $(36)_3(38)_6$

studied [82] a 1:1 mixture of **37** and **96** (R = long-chain dialkyl ether) and found evidence by IR and TEM for hydrogen-bonded assemblies appearing as strands with ca. 100 Å diameter (Fig. 23). Three models are consistent with such a structure: (1) tubes made by stacking $(37)_6(96)_6$, (2) helical tubes analogous to the schematic shown in Fig. 23, (3) zigzag ribbons of **37** · **96** packed face-to-face. Such zigzag ribbons have been a popular target for crystal engineers. In this regard, the DAD · ADA motif has been employed quite successfully. Lehn has investigated a barbituric acid · triaminotriazine system [83] and Whitesides has examined a cyanuric acid · melamine pair [79]. As noted above, cyclic structure $(38)_3(95)_3$ is favored by peripheral steric interactions with the *tert*-butylphenyl groups. When the *tert*-butyl groups are replaced with smaller groups, the 1:1 mixture crystallizes as a tape or "crinkled tape" [79]. Thus, as indicated above, the information content in the hydrogen-bonding system is insufficient to control which self-assembled structure is formed.

The DAD · ADA motif has also been used by Kunitake, Ringsdorf, and others for interfacial recognition and assembly [84, 85]. As seen in Fig. 24,

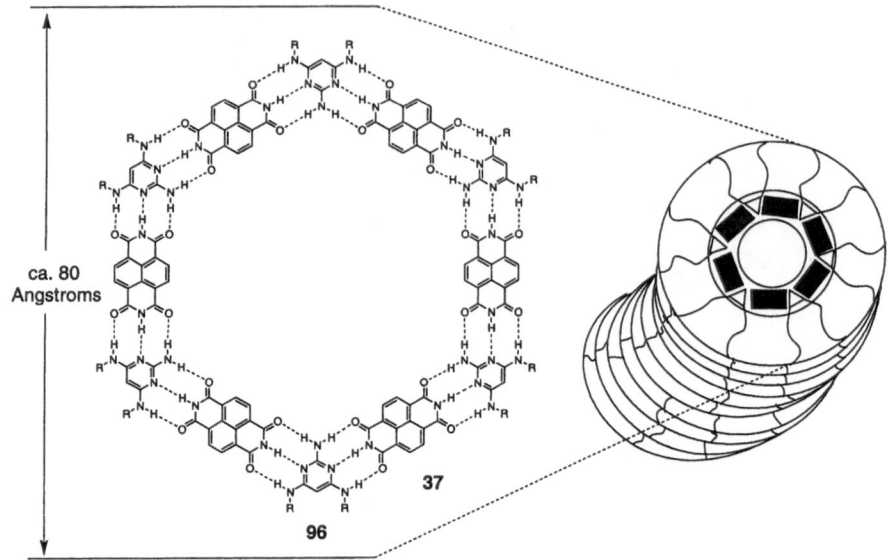

Fig. 23. One possible structure for the self-assembled material formed from a 1:1 mixture of **37** and **96**. $R = C_{12}H_{25}OC_3H_6$- or $C_4H_9(C_2H_5)CH\ C_3H_6$-

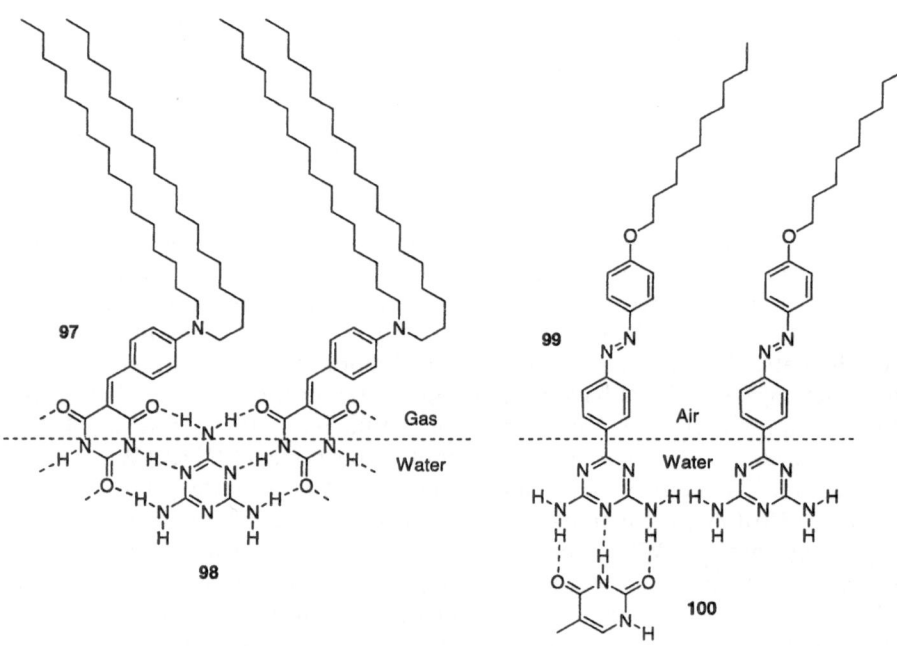

Fig. 24. Interfacial recognition and assembly at the gas–water interface using the DAD · ADA hydrogen-bonding motif

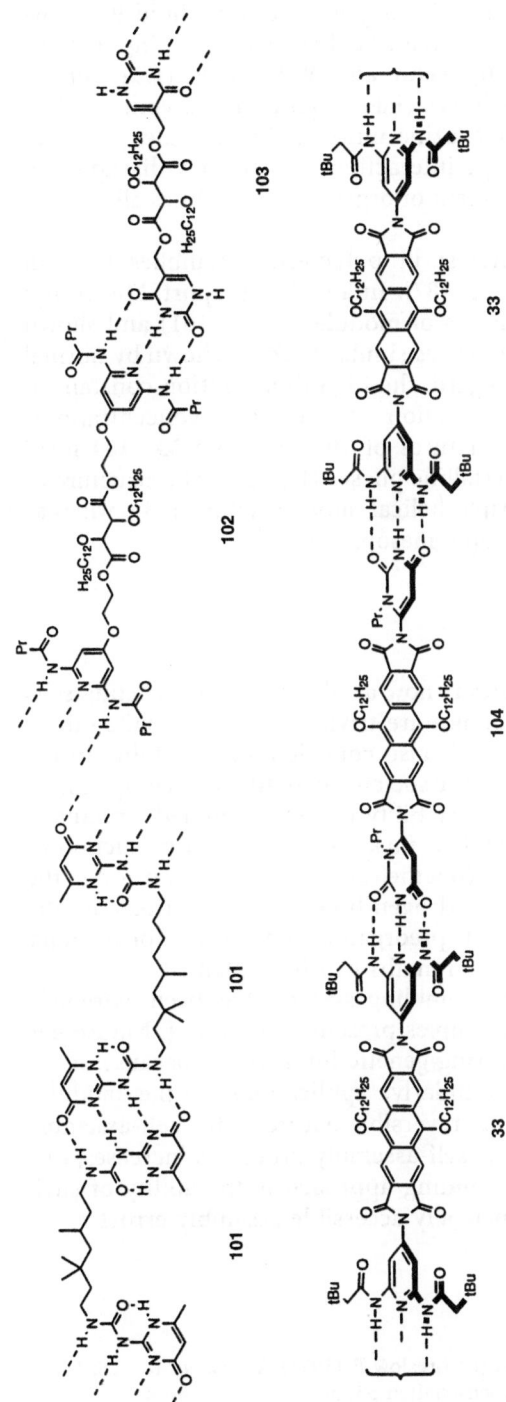

Fig. 25. Supramolecular polymers formed by hydrogen bonding in solution or mesophases

Ringsdorf reported [85] that melamine (98) organizes ditopic lipid 97 at the gas–water interface using the same polymeric hydrogen-bonding pattern found in the 1:1 crystals studied by Lehn and Whitesides (*vide supra*). Kunitake and co-workers studied the interfacial recognition of thymine 100 by a monolayer of amphiphilic diaminotriazine units at the air-water interface [85]. Normally, these base-pairing type interactions are entirely disrupted in aqueous media, but the microenvironment of organized assemblies allows this special recognition to occur [86].

Finally, there is considerable interest in polymeric assemblies both in solution and in liquid crystalline phases [87]. In a seminal report, Meijer and co-workers [49] have synthesized dimers of module 75 (e.g. 101) and shown that its solutions have rheological properties similar to those shown by normal polymer solutions (Fig. 25). In this regard, the high dimerization constant of 75 allows a high degree of polymerization at accessible concentrations. Likewise, Lehn has shown that 1:1 mixtures of 102 · 103 and 33 · 104 form supramolecular, polymeric, liquid crystalline phases (Fig. 25). The structure of 102 · 103 is believed to contain a triple helical superstructure [88], whereas rigid assembly 33 · 104 forms a lyotropic mesophase [89].

5
Conclusions

A broad set of heteroaromatic modules is now available for constructing self-assembling systems. Some of the modules are trivial to prepare while others require rather lengthy syntheses. There is also considerable variability in the strength with which these modules self-associate or bind their complement, which can be useful for fine-tuning the stability of self-assembled structures. Many factors contribute to this variability, making the a priori prediction of complex stability a daunting task. Nonetheless, when one considers the number of primary and secondary hydrogen-bonding interactions in the complexes, as well as the degree of preorganization of complementary modules, most differences in complex strengths can be explained.

The use of these modules in self-assembling systems has been extremely successful, as evidenced by the few examples presented in Sect. 4. Nature has chosen base-pairing as a means of storing genetic information because of the reversibility of hydrogen bonding. Similarly, applications of the modules described herein must benefit from the reversible nature of the self-assembly process. Thus, as the complexity of the self-assembly processes increases, one obvious advantage of the hydrogen-bonding approach is the ability of such systems to automatically edit out kinetically accessible assembly errors.

References

1. (a) Breslow R (1998) Chem Biol 5: R27; (b) Breslow R (1995) Acc Chem Res 28: 146
2. Eigen M, De Maeyer L (1966) Naturwissenschaften 53: 50
3. Lehn JM (1988) Angew Chem 27: 89

4. Zimmerman SC, Murray TJ (1994) In: Wipff G (ed) Computational approaches in supramolecular chemistry. NATO ASI Series, vol 426. Kluwer, Amsterdam, p 109
5. Kyogoku Y, Lord RC, Rich A (1967) Proc Natl Acad Sci USA 57: 250
6. Bell TW, Hou Z, Zimmerman SC, Thiessen PA (1995) Angew Chem Int Ed Engl 34: 2163 and references cited therein
7. Hamilton AD, Pant N, Muehldorf A (1988) Pure Appl Chem 60: 533
8. Rebek J Jr (1990) Angew Chem Int Ed Engl 29: 245
9. Zimmerman SC, Wu W (1989) J Am Chem Soc 111: 8054
10. Kelly TR, Maguire MP (1987) J Am Chem Soc 109: 6549
11. Kirnos MD, Khudyakov IY, Alexandrushikina NI, Vanyushin BF (1997) Nature 270: 369
12. Gryaznov S, Schultz RG (1994) Tetrahedron Lett 35: 2489 and references cited therein
13. Haaima G, Hansen HF, Christensen L, Dahl O, Nielsen PE (1997) Nucleic Acids Res 25: 4639
14. Hamilton AD, Van Engen D (1987) J Am Chem Soc 109: 5035
15. Park TK, Schroeder J, Rebek J Jr (1991) J Am Chem Soc 113: 5125
16. Schneider HJ, Juneva RK, Simova S (1989) Chem Ber 122: 1211
17. Kelly TR, Bridger GJ, Zhao C (1990) J Am Chem Soc 112: 8024
18. Murray TJ, Zimmerman SC (1992) J Am Chem Soc 114: 4010
19. Beijer FH, Sijbesma RP, Vekemans JAJM, Meijer EW, Kooijman H, Spek AL (1996) J Org Chem 61: 6371
20. Adrian JC Jr, Wilcox CS (1991) J Am Chem Soc 113: 678
21. Murray TJ, Zimmerman SC, Kolotuchin SV (1994) Tetrahedron 51: 635 and references cited therein
22. Feibush B, Figueroa A, Rosita C, Onan KD, Feibush P, Karger BL (1986) J Am Chem Soc 108: 3310
23. Hamilton AD, Van Engen D (1987) J Am Chem Soc 109: 5035
24. Leonard NJ, McCredie RS, Logue MW, Cunddall RL (1973) J Am Chem Soc 95: 2320
25. Sessler JL, Wang R (1998) J Org Chem 63: 4079
26. Hamilton AD, Little D (1990) J Chem Soc Chem Commun 297
27. Schall OF, Gokel GW (1994) J Am Chem Soc 116: 6089
28. Kotera M, Lehn J-M, Vigneron JP (1995) Tetrahedron 51: 1953
29. Prins LJ, Huskens J, de Jong F, Timmerman P, Reinhoudt DN (1999) Nature 398: 498
30. Mammen M, Simanek EE, Whitesides GM (1996) J Am Chem Soc 118: 12614
31. Berstein J, Sterns B, Shaw E, Lott WA (1947) J Am Chem Soc 69: 1151
32. (a) Berman A, Izraeli ES, Levanon H, Wang B, Sessler JL (1995) J Am Chem Soc 117: 8252; (b) Sessler JL, Wang B, Harriman A (1993) J Am Chem Soc 115: 10418
33. Roberts C, Bandaru R, Switzer C (1997) J Am Chem Soc 119: 4640
34. Kral V, Sessler JL (1995) Tetrahedron 51: 539
35. Hisatome M, Ikeda K, Kishibata S, Yamakawa K (1993) Chem Lett 1357
36. Chen L, Sakai N, Moshiri ST, Matile S (1998) Tetrahedron Lett 39: 3627
37. Fenlon EE, Murray TJ, Baloga MH, Zimmerman SC (1993) J Org Chem 58: 6625
38. Marsh A, Nolen EG, Gandinier KM, Lehn J-M (1994) Tetrahedron Lett 35: 397
39. Marsh A, Silvestri M, Lehn J-M (1996) Chem Commun 1527
40. Lehn J-M, Mascal M, DeCian A, Fischer J (1992) J Chem Soc Perkin Trans 2: 461
41. Petersen P, Wu W, Fenlon EE, Kim S, Zimmerman SC (1996) Bioorg Med Chem 4: 1107
42. Kolotuchin SV, Zimmerman SC (1998) J Am Chem Soc 120: 9092
43. Mascal M, Hext NM, Warmuth R, Moore MH, Turkenburg JP (1996) Angew Chem Int Ed Engl 35: 2204
44. Meyer H, Bossert F, Horstmann H (1978) Liebigs Ann Chem 1476
45. Bell DA, Anslyn EV (1995) Tetrahedron 51: 7161
46. Beijer FH, Kooijman H, Spek AL, Sijbesma RP, Meijer EW (1998) Angew Chem Int Ed Engl 37: 75
47. Corbin PS, Zimmerman SC (1998) J Am Chem Soc 120: 9710
48. Beijer FH, Sijbesma RP, Kooijman H, Spek AL, Meijer EW (1998) J Am Chem Soc 120: 6761

49. Sijbesma RP, Beijer FH, Brunsveld L, Folmer BJB, Hirschberg JHK, Lange RFM, Lowe JKL, Meijer EW (1997) Science 278: 1601
50. Pfleiderer M, Pfleiderer W (1992) Heterocycles 33: 905
51. Corbin PS, unpublished results
52. Sessler JL, Wang R (1998) Angew Chem Int Ed Engl 37: 1726
53. Lüning U, Kühl C (1998) Tetrahedron Lett 39: 5735
54. Corbin P, Zimmerman SC, submitted for publication
55. Jorgensen WL (1998) J Phys Chem 102: 3782
56. Hopkins HP, Alexander CJ, Ali SZ (1978) J Phys Chem 82: 1268
57. Schneider HJ, Juneva RK, Simova S (1989) Chem Ber 122: 1211. This analysis has been updated to include secondary hydrogen bonding (see [61])
58. Jorgensen WL, Pranata J (1990) J Am Chem Soc 112: 2008; Pranata J, Wierschke SG, Jorgensen WL (1991) J Am Chem Soc 113: 2810
59. Zimmerman SC, Murray TJ (1994) Tetrahedron Lett 35: 4077
60. Gardner RR, Gellman SH (1995) J Am Chem Soc 117: 10411
61. Sartorius J, Schneider HJ (1996) Eur J Chem 2: 1446
62. Cram DJ (1986) Angew Chem Int Ed Engl 25: 1039
63. Cram DJ, Trueblood KN (1985) in Vögtle F, Weber E (eds) Host guest complex chemistry, macrocycles. Springer, Berlin Heidelberg New York, p 125
64. Zimmerman SC, Mrksich M, Baloga M (1989) J Am Chem Soc 111: 8528
65. Page MI, Jencks WP (1971) Proc Natl Acad Sci USA 68: 1678
66. Searle MS, Williams DH (1992) J Am Chem Soc 114: 10690
67. Eblinger F, Schneider HJ (1998) Angew Chem Int Ed Engl 37: 826. This paper, which claims to report the first study wherein the number of free rotations within supramolecular complexes are systematically varied (see however [64]), studies the interaction between dicarboxylates and diamide. It is complicated by the fact that rigid and flexible dicarboxylates may bind the diamides in different orientations
68. Hamilton AD, Pant N (1998) J Chem Soc Chem Commun 765
69. Murray TJ, Zimmerman SC (1995) Tetrahedron Lett 36: 7627
70. Watson JD, Crick FHC (1953) Nature 171: 964
71. Douhal A, Kim SK, Zewail AH (1995) Nature 378: 260
72. Robinson H, Gao YG, Bauer C, Roberts C, Switzer C, Wang AHJ (1998) Biochemistry 37: 10897
73. Goodman MF (1995) Nature 378: 237
74. Beak P (1977) Acc Chem Res 10: 186
75. Fersht AR (1987) Trends Biochem Sci 12: 301
76. Fan E, van Araman S, Kincaid S, Hamilton AD (1993) J Am Chem Soc 115: 369; Wilcox CS, Kim EI, Romano D, Kuo LH, Burt AR, Curran DP (1995) Tetrahedron 55: 621
77. Hamilton AD, Fan E, Van Arman S, Vicent C, Tellado FG, Geib SJ (1993) Supramol Chem 1: 247
78. Drain CM, Fischer R, Nolen EG, Lehn J-M (1993) J Chem Soc Chem Commun 243
79. Zerkowski JA, Seto CT, Whitesides GM (1992) J Am Chem Soc 114: 5473
80. Seto CT, Whitesides GM (1993) J Am Chem Soc 115: 905
81. Crego Calama M, Fokken R, Nibbering NMM, Timmerman P, Reinhoudt DN (1998) J Chem Soc Chem Commun 1021
82. Kimizuka N, Kawasaki T, Hirata K, Kunitake T (1995) J Am Chem Soc 117: 6360
83. Lehn J-M, Mascal M, DeCian A, Fischer J (1990) J Chem Soc Chem Commun 479
84. Ariga K, Kunitake T (1998) Acc Chem Res 31: 371 and references cited therein
85. Bohanon TM, Denzinger S, Fink R, Paulus W, Ringsdorf H, Weck M (1995) Angew Chem 34: 58
86. Nowick JS, Cao T, Noronha G (1994) J Am Chem Soc 116: 3285
87. Zimmerman N, Moore JS, Zimmerman SC (1998) Chem Ind 604
88. Gulick-Krymicki T, Fouquey AM, Lehn J-M (1993) Proc Natl Acad Sci 90: 163
89. Kotera M, Lehn J-M, Vigneron JP (1994) J Chem Soc Chem Commun 197

Hydrogen-Bonded Liquid Crystals: Molecular Self-Assembly for Dynamically Functional Materials

Takashi Kato

Department of Chemistry and Biotechnology, Graduate School of Engineering,
The University of Tokyo, Hongo, Bunkyo-ku, Tokyo 113-8656, Japan
E-mail: kato@chiral.t.u-tokyo.ac.jp

Liquid crystals are molecular materials which combine anisotropy with a dynamic nature. This unique character can lead to the fabrication of new dynamically functional materials. A new class of liquid crystals prepared by self-assembly processes through the formation of intermolecular interactions such as hydrogen bonding would be key materials for such an approach. Supramolecular hydrogen-bonded liquid-crystalline complexes have been prepared from different and independent molecules such as carboxylic acids and pyridines. A variety of molecular structures as well as polymeric complexes have been obtained by the self-association of hydrogen-bonded components. Hydrogen-bonded complexes consisting of identical molecules such as benzoic acids, amides, and polyols also form liquid-crystalline aggregates which exhibit a variety of ordered structures. In addition to these materials which exhibit homogeneous mesophases, anisotropic liquid-crystalline gels which consist of heterogeneous structures are obtained as a new mesomorphic material. These self-assembled liquid crystals could bridge the disciplines of materials science and supramolecular chemistry.

Keywords: Hydrogen bonding, Liquid crystal, Liquid crystal polymer, Self-assembly, Self-organization, Complex, Gel, Supramolecular chemistry

1
Introduction

Liquid crystals are unique molecular materials because of their anisotropic nature and molecular dynamics [1–4]. Over the last three decades, these materials have been developed as advanced materials for electrooptical applications such as display devices. Liquid crystals also have close relationships to biomolecular systems [5]. Cell membranes form dynamic and anisotropic molecular states, which possess liquid-crystalline behavior. Recently, liquid-crystalline complexes of DNAs and liposomes have been considered as potential systems for gene therapy [6]. The design of liquid crystals by using a variety of structures and interactions may lead to wider applicability of mesomorphic materials.

The use of molecular interactions such as hydrogen bonding for the design of molecular self-assembled systems, e.g. liquid crystals [7–11], molecular solids [12], and polymeric assemblies [8, 13], has attracted attention because these materials involving non-covalent interactions have potential for dynamically functional molecular systems. Hydrogen bonding is one of the important interactions in chemical and biological processes. In nature, hydrogen bonding plays a key role in molecular recognition and molecular assembly because of its stability, dynamics, directionality, and reversibility [14, 15]. Hydrogen-bonding interactions can be expressed as $-X\cdots H-Y-$ (X, Y = O, N, etc.). The bonding energy of normal H-bonding is between 10 and 50 kJ/mol, while the energy of covalent bonding is about 500 kJ/mol and that of van der Waals interaction is about 1–0.1 kJ/mol.

In thermotropic liquid crystals, most of the mesogenic molecules consist only of covalent bonds. However, the history of liquid crystals and hydrogen bonding dates back to the early part of the 20th century when some organic acids [16] and carbohydrates [17, 18] were reported to show mesomorphism. For example, the dimer formation through hydrogen bonding between two identical benzoic acids leads to a mesogenic structure. In the middle of the 20th century the separated association of hydroxyl groups and the hydrophobic part of sugars with a long alkyl chain was also found to be mesomorphic

[17], although Fisher had observed double melting behavior [18]. However, the versatility of the use of hydrogen bonding for the design of liquid-crystalline materials has only been recognized recently [9–12].

Intermolecularly hydrogen-bonded liquid-crystalline materials are classified into three types. Liquid-crystalline molecular associates formed through intermolecular hydrogen bonding between *identical* components are shown in Fig. 1 as type A-I, a-c. The first type (A-I-a) is a type of benzoic acid dimer where molecules form complexes with well-defined closed structures [16]. Examples of the second type (A-I-b) are amide compounds that form one-dimensional H-bonded chains [17]. Polyols such as sugars [18–21], inositols [22], and diols [23, 24] are the third type (A-I-c) which show liquid-crystalline molecular association by organizing hydrogen-bonded layers separating from hydrophobic parts.

A novel type of supramolecular hydrogen-bonded liquid-crystalline materials that are conceptually different from conventional H-bonded systems such

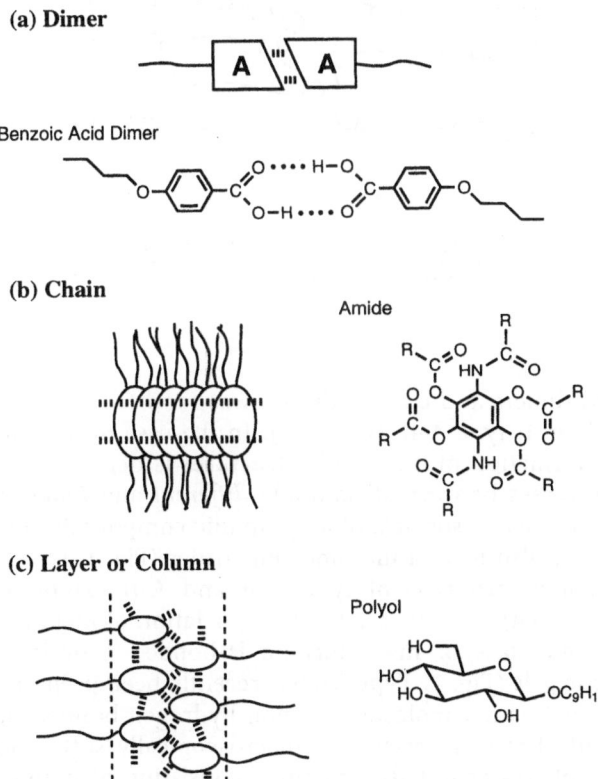

Type A-I Association of *Identical Molecules* (*Homogeneous* Phases)

(a) Dimer

Benzoic Acid Dimer

(b) Chain

Amide

(c) Layer or Column

Polyol

Fig. 1. Classification of H-bonded liquid-crystalline materials exhibiting *homogeneous* phases: Type A-I: association of identical molecules; Type A-II: association of different molecules

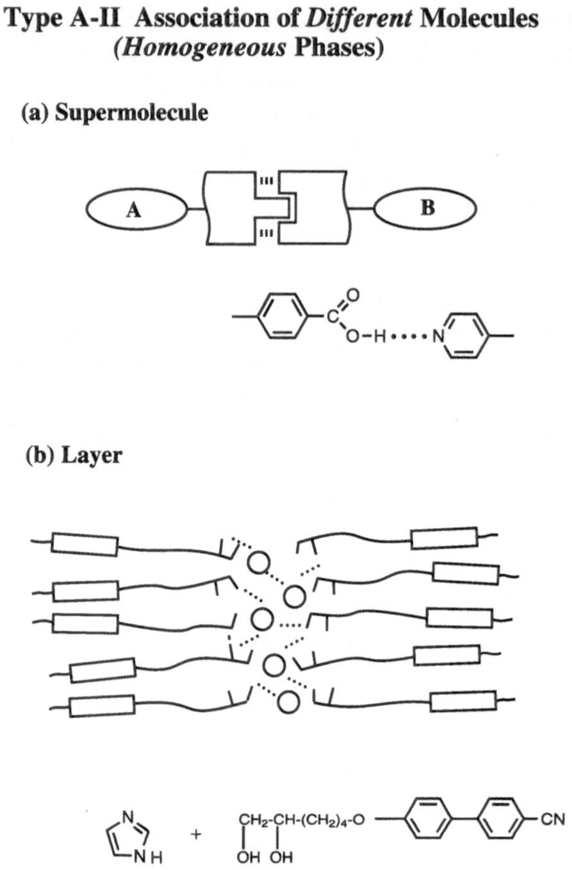

Type A-II Association of *Different* Molecules
(*Homogeneous* Phases)

(a) Supermolecule

(b) Layer

Fig. 1. (*Contd.*)

as benzoic acid dimers and carbohydrates, was reported to show mesomorphism in 1989 (Fig. 1, type A-II-a) [25–27]. In the case of the supramolecular mesogens, the formation of mesogenic structures is derived from molecular-recognition processes between structurally *different* and *independent* molecular species. Molecular associates of amphiphilic compounds and hydrophilic organic molecules also form a mesomorphic order (Fig. 1, type A-II-b) [28].

Liquid-crystalline materials of types A-I and A-II exhibit *homogeneous* (non-phase separated) mesophases by the association of identical or different molecules. A new class of mesomorphic H-bonded materials, anisotropic liquid-crystalline gels (Fig. 2, Type B), has recently been prepared by the self-aggregation of H-bonded molecules in non-hydrogen-bonded normal liquid crystals [29, 30]. These materials are macroscopically soft solids and form *heterogeneous* (phase-separated) structures consisting of liquid crystals and fibrous solids.

In this review, the design, preparation, and functionalization of hydrogen-bonded liquid-crystalline materials are described. These novel systems are

Type B Association of *Different Molecules* (*Heterogeneous* Phase)

Liquid-Crystalline Anisotropic Gel

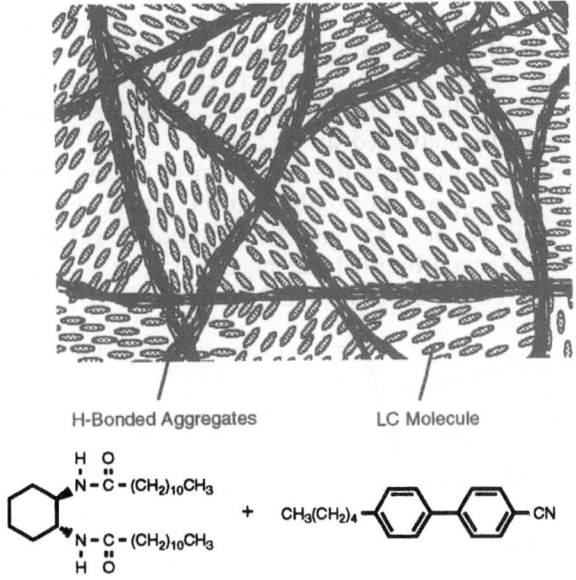

H-Bonded Aggregates   LC Molecule

Fig. 2. Classification of H-bonded liquid-crystalline materials: Type B: microphase-separated *heterogeneous* systems

now a growing area because the use of molecular interactions such as hydrogen bonding for anisotropic molecular self-organization may form a bridge between synthetic materials and biofunctional systems.

2
Supramolecular Assemblies Consisting of Different Molecules

2.1
Low-Molecular-Weight Complexes

2.1.1
Hydrogen-Bonded Mesogens

A new concept of molecular architecture for liquid-crystalline materials has been shown by the use of intermolecular hydrogen bonding that connects different molecular units at desired places [9–11].

One of the first supramolecular H-bonded mesogenic structures derived from *independent* and *different* molecular species is shown in Fig. 3 [25]. Mesogenic complex 1, which has a well-defined structure, is obtained by 1:1 association of 4-butoxybenzoic acid (2, $n = 4$) and *trans*-[4-ethoxy(benzoyl)oxy]-4′-stilbazole (3, $n = 2$) moieties through *hetero*-intermolecular hydrogen bonding between carboxylic acid and pyridyl moieties. In this molecular recognition process, the benzoic acid unit functions as the H-bonding donor and the pyridyl moiety serves as the H-bonding acceptor. The thermal stability of the mesophase is enhanced by the formation of the H-bonded mesogen. Hydrogen-bonding components of 2 ($n = 4$) and 3 exhibit nematic phases up to 147 and 213 °C, respectively. Once the 1:1 complex is formed from these molecular units, the smectic A phase appears and the nematic phase is thermally stabilized up to 238 °C.

Another representative system is the induction of mesomorphism by the self-assembly of nonmesogenic H-bonding components. Complexation of 4-methoxybenzoic acid (2, $n = 1$) and 4,4′-bipyridine (4) results in the formation of novel hydrogen-bonded mesogen 5 [:2 ($n = 1$)/4] (Fig. 4). A nematic phase is observed for 5 between 153 and 163 °C [31, 32].

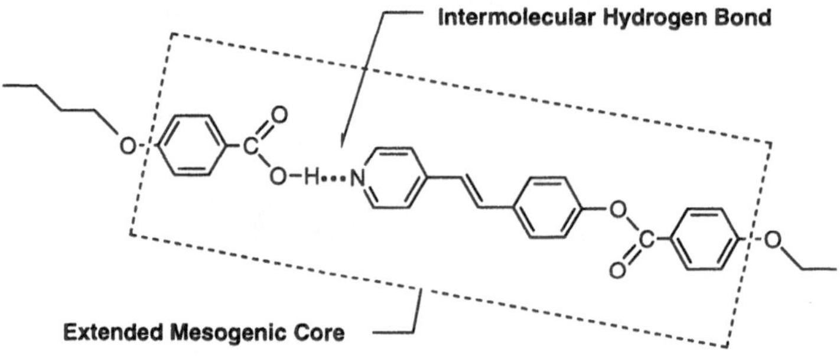

Fig. 3. Supramolecular liquid-crystalline complex 1 obtained by the formation of hydrogen bonding between benzoic acid and pyridine moieties

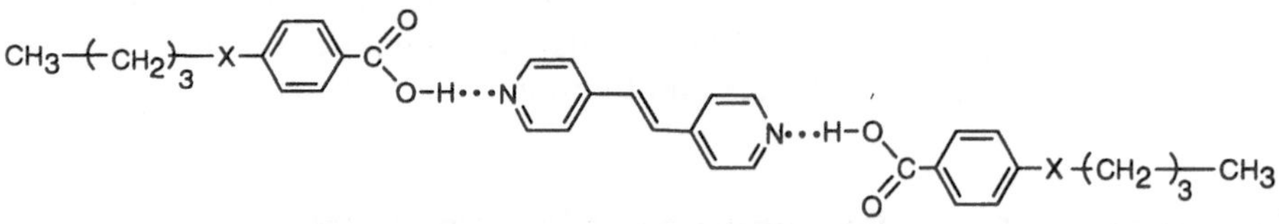

5 (n=4): -X- = -O-
6: -X- = -CH₂-

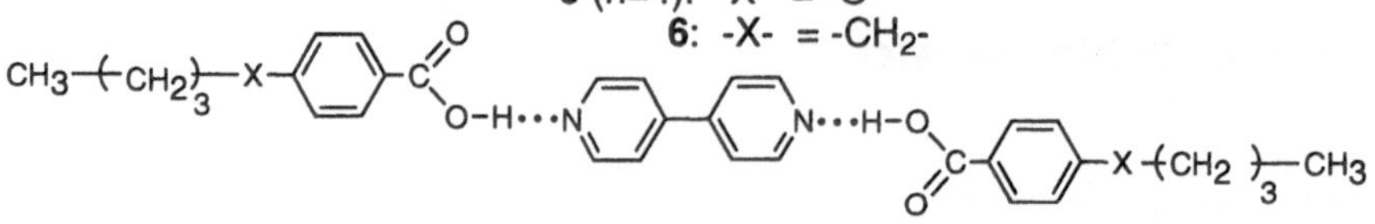

7: -X- = -O-
8: -X- = -CH₂-

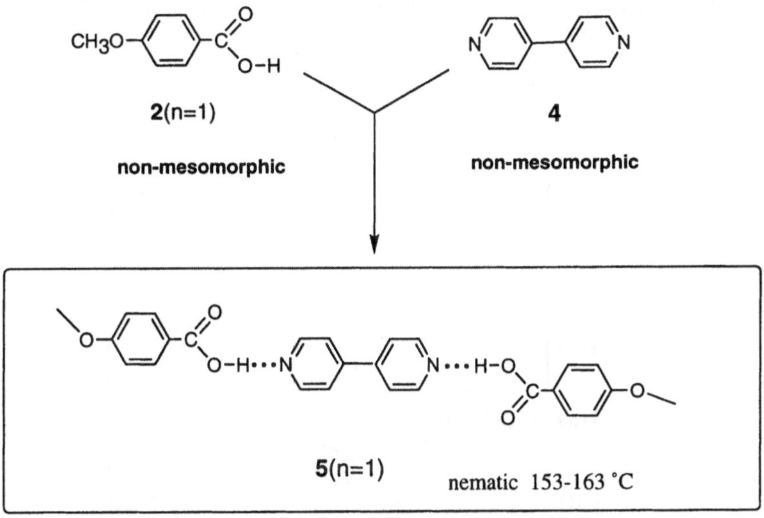

Fig. 4. Self-assembly process of benzoic acid **2** and bipyridine **4**, and induction of a mesophase by the formation of a supramolecular mesogen

It is interesting that the nonmesogenic component functions as a central core unit through the noncovalent interaction. The effect of molecular structures on liquid-crystalline behavior of the complexes based on benzoic acids and bipyridines is shown in Table 1. These complexes exhibit stable liquid crystallinity. For example, complex **5** ($n = 4$) exhibits smectic and nematic phases. When the oxygen of the alkoxy group is replaced by a methylene group, the value of T_i decreases and no nematic phase is observed for **6**. The T_i of the mesophase displayed by the complex based on *trans*-1,2-bis(4-pyridyl)ethylene **9** is higher than that of the complex based on 4,4'-bipyridine, which can be ascribed to the effect of the more planar structure of **5** that serves as the central core unit of the H-bonded mesogen.

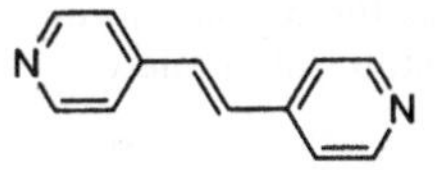

9

The relationship between the length of the alkyl end group and mesomorphic properties for complex **10** derived from *trans*-4-alkoxy-4'-stilbazole and 4-hexyloxybenzoic acid (**2**, $n = 6$) is shown in Fig. 5 [33, 34].

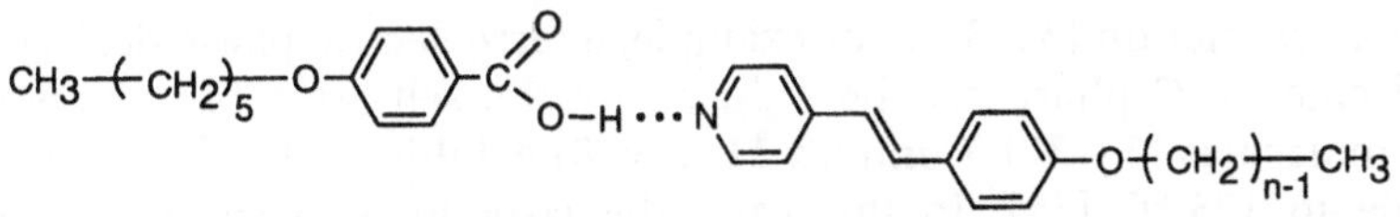

10

The type of mesophase is dependent on the alkyl chain length. Only a nematic phase is observed for the complex with $n = 1$ that consists of nonmesogenic molecules, while only smectic phases appear for the complex with long alkyl

Table 1. Liquid-crystalline behavior of supramolecular 2:1 hydrogen-bonded complexes derived from 4-butoxybenzoic acid and bipyridines[a]

Complex	Cr		S_A		N		I
5	–	146	–	168	–	178	–
6	–	132	–			167	–
7	–	134	–	150	–	159	–
8	–	110	–			153	–

[a] Cr: crystalline; S_A: smectic A; N: nematic; I: isotropic.

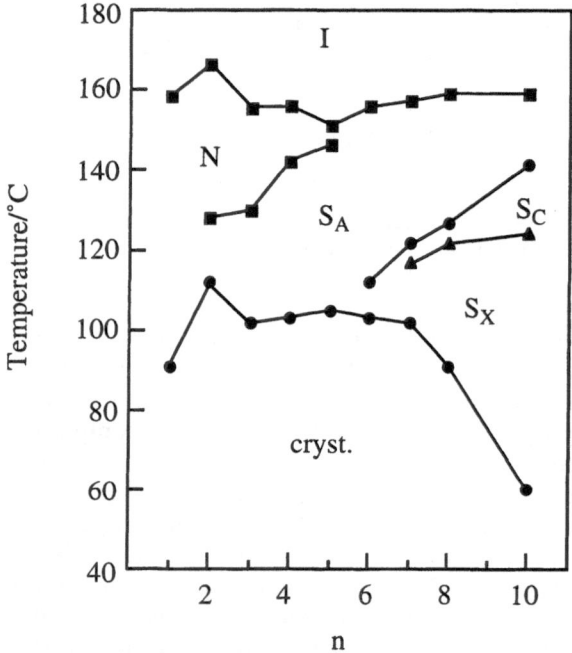

Fig. 5. Plot of transition temperatures for supramolecular liquid-crystalline complex **10** as a function of chain length (*n*) of the stilbazole moiety

chains (*n* = 6–8, 10). However, T_i does not change significantly as the chain length increases. A similar trend is observed for a series of complexes of benzoic acids and bipyridines [31].

The use of chiral H-bonding components yields stable chiral phases by molecular association [35–37]. For example, a ferroelectric phase derived from a chiral smectic C phase can be induced by the self-assembly of nonferroelectric molecules [35–37]. Complex **11** (*n* = 8) exhibits a chiral smectic phase from 109 to 123 °C [35]. In this case, the H-bonded mesogen consists of nonmesomorphic chiral benzoic acid and a stilbazole that shows a very narrow temperature range of ordered smectic phases. The value of spontaneous polarization for **11** at 115 °C is 33.0 nC/cm². When the oxygen in the alkoxy group is replaced by an oxycarbonyl linkage, the S_C^* phase appears between 100 and 125 °C for complex **12** [36]. Complexes containing a chiral stilbazole also exhibit ferroelectric behavior [37]. An antiferroelectric phase has recently been observed for an H-bonded chiral complex consisting of four aromatic rings [38].

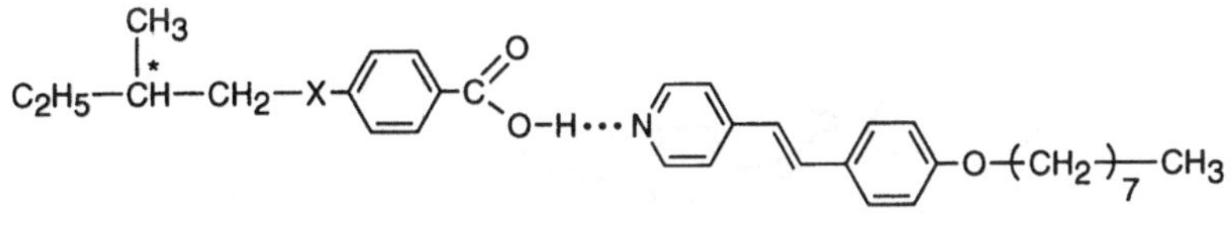

11: -X- = -O-

12: -X- = -O-C-
                (O)

Room-temperature liquid-crystalline (LC) phases have been achieved for two-ring complexes **13** and **14** [39–41]. Complexes **13** ($n = 8$–12) show nematic phases below 50 °C. In contrast, complexes **14** exhibit only room-temperature smectic phases. The nematic compounds show twisted nematic (TN) molecular alignment and they are responsive to electric fields [40, 41]. The response of a room-temperature LC complex to applied electric fields on modified surfaces has been examined by time-resolved infrared (IR) spectroscopy [41]. The response of the interfacial complex is smoothly transferred to the complex in the bulk layer due to the dynamic behavior of the hydrogen bonding. The molecular dynamics of the hydrogen-bonded systems are interesting because these mesogens are expected to be *softer* than conventional mesogens [41].

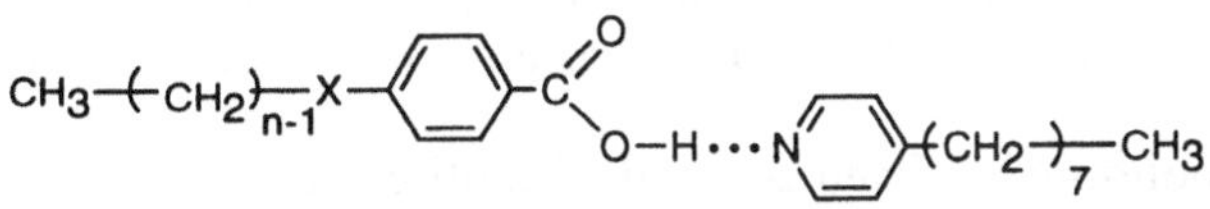

13: X= -O-
14: X= -CH₂-

The mesomorphic properties of hydrogen-bonded mesogens have been compared [40] with those of covalently bonded compounds **16** and **17** [42] (Fig. 6 and Table 2). The melting and isotropization temperatures of the complexes are lower than those of ester compounds, which leads to wider and lower mesomorphic temperature ranges [40]. Moreover, smectic mesophases are observed for fluorinated H-bonded complex **15**.

Hydrogen bonding between carboxylic acid and pyridine has been widely applied for the formation of a variety of mesogenic structures [43–54]. Components of dicarboxylic acid have been used for the complexation [43–45]. A twin mesogenic structure shows an odd-even effect on transition

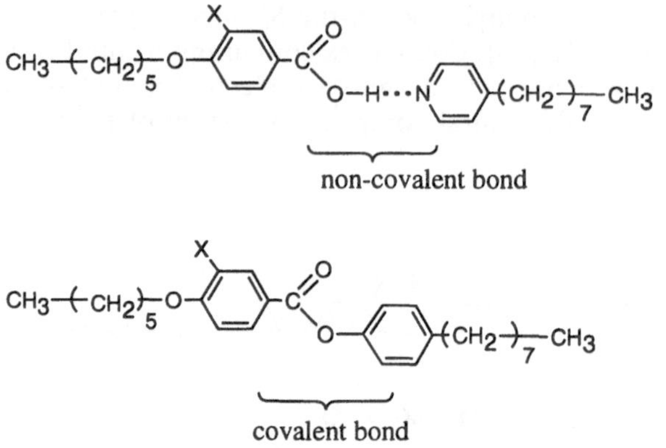

Fig. 6. Structures of hydrogen-bonded complexes and covalently bonded aromatic esters

Table 2. Comparison of transition temperatures of hydrogen-bonded liquid-crystalline complexes and normal liquid crystals[a]

Complex	Cr		S_C		S_A		N		I
13	–	38					–	48	–
15	–	28	–	31	–	33	–	39	–
16	–	43			–	61	–		
17	–	57					–	(45)	–

[a] Cr: crystalline; S_C: smectic C; S_A: smectic A; N: nematic; I: isotropic.

temperatures [42]. A stilbazole calix[4]arene and benzoic acids have been complexed for the formation of a mesomorphic assembly by molecular stacking [46].

Hydrogen bonding has been coupled with other molecular interactions. The incorporation of perfluorinated alkyl chains into the end group of the complexes leads to stabilization of the mesophase [47]. The combination of an ionic interaction and H-bonding has been introduced into complex structures [48].

The interaction between phenol and pyridine has also been used to form mesomorphic complexes **18–20** [55–57]. A 1:1 mixture of 4-cyanophenol and a 4-alkoxy-4′-stilbazole shows a monotropic nematic phase, while the stilbazole exhibits a very narrow temperature range of ordered smectic phases [58] and 4-cyanophenol is nonmesogenic. Complexes **19**, based on 3-cyanophenol, show more thermally stable mesophases than those based on 4-cyanophenol because molecular linearity is well retained for the meta-substituted system [56]. Complexes **20**, derived from 2,4-dinitrophenol and a stilbazole, show smectic A phases [57]. Induction of smectic phases has also been observed for a mixture of a bipyridine compound and 4-alkoxyphenol [59].

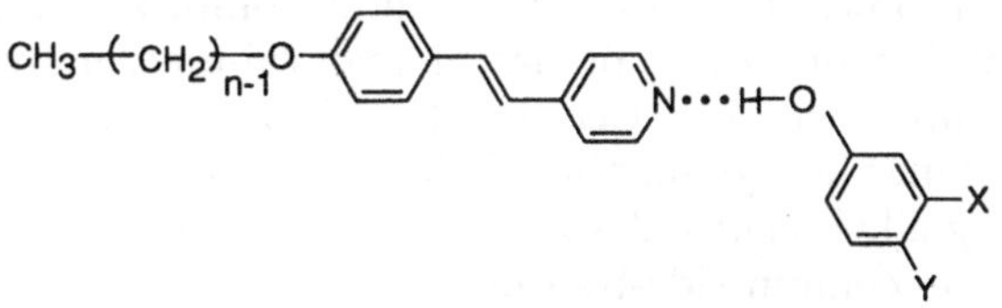

18: X=H, Y=CN
19: X=CN, Y=H

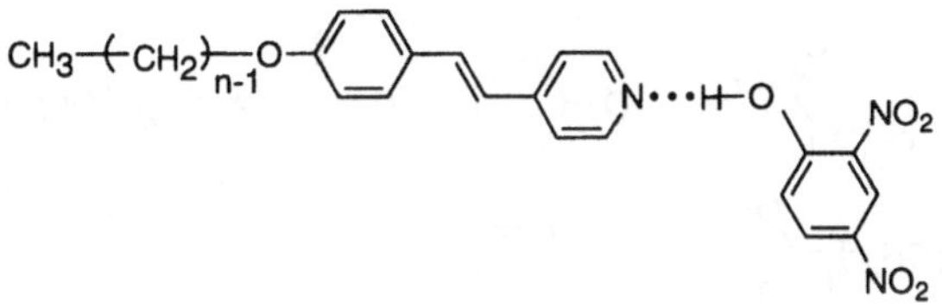

20

Double hydrogen bonding has been used for molecular recognition and organization [60]. Doubly H-bonded supramolecular liquid-crystalline complexes based on 2,6-bis(acylamino)pyridines show unique mesomorphic behavior. Complexes **21** exhibit monotropic crystal B phases [61, 62].

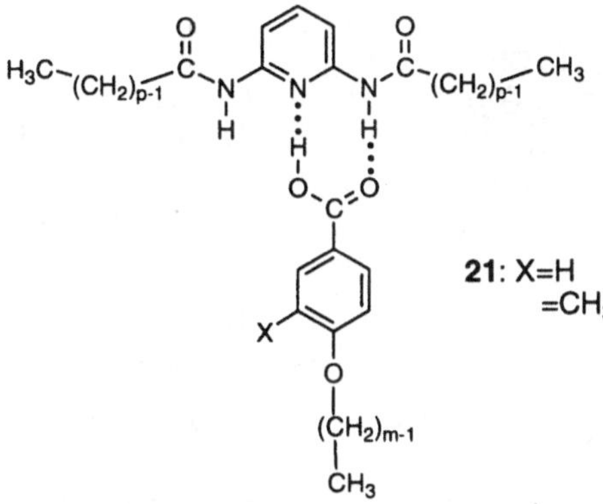

21: X=H
=CH$_3$

As this supramolecular mesogen does not form a simple rod or disc shape, the molecular arrangements of calamitic and columnar mesophases can be tuned when appropriate substituents are introduced into either of the components [61, 63]. Triazines and carboxylic acids are complexed to form doubly H-bonded mesogens which display columnar mesophases, e.g. **22** [64, 65].

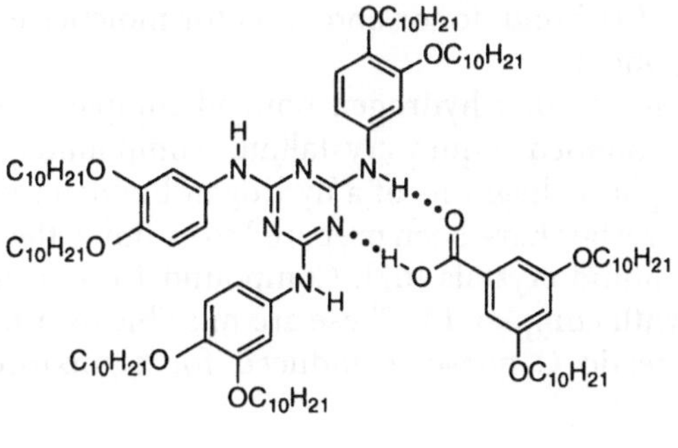

22

Triply hydrogen-bonded 1:1 liquid-crystalline complexes **23** are formed by molecular-recognition processes of uracyl and 2,6-diaminopyridine derivatives [27]. The complex with R = CH$_3$, k = l = 10, m = 11, and n = 16 exhibits a metastable columnar hexagonal phase below 72 °C. In this case, the complex contains four long alkyl chains. An X-ray study has shown that two complex molecules form one column side-by-side.

23

The combination of a host-guest interaction and an intermolecular hydrogen bond induces a bowlic columnar mesophase for calix[4]arene **24** and diformamide **25** [66]. In this case, compound **25** in the core of **24** binds these two calix[4]arenes through a noncovalent interaction.

24

25

As the H-bonded mesogenic core consists of *independent* and *different* molecules, nonstoichiometric mixtures can be prepared from H-bond donor and acceptor compounds [31, 32]. Figure 7 shows the phase diagram of 4-butoxybenzoic acid **2** (n = 4) and *trans*-1,2-bis(4-pyridyl)ethylene **9**. Significant positive deviation is seen for the isotropization temperature. The isotropization temperature becomes the highest at the point where an equimolar amount of H-bond donor and acceptor moieties exists and behaves as one single component.

The question as to whether hydrogen-bonded complexes are miscible with normal covalently bonded liquid-crystalline compounds arises for their application. Binary phase diagrams of a hydrogen-bonded mesogenic complex with normal liquid crystals have been prepared to examine the miscibility of two different classes of liquid crystals [67]. Compound **16** or 4'-heptyl-4-cyanobiphenyl was mixed with complex **14**. These are miscible over the whole range of composition. A smectic C phase is induced for a mixture of **16** and the cyanobiphenyl.

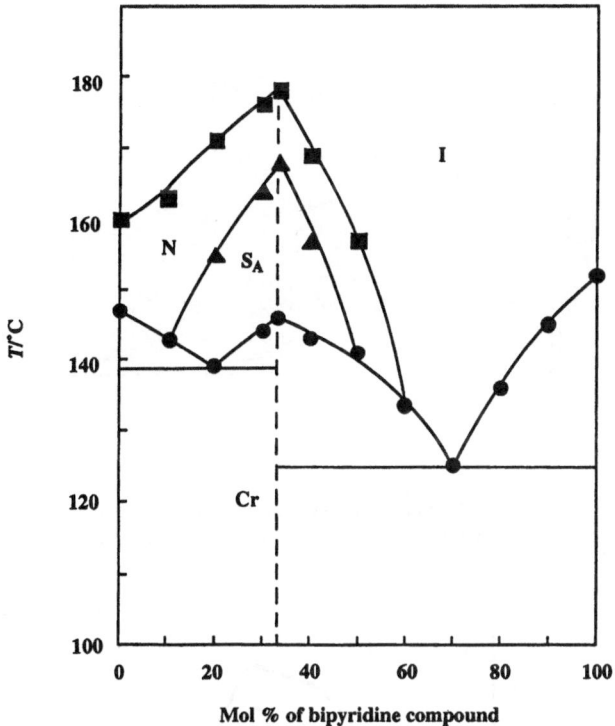

Fig. 7. Binary phase diagram for mixtures of hydrogen-bonding components of 4-but-oxybenzoic acid (2, *n* = 4) and *trans*-1,2-bis(4-pyridyl)ethylene 9

2.1.2
Stability of Hydrogen Bonding

The stability of hydrogen bonding leading to well-defined molecular structures has been examined. The energies of the hydrogen bonding between pyridines and carboxylic acids are estimated to be about 45 kJ/mol [31, 34, 68–70]. The infrared (IR) spectra of the pyridine/carboxylic acid complexes show that the O-H stretching band appears at ca. 2500 cm^{-1}, while carboxylic acid dimers show the O-H band at 3000 cm^{-1}. Variable-temperature IR spectra have been obtained for the H-bonded complexes [31, 34]. Figure 8 shows the IR spectra of complex 5 (n = 4) in the range 1500–1800 cm^{-1} between 100 and 180 °C [31]. The spectra show that the stability of the H-bond is not simply a function of temperature. The band abruptly changes at the phase transition temperature. Curve fitting has been performed for the carbonyl bands. Figure 9 shows the results of the bands at 120 °C (crystalline), 140 °C (smectic A) and 175 °C (isotropic) [31]. In the crystalline state (Fig. 9A), the band is separated into two bands at 1686 and 1703 cm^{-1}, which indicates a double minimum potential of this hydrogen bonding. The single crystal structure of an H-bonded complex analyzed by X-ray spectroscopy indicates that an equimolar complex quantitatively forms in the crystalline state [50]. The fraction of the

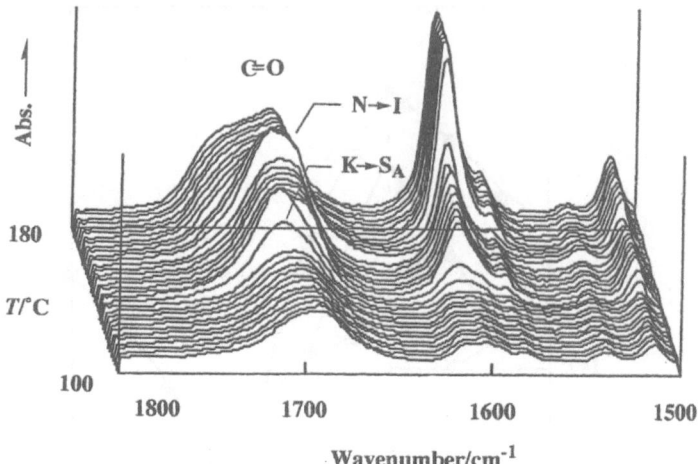

Fig. 8. Variable-temperature IR spectra of supramolecular hydrogen-bonded complex **5** ($n = 4$) consisting of 4-butoxybenzoic acid **2** and bipyridine **4** in the range 1500–1800 cm^{-1} from 100–180 °C

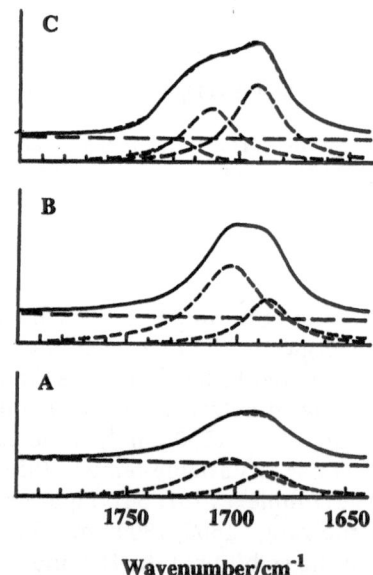

Fig. 9A–C. Curve fitting results of the carbonyl bands of supramolecular hydrogen-bonded complex **5** ($n = 4$). A 120 °C (crystalline); B 140 °C (smectic); and C 175 °C (isotropic)

band at 1703 cm^{-1} increases in the liquid-crystalline state (Fig. 9B). No carbonyl band due to the dissociation of the H-bonding is observed in the crystalline and liquid-crystalline states. Once the temperature of the complex has reached the isotropization temperature at 159 °C, the carbonyl band free of the hydrogen bonding appears as a shoulder peak at 1728 cm^{-1} (Figs. 8 and 9C). In contrast, for simple thermotropic benzoic acid dimers, the dissociated monomer is observed both in the crystal and liquid-crystalline phases [71].

The combination of the stabilization effect by molecular orientation in liquid crystals, the geometry effect, and the relatively strong bonding energy of the carboxylic acid/pyridine interaction, leads to the formation of supramolecular liquid-crystalline materials which exhibit stable mesomorphic behavior. The stability and molecular orientation for H-bonded systems have been discussed theoretically [72–74].

A stabilizing effect by the molecular orientation is observed for complex 20, which exhibits a smectic A phase. Variable-temperature electronic spectra of the complex show that the ionic hydrogen-bonded state by proton transfer is stabilized in the mesophase [57].

2.1.3
Amphiphilic Assemblies

The molecular self-organization of an amphiphilic compound with imidazole induces smectic layer structures [28]. Mesogenic compounds 26 and 27 contain dihydroxy and monohydroxy moieties at the extremity of their alkyl terminal groups, respectively. These compounds show only nematic phases. However, a mixture of compound 26 and imidazole forms a self-organized smectic A structure over a wide range around room temperature. A molecular self-assembly of type A-II-b (Fig. 1) is induced for the mixture. No induction of a smectic phase is seen for 27 and imidazole 28.

26

27

28

2.2
Polymeric Complexes

2.2.1
Design of Polymeric Structures

A wide variety of polymeric liquid-crystalline structures have been prepared by the synthetic formation of covalent bonding [3, 75]. Recently, the use of noncovalent interactions has been recognized to be a versatile approach to the design of functional polymeric materials such as mesogenic polymers. Dynamic functions such as photo and electric properties, molecular sensing, molecular information, and catalytic activities, can be introduced into these systems because biomacromolecules, such as nucleic acids, polypeptides and polysaccharides, all have H-bonding functional groups that play a critical role in biological processes and the formation of supramolecular structures.

The approach to the construction of well-defined mesogenic polymer structures through the formation of *hetero*-intermolecular hydrogen bonding between *different* molecular species was first demonstrated by Kato and Fréchet who in 1989 prepared a supramolecular liquid-crystalline side-chain polymer [26]. Lehn and co-workers reported polymeric main-chain mesogenic polymers which were obtained by the complexation of two bifunctional moieties [76]. Since then, a wide variety of polymeric complexes of side-chain, main-chain, combined, and network structures (Fig. 10) have been prepared by the formation of hydrogen bonding [7–11]. Hydrogen bonding has also been shown to be useful for liquid-crystalline polymer blends by the increase in miscibility [77]. In the following sections, supramolecular polymers and their related systems are described.

2.2.2
Side-Chain Type

An initial approach to supramolecular H-bonded mesogenic polymer complexes involves a polyacrylate with 4-oxybenzoic acid moieties via a hexamethylene spacer **29** [26]. The 1:1 complexation of the side chain of the polymer and stilbazole **3** ($n = 2$) (nematic, 168–216 °C) results in the formation of an extended supramolecular mesogen in the side chain (Fig. 11). Side-chain polymer complex **30** exhibits a nematic phase up to 252 °C, which shows that a significantly stabilized mesophase is achieved by the complexation of two different components. Liquid-crystalline properties have been examined for the series of complexes formed between polyacrylates and *trans*-4-alkoxy-4'-stilbazoles [33, 78]. Figure 12 shows transition temperatures against the carbon number of the alkyl chain for the series of complexes **31** [33]. They exhibit thermally stable smectic liquid-crystalline phases. For example, smectic E, B, and A phases are observed until 192 °C after the glass transition at 38 °C for the complex with m = 6 [78a].

The polymeric complexes derived from 4-nitro- and cyanostilbazoles also show a smectic A phase up to about 200 °C [78a]. Polysiloxane complexes **32** also exhibit thermally stable smectic A or C mesophases [79–81]. These carboxyl-functionalized polymers and stilbazoles are miscible in a whole range of composition and show stable mesomorphic behavior [26, 79]. The introduction of the chiral stilbazole for the formation of a mesogenic complex leads to the induction of ferroelectricity [80]. Polymeric complex **33** exhibits a chiral smectic C phase, while no ferroelectricity is observed for each of single components. The value of spontaneous polarization for **33** ($x = 0.43$, $n = 5$) is 21.0 nC/cm^2 at 112 °C. The hydrogen bonding between the carboxylic acid and

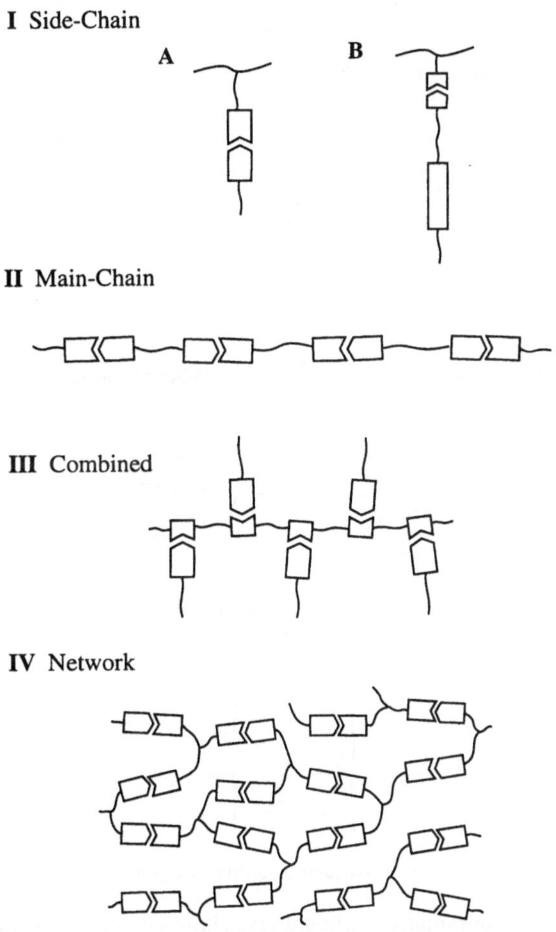

I Side-Chain

II Main-Chain

III Combined

IV Network

Fig. 10. Schematic illustration of supramolecular liquid-crystalline polymeric complexes: *I* side-chain type; *II* main-chain type; *III* combined type; and *IV* network type

the *N*-oxide moieties is also useful for the formation of mesogenic structures in the side chain of polymers **34** [82].

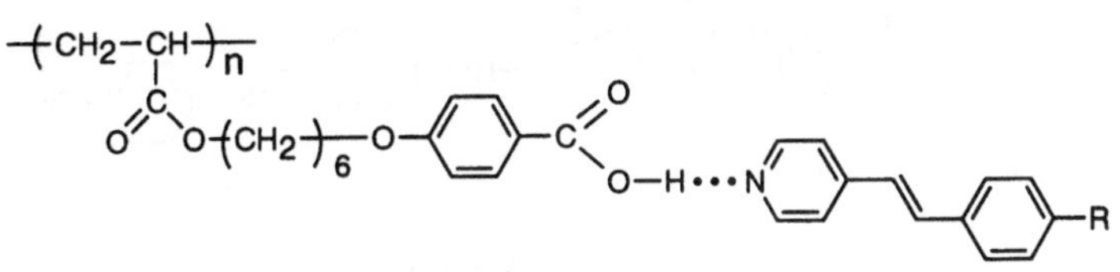

31 R= $-O-(CH_2)_{m-1}CH_3$

$-CN$

$-NO_2$

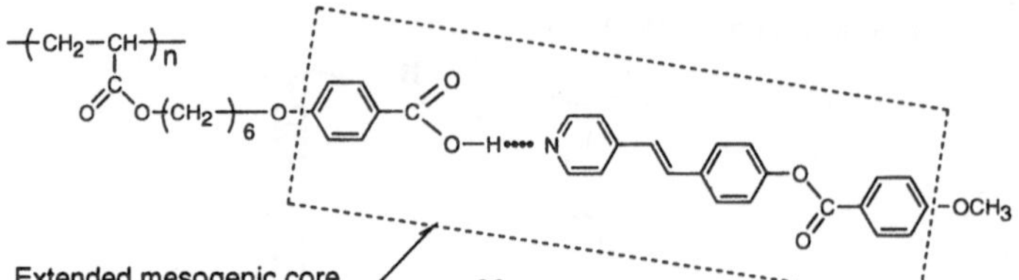

Extended mesogenic core
in polymer side chain

30

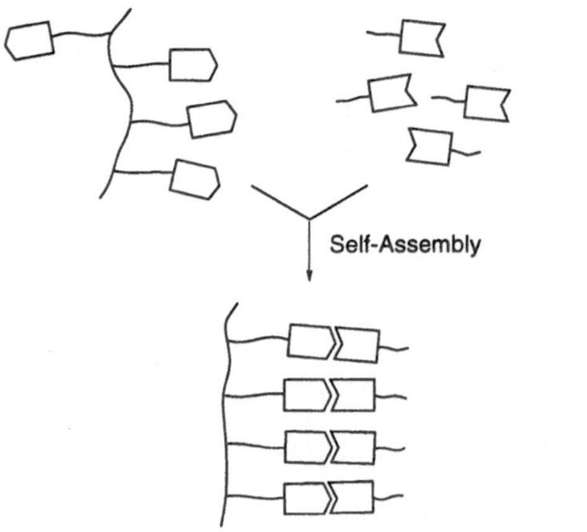

H-Bond donor   H-Bond acceptor

Self-Assembly

Surpamolecular Side-Chain Polymer

Fig. 11. Self-assembly of supramolecular liquid-crystalline side-chain polymer **30** consisting of polymer **29** and stilbazole **3**

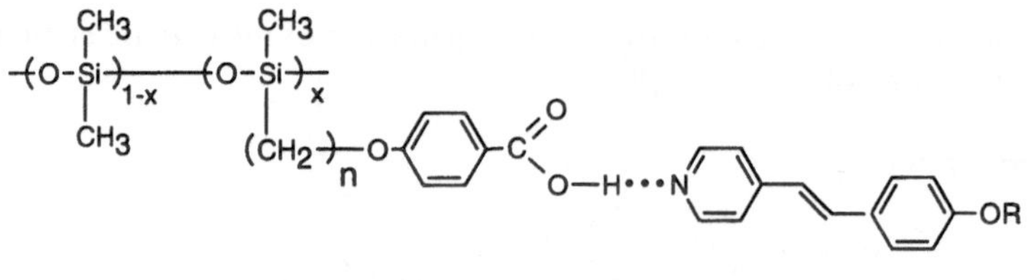

32: R= -CH$_2$CH$_3$

-C$_6$H$_4$OCH$_3$

33 : R= -CH$_2$-CH-CH$_3$

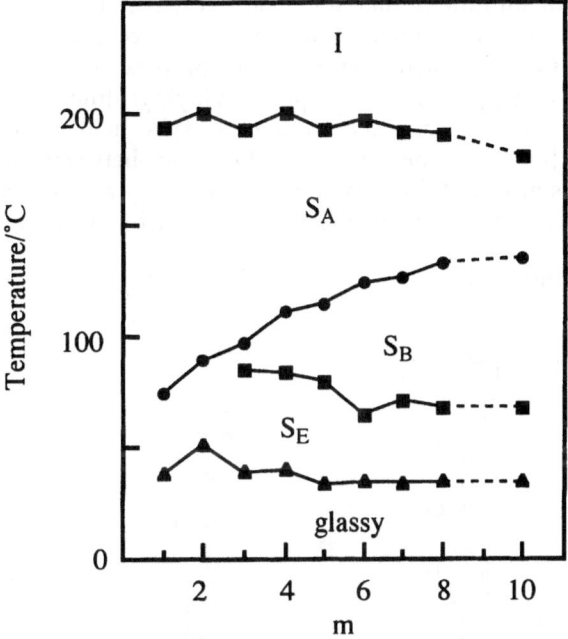

Fig. 12. Plot of transition temperatures for supramolecular liquid-crystalline complex **31** as a function of chain length (m) of the stilbazole moiety

34

Doubly hydrogen-bonded supramolecular polymers **35** and **36** consisting of carboxylic acid and diaminopyridine moieties exhibit monotropic liquid-crystalline behavior [63a, 83]. A texture of dendritic growth characteristic of columnar phases is seen for the complex [63a].

35: R=H

36: R=CH₃

Another type of supramolecular side-chain polymers (shown in Fig. 10B) is obtained by the complexation of functionalized mesogenic molecules with polymer backbones. The hydrogen-bonded complex of **37** is obtained by mixing poly(4-vinylpyridine) or poly(4-vinylpyridine-*co*-styrene) with mesogenic compounds terminated by a carboxylic acid moiety through a flexible spacer [84–88]. The hydrogen bonding between imidazole and carboxylic acid is also useful for the formation of supramolecular complexes. Polymeric complex **38** exhibits a smectic A phase from 10 to 65 °C [89].

The interactions of carboxylic acid/amine [90, 91], phenol/amine [92], and phenol/pyridine [93] are also used for the formation of side-chain complexes such as **39** and **40**.

The complexation of amphiphilic molecules with functionalized polymers forms layered smectic structures [94–97]. Polymeric complexes **41** consisting of poly(vinylpyridine) and an alkylphenol also form layered organized mesophases [94]. The incorporation of this structure into block copolymers with polystyrene results in the self-organization in two length scales, that is, "block copolymer length" and "nanoscale length" [95, 96].

41

Moreover, the hierarchical structure is formed by hydrogen bonding from the mixture (**42**) of poly[(styrene)-*block*-(4-vinylpyridine)] and methanesulfonic acid, which exhibits temperature-dependent electrical conductivity by the change of microphase-separated structure (Fig. 13) [96].

Inclusion complexes based on hydroxyl-functionalized polystyrene exhibit liquid-crystalline behavior. In this case, hydrogen bonding may play a key role for the complexation and the induction of mesophases [98].

2.2.3
Main-Chain Type

Liquid-crystalline supramolecular polymers **43** have been prepared by the association of bifunctional components through the formation of triple hydrogen bonding [76]. These bifunctional species have been derived from L-(+)-, D-(−)-, and *meso*-(M)-tartaric acid, respectively. The liquid-crystalline phases of the polymeric complexes consisting of chiral species are more thermally stable than those of polymers consisting of meso components. X-ray patterns and electron micrographs show that a triple helix structure is formed by the polymeric complex derived from L-(+)-tartaric acid, while the complex consisting of *meso*-tartaric acid forms a twofold structure [76, 99]. The hydrogen bonding between carboxylic acids and pyridines can also be used for the formation of polymeric structures such as **44** [100–103]. The polymeric association of bifunctional components induces mesophases. For example, **44** (A = -CH$_2$=CH$_2$-) shows nematic and smectic phases. However, in this case, no rheological behavior characteristic of a polymer in solution is observed for these main-chain complexes.

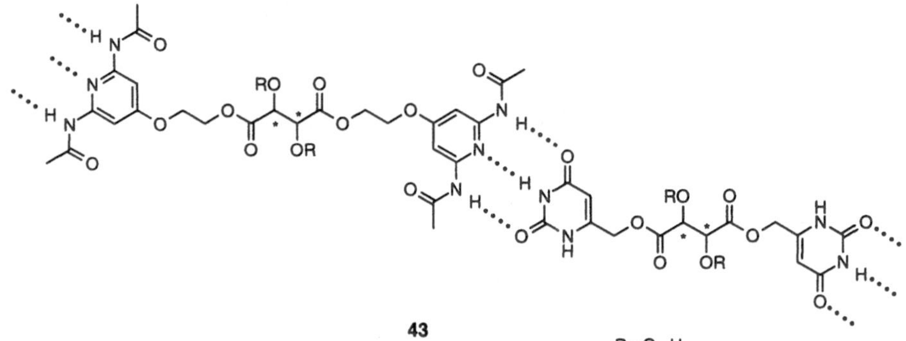

43

R= C$_{12}$H$_{25}$

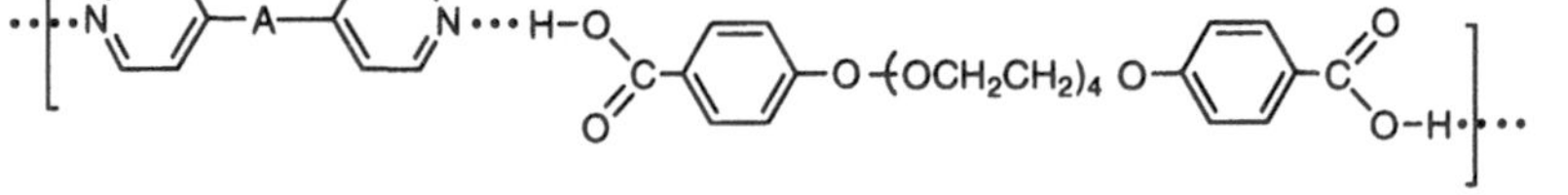

44   A =   —CH$_2$=CH$_2$ —

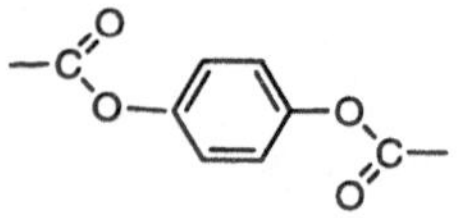

2.2.4
Combined Type

The incorporation of a hydrogen-bonding component, a 2,6-diaminopyridine moiety, into polyamide backbones enables the polymers to recognize molecules on their backbone through H-bonding [62, 104–107]. Polyamides containing the pyridine moiety and benzoic acid derivatives form liquid-crystalline complexes **45**. In particular, a supramolecular polyamide containing 4-[4'-(octyloxy)phenyl]benzoic acid results in the formation of a significantly stabilized mesophase up to above 350 °C [104]. The X-ray diffraction pattern of the complex suggests that the supramolecular polyamide is arranged as shown in Fig. 14.

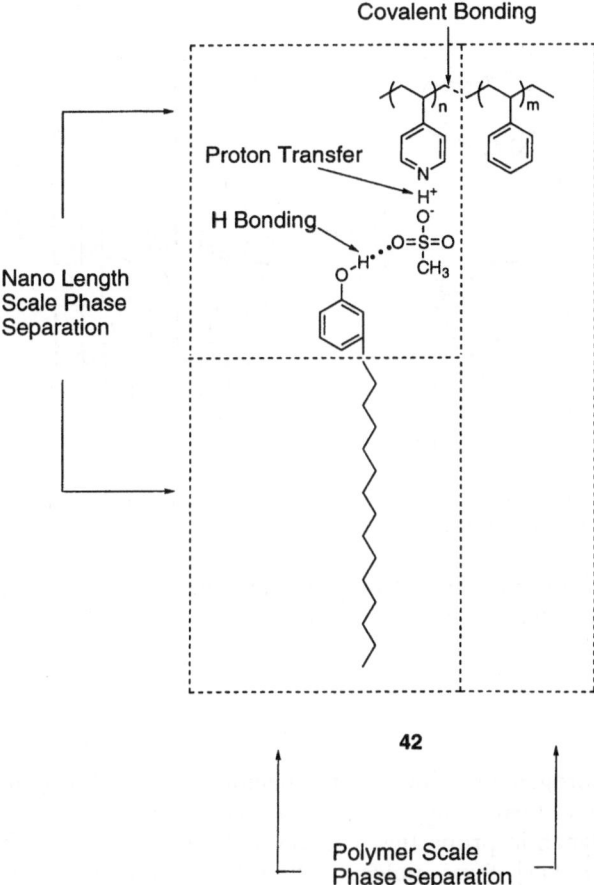

42

Fig. 13. Hierarchical structure of a supramolecular complex based on a block copolymer

45

R: $-O-(CH_2)_{\overline{m-1}}CH_3$

$-\!\!\!\!\bigcirc\!\!\!\!-O-(CH_2)_{\overline{7}}CH_3$

X: -Cl, -NO₂, -OCH₃, -CH₃,

It is of interest that supramolecular polyamides incorporating 4-alkoxy-benzoic acids with polar and nonpolar groups at the 3 position exhibit meso-phases until ca. 200 °C, while no stable complexation behavior is observed when the simple 4-alkoxybenzoic acid is used for complexation [62, 105, 106].

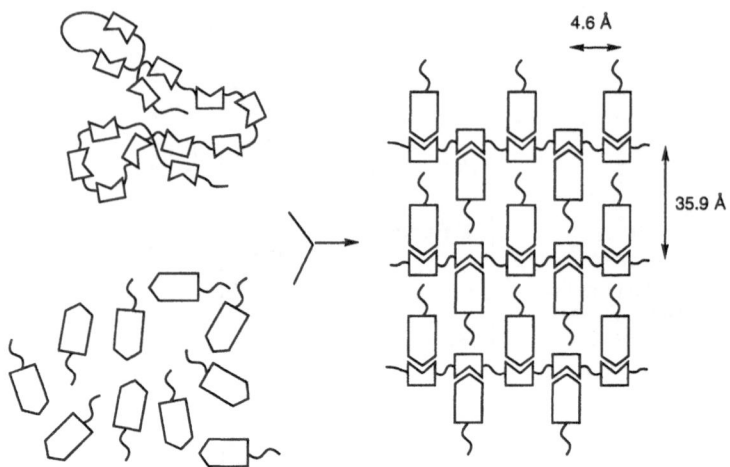

Fig. 14. Self-assembly of supramolecular polyamide **45** through a molecular recognition process on the polymer backbone

Steric effects of the substituents may stabilize the molecular packing of these supramolecular polyamides.

2.2.5
Network Type

The use of hydrogen bonding for mesogenic network formation has been expected for the formation of dynamic mesomorphic states due to the induction of dynamic properties, e.g. reversibility and fast exchange of the H-bonding [108–111]. Supramolecular liquid-crystalline networks have been prepared by the self-assembly of a polyacrylate with a benzoic acid moiety and 4,4'-bipyridine [108]. Complexes **46** form dynamically cross-linked structures obtained by the complexation of a carboxyl-functionalized polyacrylate and a mixture of a stilbazole, which has the single H-bond acceptor, and a bipyridine, which has the two acceptors.

Liquid-crystalline behavior has been examined for this system retaining the 1:1 stoichiometry of the carboxylic acid and pyridyl groups. Interestingly, the 1:1 complex of the polymer and the bipyridine exhibits a smectic A phase from 85 to 205 °C. The anisotropic melt of the complex ($x = 1.0$) is viscous, but exhibits shear flow between glass plates. The X-ray diffraction pattern for the layer spacing of the smectic A phase is 39.8 Å, which agrees with the molecular modeling of the length of the fully stretched hydrogen-bonded mesogen.

The transition between the liquid-crystalline phase and the isotropic state is thermally reversible due to the association and dissociation of hydrogen bonds. Figure 15 presents the thermally reversible formation of the supramolecular cross-linked structures. If these cross-linked structures are prepared only by covalent bonding, no such dynamic reversible behavior is observed.

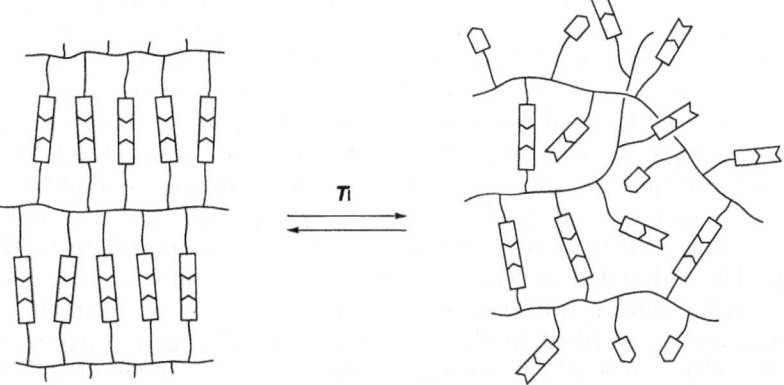

Fig. 15. Reversible phase transition of supramolecular liquid-crystalline network **46**

The hydrogen bonding between carboxylic acid and imidazole is also useful. A complex derived from poly(acrylic acid) and bis(imidazolyl)biphenyls forms a smectic A phase [89].

46

Supramolecular mesogenic networks have also been prepared by self-assembly involving multifunctional low-molecular-weight components [109–112]. Network complexes have been prepared from trifunctional H-bonding donors **47** and **48**, and bifunctional H-bonding acceptors **4** and **9**, maintaining the 1:1 stoichiometry of the carboxylic acid and pyridine moieties. Complexes **47/4** and **47/9** exhibit monotropic smectic A phases, while enantiotropic nematic phases are seen for complexes **48/4** and **48/9**. It should be noted that all single H-bonding components are nonmesomorphic. A nematic phase is observed for complex **48/4** from 200 to 87 °C on cooling. The H-bonded networks based on **47** show only smectic phases, while nematic phases are seen for the networks derived from **48**. This difference can be ascribed to the favored conformation of the trifunctional molecules (Figs. 16 and 17). Molecular modeling of compound **47** shows that one of the spacers lies parallel to the other, thereby, the complexes based on **47** exhibit smectic phases, as shown in Fig. 16. In contrast, the complexation of **48** that forms an asymmetric molecular conformation with **4** and **9** results in the induction of a nematic phase (Fig. 17).

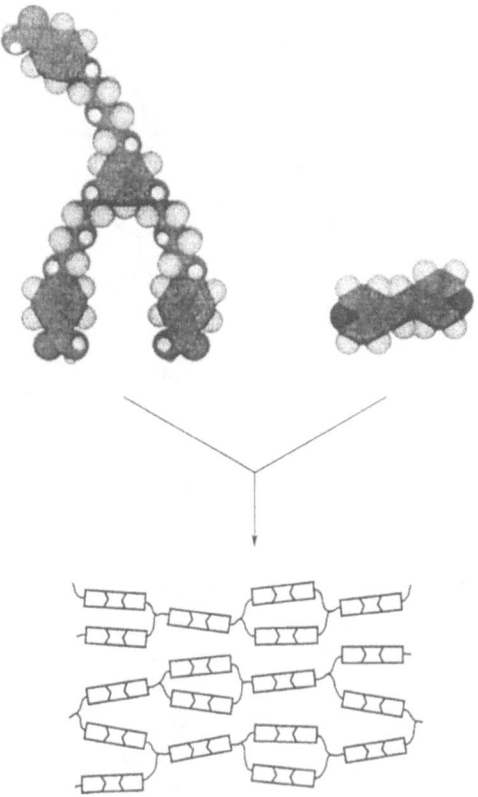

Fig. 16. Smectic network formation by supramolecular self-assembly of trifunctional compound **47** and bifunctional bipyridine **4** or **9**

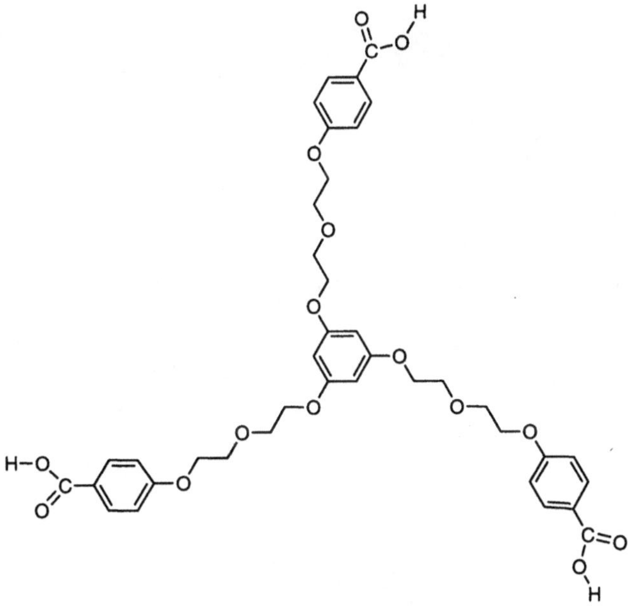

47

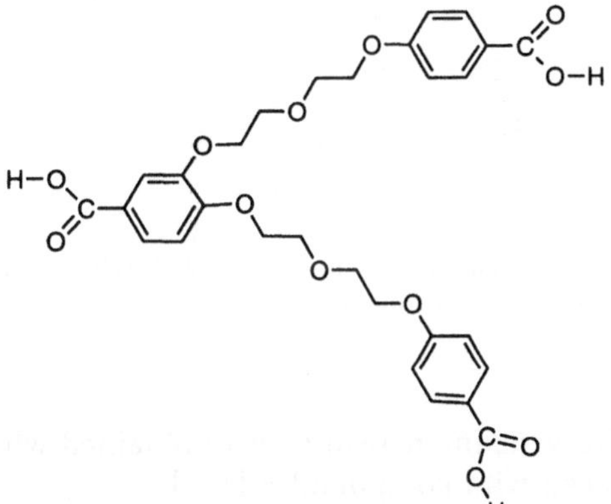

48

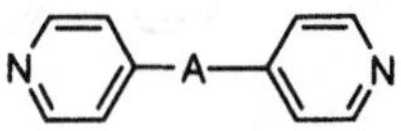

4 :   A = none

9 :   A = —CH₂=CH₂ —

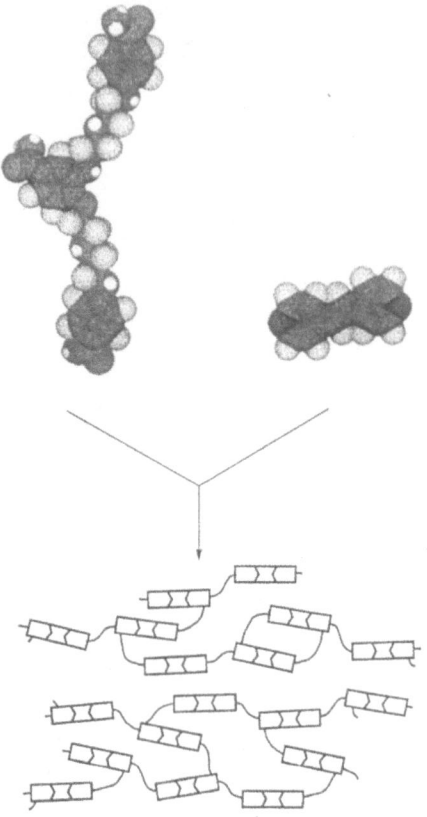

Fig. 17. Nematic network formation of supramolecular self-assembly of trifunctional compound **48** and bifunctional bipyridine **4** or **9**

Cholesteric liquid-crystalline networks were obtained when chiral building block **49** was complexed with compound **4** [111].

$$\text{(structure of compound 49)}$$

49

Hydrogen-bonded networks have been prepared from compound **50**, which has four pyridine units, with dicarboxylic acid **51**. The complex forms network and polymeric structures. Fibrous solids are drawn from the melt state of the complexes [112].

The formation of hydrogen-bonded networks can also be obtained by the reaction of bifunctionalized polymerizable complex **52** [113].

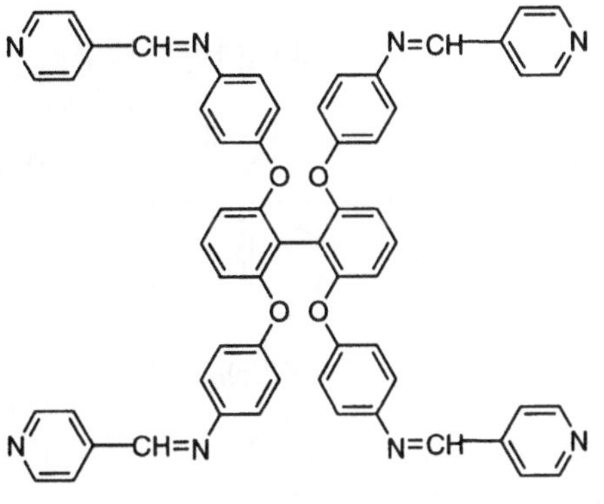

50

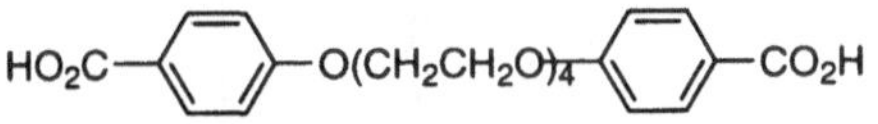

51

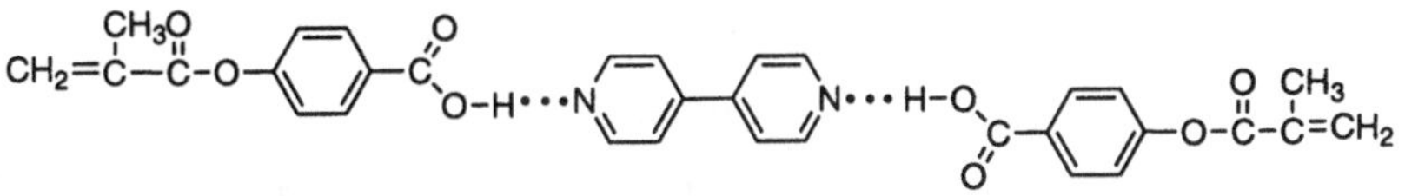

52

The formation of covalent cross-linking in liquid-crystalline polymers is useful to lock in the molecular order in solid states, which leads to static functionality [114, 115]. For hydrogen-bonded mesogenic networks, the use of the dynamic nature of H-bonding induces dynamic properties [108, 109]. It is important to choose an appropriate method of cross-linking for the purpose of the use of network materials.

2.2.6
Host-Guest Type

Liquid-crystalline host-guest functional polymers have shown to be useful for electrooptic materials [116, 117]. The introduction of noncovalent interactions between mesogenic polymer matrices and functional molecules has the advantage of producing a host-guest liquid-crystalline system (Fig. 18) [84]. A hydrogen-bonded photoresponsive nematic material 53 has been prepared from a mesogenic side-chain copolymer and a photofunctional guest compound [84]. This complex exhibits isothermal nematic-isotropic transitions by *cis-trans* isomerization of the azobenzene. In this host-guest material, miscibility of the guest is enhanced by the interaction compared with simple mixture materials and the synthesis is easier than that of copolymeric materials.

2.2.7
Blends

The blend of two different polymers is a versatile way to develop new polymeric materials with desirable properties. The use of hydrogen bonding

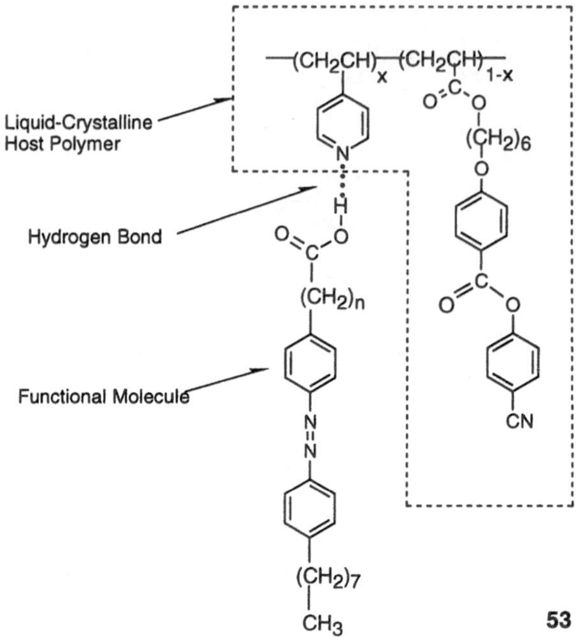

Fig. 18. Supramolecular hydrogen-bonded host-guest functional system

has become important to increase the miscibility of polymers (Fig. 19) [118]. It is also effective to control the self-organized structures of the resultant blends. However, in the case of liquid-crystalline main-chain polymers, miscibility with conventional amorphous polymers is poor. Hydrogen-bonded mesomorphic blends have been prepared by mixing a main-chain liquid-crystalline polyester and a nonmesogenic vinyl polymer [77]. Polyester **54** was designed and prepared as a thermotropic polyester capable of the formation of stronger hydrogen bonding with acids. This polymer exhibits a glass transition

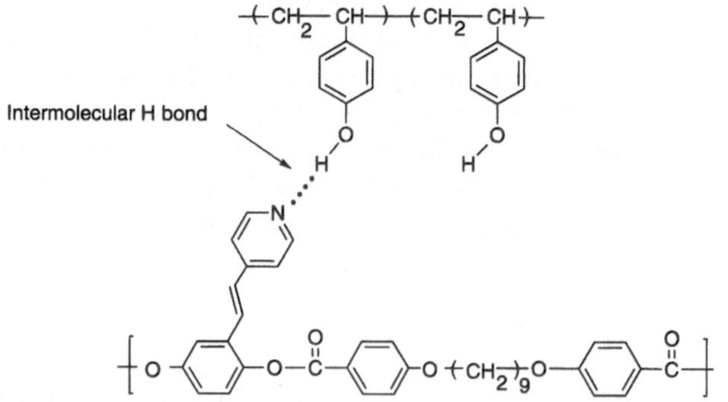

Fig. 19. Hydrogen-bonded liquid-crystalline polymer blend

temperature at 88 °C and a nematic phase up to 175 °C, while poly-(4-vinylphenol) is nonmesomorphic. A homogeneous viscous nematic state with no phase separation is visually observed for the 1:1 blend from 96 to 150 °C. IR measurements of the blends indicate the existence of an H-bond between the pyridyl group and the phenol unit.

2.2.8
Related Nonmesomorphic Materials

Polymeric association of *different* molecules through multiple hydrogen bonding has been used for the formation of non-liquid-crystalline bulk solids [12], fibrous solids for gelation in solvents [119], and as monolayers [120]. Simpler H-bonding such as the interaction between carboxylic acid and pyridine has also been shown to be useful for one- or two-dimensional aggregates in solid states [121–124]. For example, a one-dimensional polymeric complex from an A-B type monomer is formed in crystalline solids [121].

3
Molecular Assemblies of Identical Molecules

3.1
Closed Structures

3.1.1
Rod-like Complexes

The dimers of benzoic and cinnamic acid were probably the first hydrogen-bonded liquid crystals [16]. Since this report early in the 20th century, a wide variety of aromatic acid dimers such as **55** have provided information on the structure and properties of liquid crystals [125–127]. A notable example is compound **56** which shows an optically isotropic smectic D phase [127].

To examine the stability of mesogenic carboxylic acid dimers, variable temperature IR spectra have been obtained for 4-alkylbenzoic acids [71]. The monomeric molecule of a benzoic acid appears once the compound melts to a nematic state. For 4-pentylbenzoic acid **55** (X = CH_2, $n = 4$) (nematic: 88–127 °C), the fraction of the carbonyl peak due to the monomer becomes 6.2 mol% at T_i. In the isotropic state, a steady increase in the free monomeric component is observed as the temperature increases. No monomeric species is observed for carboxylic acid/pyridine complexes, which are formed by more stable H-bonding [31, 34].

$$CH_3-(CH_2)_{n-1}-X--C\genfrac{}{}{0pt}{}{O\cdots H-O}{O-H\cdots O}C--X-(CH_2)_{n-1}-CH_3$$

55   X = —CH_2— or
                —O—

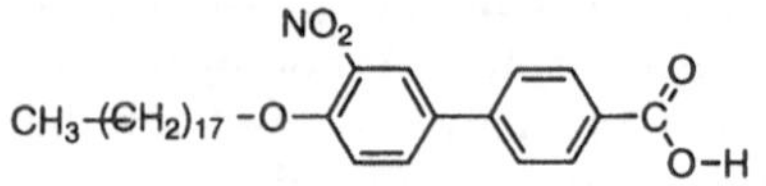

CH₃–(CH₂)₁₇–O–[benzene ring with NO₂]–[benzene ring]–C(=O)–O–H

56

The mesogenic structure of a benzoic acid dimer has been introduced as a noncovalent cross-linker for polysiloxanes [79]. Polymer 57 exhibits a smectic C phase due to the dynamics of H-bonding. In contrast, mesomorphic order is locked in the solid state of poly[(4-acryloyl)benzoic acid] by polymerization in its mesophase [128]. No liquid-crystalline state is observed for this material because of the lack of flexibility of the structures. Main-chain-type polymeric liquid-crystal associates are formed from carboxyl-bifunctionalized aromatic compounds [129].

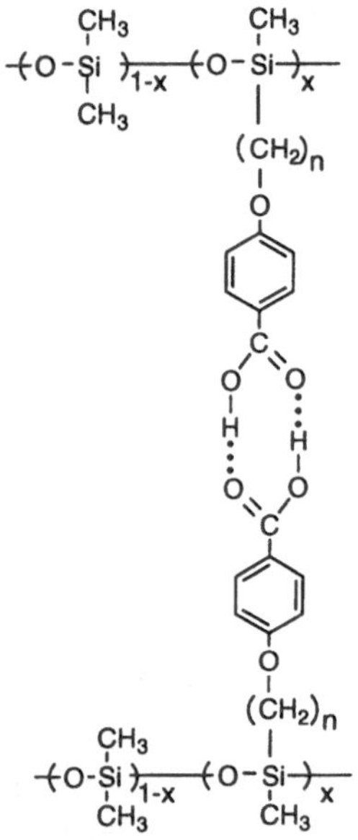

57

The interaction of carboxylic acids affects the molecular order of side-chain polymer liquid crystals [130]. Copolymerization of an acrylate containing a cyanobiphenyl unit through butyl group and acrylic acid leads to the induction of a smectic A phase. The intra- and intermolecular H-bonds of carboxylic groups stabilize layered structures.

Dimerization of carboxylic acids is also used for the formation of nonmesomorphic molecular assemblies in solid states such as one-dimensional polymeric structures [131, 132], two- or three-dimensional networks [133], and dendritic materials [134]. Dimer formation due to quadruple H-bonding has been designed to obtain highly stable molecular association [135]. The quadruply H-bonded polymer associates from bifunctional compounds show polymer-like behavior in the solution state due to large association constants [135].

3.1.2
Disk-like Complexes

Pyridone forms a hydrogen-bonded dimer with a closed structure [136]. This dimerization has been used for the design of crystalline assemblies such as three-dimensional networks [136b]. This hydrogen bonding has been used for the formation of discotic liquid-crystalline assemblies [137]. Disk-like complex **58** with dodecyl chains displays a hexagonal columnar phase (D_h) between 88 and 108 °C. Columnar mesophases are also exhibited by benzamide dimers [138].

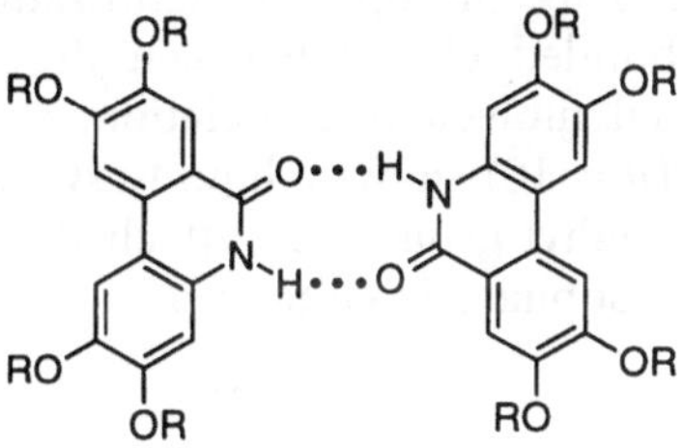

58   R = $C_{12}H_{25}$

Trimeric H-bonded disk **59** from phthalhydrazides self-assembles into columnar phases [139].

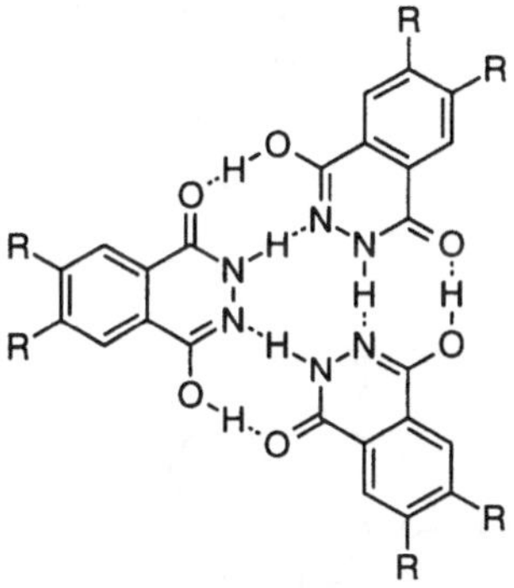

59

R: C_nH_{2n+1} n = 6, 10, 14

3.2
Chains

The amide group is important in nature because proteins and peptides consist of amide linkages which connect amino acids. They are capable of forming an H-bonding chain, which is useful for molecular self-organization. For thermotropic liquid crystals, the number of amide-containing compounds is limited because the intermolecular H-bonding chain often raises the melting temperatures of the compounds and reduces liquid crystallinity [4]. In contrast, numerous mesogenic compounds containing ester linkages have been prepared [4]. However, the appropriate design of the use of H-bonding by amide groups leads to the formation of stable mesomorphic molecular assemblies. A few examples of amide-containing liquid crystals and their related systems are described in this section.

Compound **60** was first synthesized as a discotic liquid crystal in 1977 [140]. A wide range of mesomorphic temperature is seen until about 200 °C for compound **61** where the ester linkage at the 1 and 4 position of **60** is replaced by the amide linkages [141], while the mesomorphic temperature range of **60** is narrower below 100 °C. This mesophase stabilization is ascribed to the formation of hydrogen-bonded chains between the amide linkages, which induces a one-dimensional molecular association, as shown in Fig. 20 [24, 141–144]. Compound **62** ($n = 10$) shows a liquid-crystalline phase from 130 to 203 °C [24, 142]. When methyl groups are attached to the benzene ring, the thermal stability of the mesophase is enhanced.

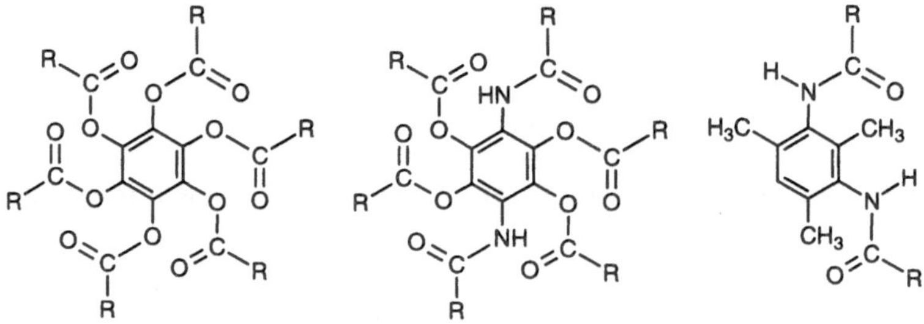

60   **61**   **62**

R: C_nH_{2n+1}

H Bonding

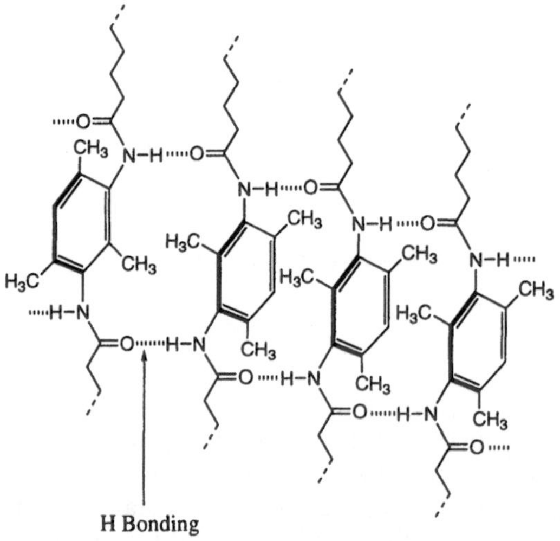

Fig. 20. Molecular assembly of a benzene-diamide compound by formation of intermolecular hydrogen bonding chains

Taper-shaped bis-amide compounds **63** show hexagonal columnar meso-phases due to chain hydrogen bonding [145]. The effect of amide groups has also been examined for supramolecular doubly H-bonded complexes [146]. For hydrazine compounds, the mesomorphism was reported as early as 1977 [147].

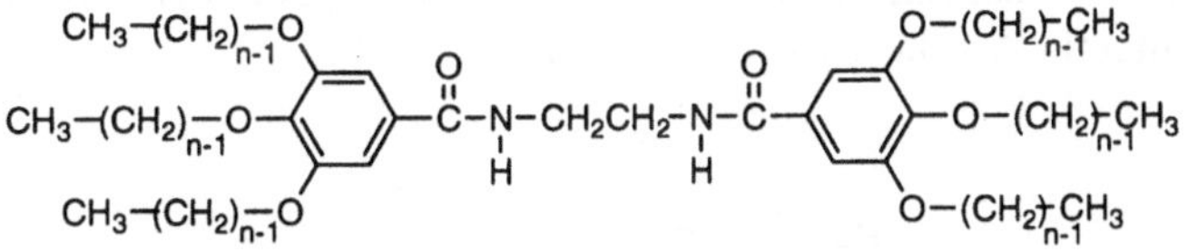

63

Compound **64** exhibits only a smectic C phase [148]. In this case, the alkylamino group introduced into this structure induces the association of H-bonded molecular chains and a smectic C arrangement, as shown in Fig. 21.

3.3
Layers or Columns

Carbohydrates are natural compounds which have hydroxyl groups capable of hydrogen bonding. Amphiphilic sugars with long alkyl chains form layered or columnar mesomorphic structures due to microphase segregation, as shown in Fig. 1 (Type A-I-c) [17–20, 149–153]. Alkylglucosides are representative

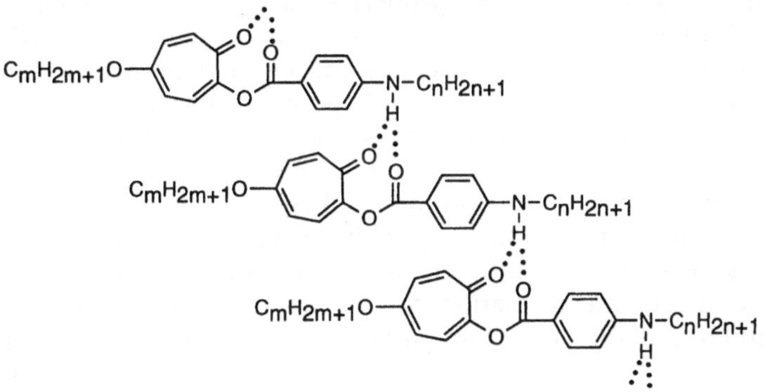

64

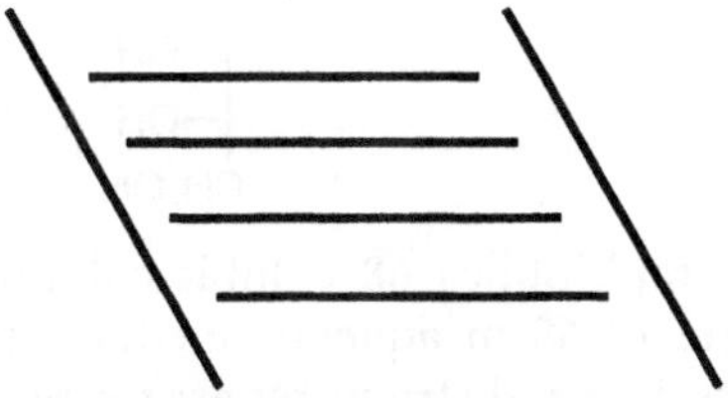

Fig. 21. Smectic C molecular arrangement through the formation of lateral hydrogen bonding between a tropone compound containing an N-H group

examples of liquid-crystalline thermotropic carbohydrates. For example, octyl-
β-D-glucopyranoside (**65**, $n = 8$) exhibits a smectic phase between 69 and
110 °C [19, 20]. Smectic A phases are also observed for manno- and
galactopyranosides. Di- and trisaccharides with long alkyl chains show more
thermally stable smectic phases due to the increase in hydrogen bonds [149].
Mesophases are observed until 200 °C. For example, dodecyl-β-D-maltoside **66**
exhibits a smectic A phase from 80 to 244 °C.

65

66

Sugars incorporating a 1,4-phenylene or 1,4-cyclohexylene unit in the alkyl
part display a mesophase [150]. Sugars with open-chain structures also show
smectic phases [151]. The X-ray diffraction patterns of the mesophases of
amphiphilic sugars have shown that the phase is a bimolecular smectic A
phase [19]. These thermotropic properties are due to the cooperation of
intermolecular hydrogen bonding between hydroxyl groups and hydrophobic
interactions of long alkyl chains, as shown in Fig. 1 (type A-I-c). The smectic
phases are induced by the formation of the micro-phase-segregated layered
structure of the hydrogen-bonded and hydrophobic regions. The smectic A
phase displayed by the sugar amphiphilic molecules is not miscible with
normal calamitic smectic A compounds [20]. Carbohydrates with dialkyl
dithioacetal groups show liquid-crystalline columnar phases [153]. As the
molecule has two flexible chains, the association of the central sugar part by
H-bonds results in the formation of a columnar structure. Compound **67**
($n = 8$) exhibits a hexagonal columnar phase between 36 and 85 °C on heating.

67

Bolaform sugar amphiphiles **68** exhibit columnar mesophases [154]. The
aggregated structures of **68** in aqueous media depend on the stereochemical
factors of the spacers. Layered structures are formed in crystalline states by self-
assembly of glucopyranose-based bolaamphiphiles due to the cooperation of the
hydrogen-bonding interactions of hydroxyl groups and amide linkages [155].

68

A mesophase is formed for inositol derivatives **69** with alkoxyl and hydroxyl groups [22, 156]. They exhibit columnar mesophases.

69

$$-X: \quad -OH \quad or \quad -OC_nH_{2n+1}$$

The structure of the molecular association can be controlled by the substitution patterns of the hydroxyl groups on the cyclohexane rings, as shown in Fig. 22. When all the hydroxyl groups are modified by alkoxyl chains, the compound forms one column by one molecule (Fig. 22A), while the molecules that have two hydroxyl groups form one column by two molecules by hydrogen bonding (Fig. 22B). In the case of inositols with four hydroxyl groups, five molecules behave as one molecular discotic component (Fig. 22C).

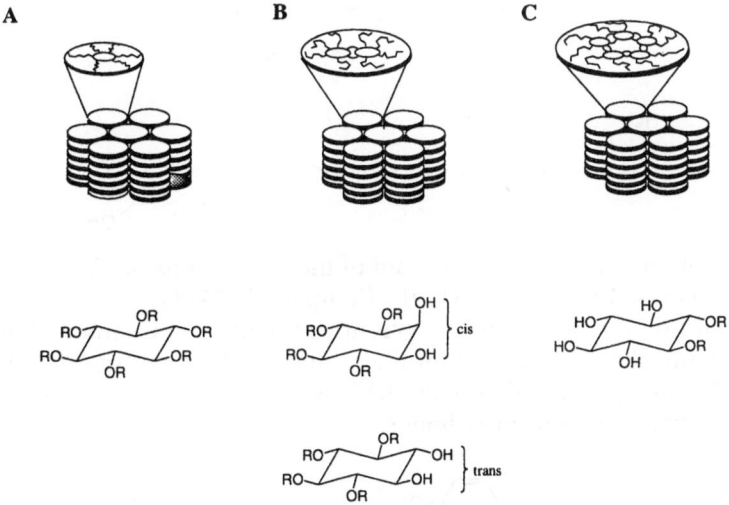

Fig. 22. Control of molecular association of inositol derivatives by the substitution pattern of hydroxyl groups

Dihydroxycyclohexyl compounds **70** and **71** containing two and three alkyl chains, respectively, have been shown to exhibit columnar mesophases [157, 158]. In the liquid-crystalline behavior for the diol mesogens, intermolecular hydrogen bonding plays a determining factor for the formation of meso-morphic hexagonal discotic aggregates.

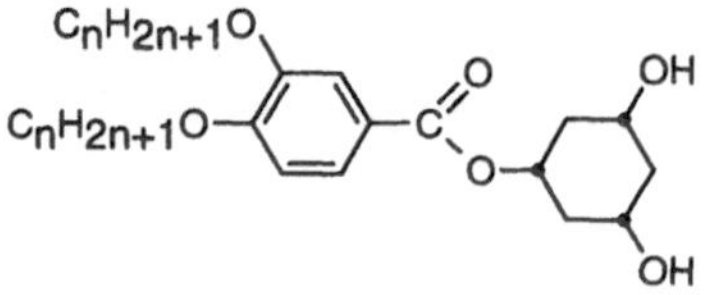

70

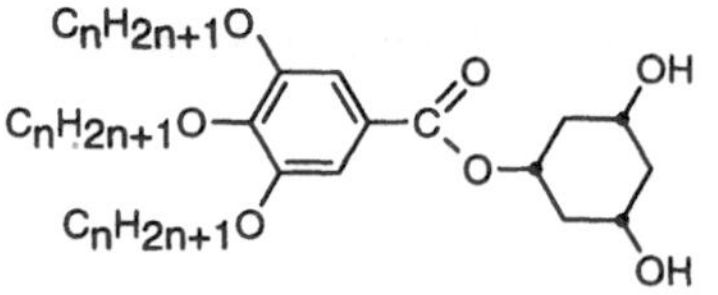

71

A diol compound, cyclohexylpropanediol **72**, which consists of a single ring, exhibits a smectic phase due to the formation of hydrogen-bonded layers [22]. Simple alkyl compounds **73** and **74** containing 1,2-diol moieties at the extremity at their ends also show thermotropic liquid crystallinity [23, 159]. These smectic phases are completely immiscible with conventional liquid crystals. These results suggest that the 1,2-diol groups are capable of forming stable hydrogen-bonded networks and induce layered liquid-crystalline molecular association.

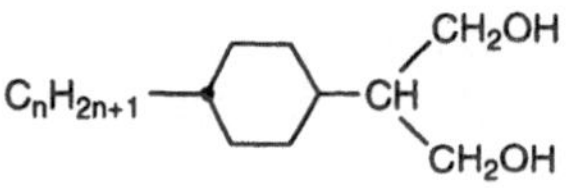

72

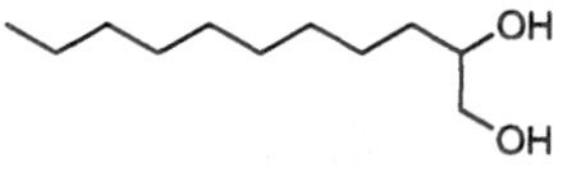

73

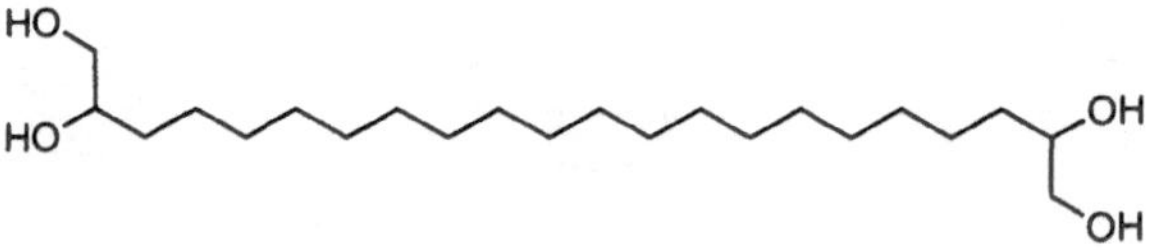

74

The effect of the diol groups at the end of the alkyl chains has been compared for compounds **75** and **76** [160]. Compound **75** has a structure that incorporates diol moieties into its parent compound **76**. Compound **76** melts into a nematic phase up to 79 °C, while compound **75** shows smectic phases to 165 °C. The mesophase is thermally stabilized at 86 °C by the incorporation of the diol groups, which form H-bonds.

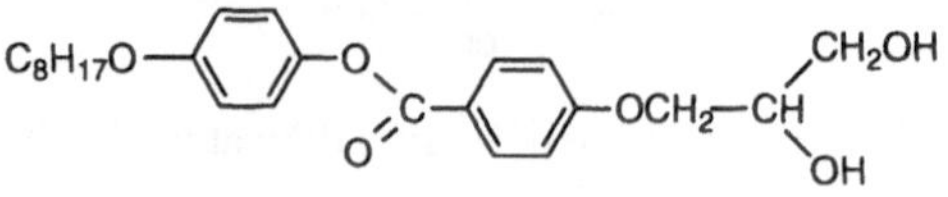

75

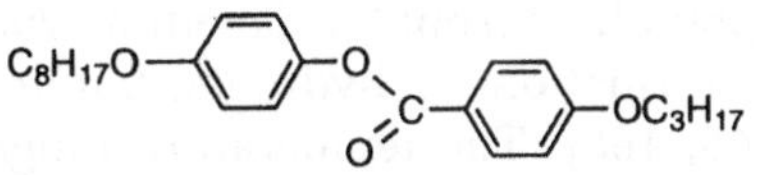

$$C_8H_{17}O-\text{(structure)}-OC_3H_{17}$$

76

The effect of the number of the hydroxyl groups in alkyl terminal chains has been examined for cyanobiphenyl compound **77** [161]. With an increase in the number of OH groups, the thermal stability of the smectic phases is enhanced. This is due to an increase in the number of hydrogen bonds which stabilize the aggregated structures of molecules. The effect of the hydroxyl group has also been studied for the laterally attached group of terphenyl compounds [162, 163]. Lateral fixing of molecules through hydrogen bonding results in the stabilization of mesophases [162]. Moreover, the attachment of hydrophilic dihydroxy groups to hydrophobic rigid rod mesogenic cores produces new mesogenic block molecules [163].

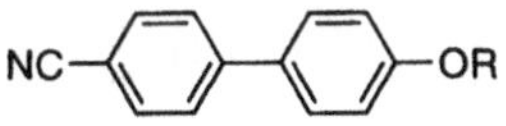

$$NC-\text{(structure)}-OR$$

77

R:

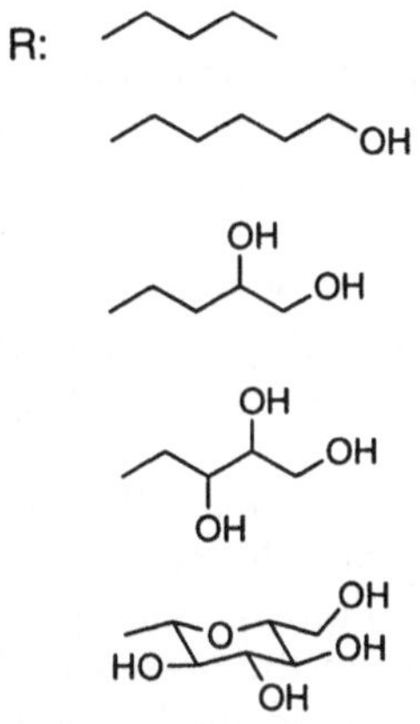

Columnar mesomorphic behavior is observed for compound **78** because the associated lateral groups disturb the formation of smectic layers and the hydrophilic lateral chain and the hydrophobic core unit form microphase-separated structures. A variety of block molecules capable of forming mesophases have been prepared and examined in relation to phase-separated association [164].

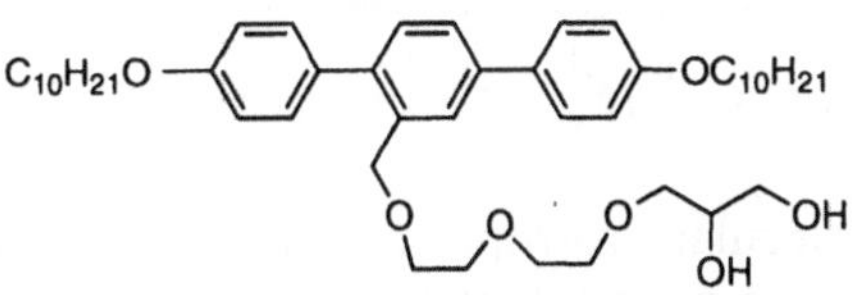

$$C_{10}H_{21}O-\text{(structure)}-OC_{10}H_{21}$$

78

An inorganic compound, diisobutylsilanediol **79**, shows a columnar mesophase due to the formation of hydrogen-bonded columnar networks based on the dimer [165, 166]. The temperature range of the mesophase is between 88 and 99 °C. Although this compound was synthesized in 1952, the liquid-crystalline behavior of the compound was not established until later. Analogous compounds, dihydroxytetraalkyldisiloxanes, show thermotropic columnar phases [167].

79

4
Lyotropic Assemblies

Self-assembly and molecular recognition of some biomolecules lead to lyotropic liquid-crystalline behavior with water [5]. These molecular complexes derived from biomolecules could be more important in view of biofunction. Nucleic acids such as DNA and RNA, which self-organize supramolecular complexes, show lyotropic mesophases [5]. Complexes of nucleic acid and cationic amphiliphic molecules form mesomorphic molecular arrangements, which can be used for drug delivery such as gene therapy [6].

Interesting examples of biomolecules that show lyotropic liquid crystallinity by association of *identical* molecules are guanosines and folic acids [168–171]. These molecules exhibit columnar mesophases due to the formation of hydrogen-bonded complexes in the presence of water. For example, folic acid **80** sodium salt exhibits cholesteric and hexagonal phases with water due to the formation of a tetramer, which stacks as shown in Fig. 23. The introduction of long alkyl chains into the guanosines leads to the assembly of ribbon-like structures in organic solvents [172].

80

Synthetic supramolecular lyotropic polymer **81** has been prepared from bifunctional molecular components through triply hydrogen bonds [173]. The

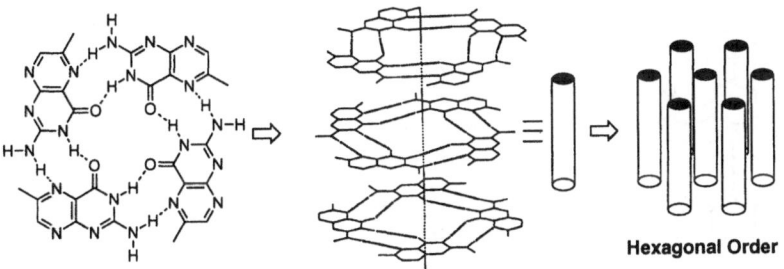

Fig. 23. Tetramar formation and induction of a mesophase of folic acids

polymer shows liquid-crystalline behavior with an organic solvent whereas it melts with decomposition at high temperature due to its high rigidity.

(Structure Next Page)

5
Anisotropic Gels

In the cases of the hydrogen-bonded materials described in the previous sections, single *homogeneous* liquid-crystalline phases without phase separation are displayed by the formation of intermolecular hydrogen bonds. Here, liquid-crystalline physical gels, anisotropic functional materials with *heterogeneous* self-organized structures (Type B in Fig. 2), are discussed.

Recently, physical gels of common organic solvents have been prepared by the self-assembly of low-molecular-weight compounds [174–177]. In most cases, hydrogen-bonded compounds that can form fibrous solids which disperse in solvents have been used for the gelation. Anisotropic solvents such as thermotropic liquid crystals were expected to be gelled by these H-bonded molecules. Such physical gelation has been successfully achieved by simple self-association processes of hydrogen-bonded molecules [29, 30, 178–180]. In contrast, the formation of covalently bonded polymer networks is needed for the preparation of chemical anisotropic gels [114, 181].

Gelling agents, *trans*-(1R,2R)-bis(dodecanoylamino)cyclohexane **82** [176] and amino acid derivatives **83** and **84** [177], were chosen for the gelation of liquid crystals. Room-temperature liquid crystal **85** (nematic: 24–35 °C), which is a non-hydrogen-bonded anisotropic solvent, is efficiently gelled by these hydrogen-bonded molecules. Figure 24 presents the phase-transition behavior of a mixture of **82** and **85** as a function of the mol% of **82**. Three thermoreversible states, isotropic liquid, *normal* (isotropic) gel, and liquid-crystalline (anisotropic) gel states, have been achieved for the mixtures. The sol-gel transition temperatures from isotropic liquids to gel states increase with the increase in the mol% of **82**. IR spectra of the mixture show that the amide group is involved in the formation of hydrogen bonding (Fig. 25) and the association and dissociation of the H-bonding leads to the reversible sol-gel transition, as depicted in Fig. 26.

81

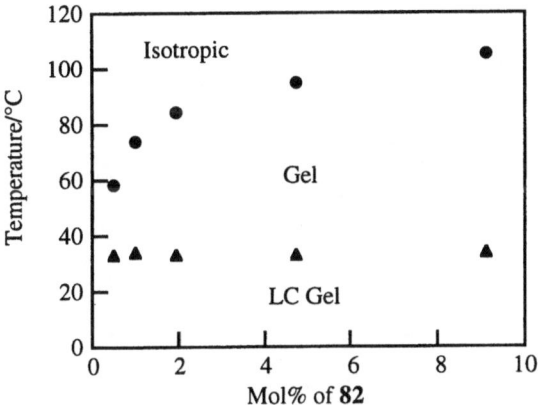

Fig. 24. Phase transition behavior of the mixture of a room-temperature nematic liquid crystal and a gelling agent

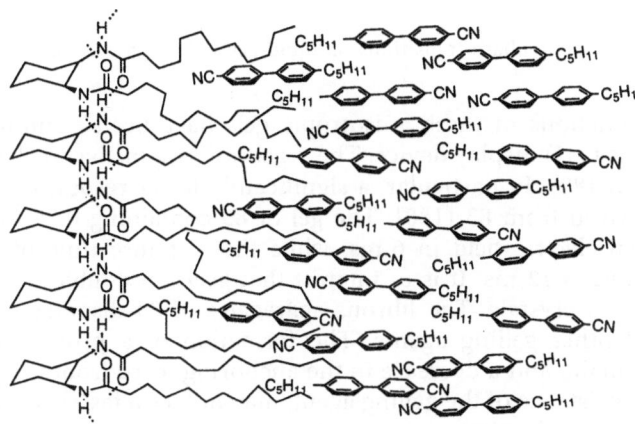

Fig. 25. Structure of liquid-crystalline gels consisting of a liquid crystal and hydrogen-bonded gelling agent which are micro-phase separated

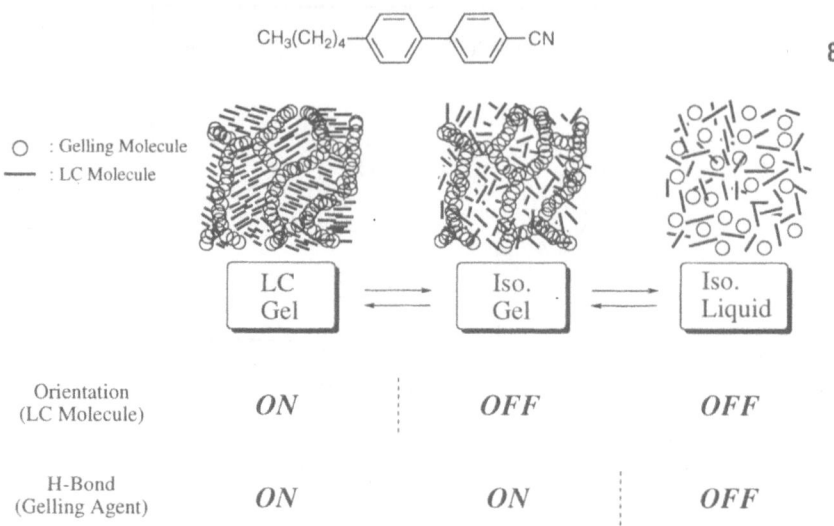

Fig. 26. Thermoreversible phase transitions of liquid-crystalline physical gels

Electrooptic functions of these anisotropic gels have been examined with a twisted nematic (TN) display device. These gels are responsive to electric fields in TN cells [29, 180]. In particular, a significantly faster response is observed for the gel derived from **83** [180]. The gel of **85** containing 0.25 mol% of **83** responds to an electric field in 6 ms, while the response time of the single component of **85** is 12 ms. It is of interest that gelation doubles the response rate of the liquid crystals. The fibrous aggregates of **83** disperse more finely than those of other gelling agents. The formation of a proper size of the mesophase domain, and a decrease in the anchoring force from the polyimide surface by the existence of the gelling agent, may induce a faster response from the liquid-crystal molecules.

Smectic liquid-crystalline physical gels are also obtained by the self-aggregation of hydrogen-bonded molecules [30]. The phase diagram for a mixture of a smectic liquid crystal (transition temperatures: isotropic 102 °C; smectic A 78 °C; smectic C 58 °C) and gelling agent **83** is presented in Fig. 27. Interestingly, the aggregates of **83** are observed to develop in the direction perpendicular to the molecular long axis between the smectic molecular layers (Fig. 28). In contrast to the case shown in Fig. 26, the sol-gel transition occurs in the liquid-crystalline state, which results in an oriented anisotropic aggregation of the hydrogen-bonded molecules, as shown in Fig. 29.

These liquid-crystalline gels are a new class of materials that combine the properties of physical gels, which form *soft solids*, with liquid crystals, which possess dynamic molecular anisotropy. Moreover, these gels are a novel type of composites, which have anisotropic *heterogeneous* structures. These soft materials will be widely applicable to functional systems responsive to various external stimuli.

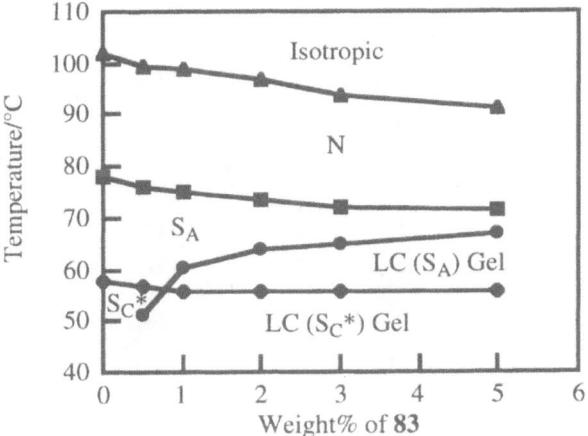

Fig. 27. Phase behavior of a mixture of liquid crystal SCE8 and **83** on cooling

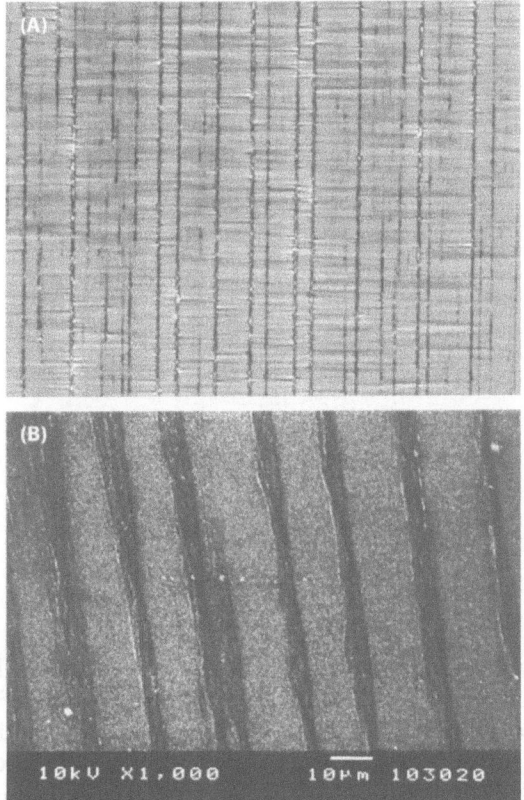

Fig. 28. A Optical polarized photomicrograph of a mixture of liquid crystal SCE8 and amino acid derivative **83** at 60 °C in a parallel rubbed cell; **B** SEM image of oriented fiber bundles made of **83** after the extraction of SCE8

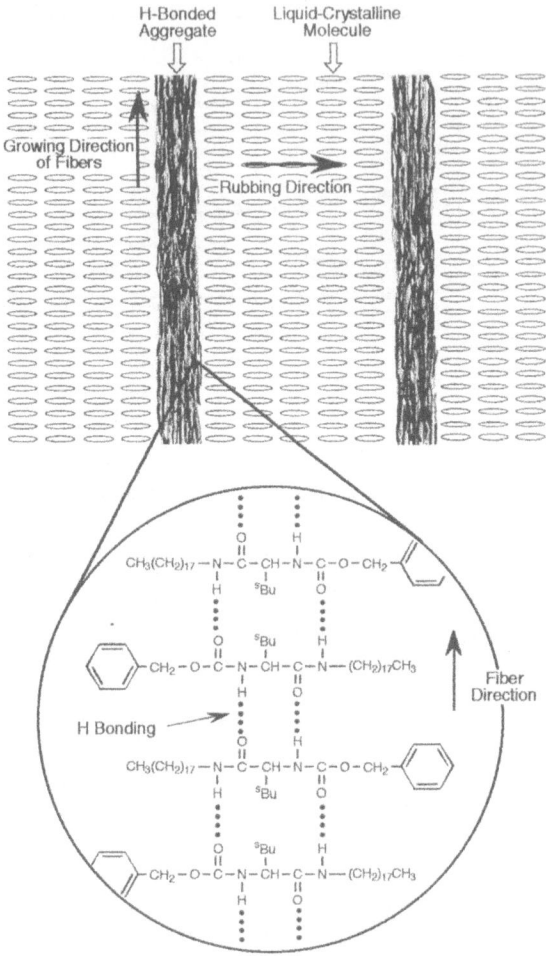

Fig. 29. Schematic representation of the anisotropic aggregation structure of **83** in a smectic liquid crystal

6
Conclusions

In addition to applications for electrooptical displays and high-strength fibers [1, 2, 75], liquid crystals should have greater potential for a variety of functional materials because it is only these materials that possess dynamic properties and anisotropic structures. Such a unique nature is similar to biomolecular systems, which exhibit highly dynamically functional processes and consist of numerous self-assembled molecules through a variety of molecular interactions. The versatility of hydrogen bonding for the anisotropic self-organization of functional synthetic materials has been demonstrated here. Moreover, intermolecular interactions such as ionic [182], ion–dipolar [183], and charge-transfer [184] as well as hydrogen bonding are also reported to induce liquid crystallinity. The author believes that for future materials the use of molecular interactions and the design of hierarchical structures will be very important for the fabrication of highly and dynamically functional materials.

7
References

1. Gray GW, Demus D, Spiess HW, Goodby JW, Vill V (eds) (1998) Handbook of liquid crystals. Wiley-VCH, Weinheim
2. (a) Gray GW (ed) (1987) Thermotropic liquid crystals. Wiley, Chichester; (b) Gray GW and Goodby JW (1984) Smectic liquid crystals. Leonard Hill, Glasgow; (c) Collings PJ, Hird M (1997) Introduction to liquid crystals. Taylor & Francis, London; (d) Goodby JW, Mehl GH, Saez IM, Tuffin RP, Mackenzie G, Auzély-Velty R, Benvegnu T, Plusquellec D (1998) Chem Commun 2057
3. (a) Percec V, Tomazos D (1992) Molecular engineering of liquid crystalline polymers. In: Allen G (ed) Comprehensive polymer science, 1st supplement. Pergamon, Oxford, p 300; (b) McArdle CB (ed) (1989) Side chain liquid crystal polymers. Blackie, Glasgow
4. Vill V, Liq Cryst – Database of liquid crystal compounds for personal computers
5. Hoffman S (1998) In: Gray GW, Demus D, Spiess HW, Goodby JW, Vill V (eds) Handbook of liquid crystals, vol 3. Wiley-VCH, Weinheim, p 393
6. For example: Radler JO, Koltover I, Salditt T, Safinya CR (1997) Science 275: 810
7. Kato T (1998) In: Gray GW, Demus D, Spiess HW, Goodby JW, Vill V (eds) Handbook of liquid crystals, vol 2B. Wiley-VCH, Weinheim, p 969
8. Kato T, Fréchet JMJ (1995) Macromol Symp 98: 311
9. (a) Kato T, Fréchet JMJ (1996) In: The polymeric materials encyclopedia, synthesis, properties and applications. p 8158; (b) Kato T (1996) Supramolecular Science 3: 53; (c) Kato T (1993) Hyomen (Surface) 31: 221; (d) Kato T (1997) Kobunshi Ronbunshu 54: 855
10. (a) Lehn J-M (1993) Makromol Chem Macromol Symp 69: 1; (b) Lehn J-M (1995) Supramolecular chemistry. VCH, Weinheim
11. Paleos CM, Tsiourvas D (1995) Angew Chem Int Ed Engl 34: 1696
12. (a) Special Issue: Structure and chemistry of the organic solid state (1994) Chem Mater 6, No 8; (b) Etter MC (1990) Acc Chem Res 23: 120; (c) MacDonald JC, Whitesides GM (1995) Chem Rev 34: 1555; (d) Lawrence DS, Jiang T, Levett M (1995) Chem Rev 95: 2229
13. (a) Zimmerman N, Moore JS, Zimmerman SC (1998) Chem Ind 604; (b) Muthukumar M, Ober CK, Thomas EL (1997) Science 277: 1225; (c) Ober CK, Wegner G (1997) Adv Mater 9: 17; (d) Stupp SI, LeBonheur V, Walker K, Li LS, Huggins KE, Keser M, Amstutz A (1997) Science 276: 384; (e) Stupp SI, Keser M, Tew GN (1998) Polymer 39: 4505
14. Joesten MD, Schaad LJ (1974) Hydrogen bonding. Dekker, New York
15. Jeffrey GA (1997) An introduction to hydrogen bonding. Oxford Press, Oxford
16. (a) de Kock AC (1904) Z Phys Chem 48: 129; (b) Vorlander D (1908) Ber Dtsch Chem Ges 41: 2033; (c) Bradfield AE, Jones B (1929) J Chem Soc 2660; (d) Weygand C, Gabler RZ (1940) Z Phys Chem B46: 270
17. Matsunaga Y, Terada M (1986) Mol Cryst Liq Cryst 141: 321
18. Noller CR, Rockwell WC (1938) J Am Chem Soc 60: 2076
19. Fischer E, Helferich B (1911) Justus Liebigs Ann Chem 383: 68
20. (a) Jeffrey GA (1986) Acc Chem Res 19: 168; (b) Jeffrey GA, Wingert LM (1992) Liq Cryst 14: 179; (c) Jeffrey GA (1984) Mol Cryst Liq Cryst 110: 221
21. Goodby JW (1984) Mol Cryst Liq Cryst 110: 205
22. Praefcke K, Marquardt P, Kohne B, Stephan W (1991) J Carbohydr Chem 10: 539

23. Diele S, Madicke A, Geissler E, Meinel D, Demus D, Sackmann H (1989) Mol Cryst Liq Cryst 166: 131
24. Hentrich F, Diele S, Tschierske C (1994) Liq Cryst 17: 827
25. Kato T, Fréchet JMJ (1989) J Am Chem Soc 111: 8533
26. Kato T, Fréchet JMJ (1989) Macromolecules 22: 3818; Erratum, ibid (1990) 23: 360
27. Brienne MJ, Gabard J, Lehn J-M, Stibor I (1989) J Chem Soc Chem Commun 1868
28. Kato T, Kawakami T (1997) Chem Lett 211
29. Kato T, Kutsuna T, Hanabusa K, Ukon M (1998) Adv Mater 10: 606
30. Mizoshita N, Kutsuna T, Hanabusa K, Kato T (1999) Chem Commun 781
31. Kato T, Fréchet JMJ, Wilson PG, Saito T, Uryu T, Fujishima A, Jin C, Kaneuchi F (1993) Chem Mater 5: 1094
32. Kato T, Wilson PG, Fujishima A, Fréchet JMJ (1990) Chem Lett 2003
33. Kato T, Kihara H, Uryu T, Fujishima A, Fréchet JMJ (1992) Macromolecules 25: 6836
34. Kato T, Uryu T, Kaneuchi F, Jin C, Fréchet JMJ (1993) Liq Cryst 14: 1311
35. Kato T, Kihara H, Uryu T, Ujiie S, Iimura K, Fréchet JMJ, Kumar U (1993) Ferroelectrics 148: 161
36. Yu LJ (1993) Liq Cryst 14: 1303
37. Kihara H, Kato T, Uryu T, Ujiie S, Kumar U, Fréchet JMJ, Bruce DW, Price DJ (1996) Liq Cryst 21: 25
38. Walba DM, Korblova E, Shao E, Maclennan JE, Link DR, Clark NA (1998) Proceedings of the Fifth International Display Workshops, p 97
39. Fukumasa M, Kato T, Uryu T, Fréchet JMJ (1993) Chem Lett 65
40. Kato T, Fukumasa M, Fréchet JMJ (1995) Chem Mater 7: 368
41. Machida S, Urano T, Sano K, Kato T (1997) Langmuir 13: 576
42. (a) Steinstrasser RZ (1972) Naturforsch 276: 774; (b) Neubert ME, Jirousek MR, Hanlon CA (1986) Mol Cryst Liq Cryst 133: 223
43. Kato T, Fujishima A, Fréchet JMJ (1990) Chem Lett 919
44. Kato T, Adachi H, Fujishima A, Fréchet JMJ (1992) Chem Lett 265
45. Willis K, Luckhurst JE, Price DJ, Fréchet JMJ, Kihara H, Kato T, Ungar G, Bruce DW (1996) Liq Cryst 21: 585
46. (a) Koh KN, Araki K, Shinkai S (1994) Tetrahedron Lett 35: 8255; (b) Koh KN, Arai K, Komori T, Shinkai S (1995) Tetrahedron Lett 36: 5191
47. Bernhardt H, Weissflog W, Kresse H (1997) Chem Lett 151
48. (a) Kato T, Saito G, Uryu T, Ujiie S, Iimura K, Fréchet JMJ (1993) Abst Jpn Liq Cryst Conf p 106; (b) Bernhardt H, Weissflog W, Kresse H (1998) Liq Cryst 24: 895
49. Deschenaux R, Monnet F, Serrano E, Turpin F (1998) Helv Chim Acta 81: 2072
50. Price DJ, Adams H, Bruce DW (1996) Mol Cryst Liq Cryst 289: 127
51. Kresse H, Szulzewsky I, Diele S, Paschke R (1994) Mol Cryst Liq Cryst 238: 13
52. Sideratou Z, Paleos CM, Skoulios A (1995) Mol Cryst Liq Cryst 265: 19
53. Tian Y, Xu X, Zhao Y, Tang X, Li T (1997) Liq Cryst 22: 87
54. Lin HC, Ko CW, Guo K, Cheng TW (1999) Liq Cryst 26: 613
55. Bruce DW, Price DJ (1994) Adv Mater Opt Electron 4: 273
56. Willis K, Price DJ, Adams H, Ungar G, Bruce DW (1995) J Mater Chem 5: 2195
57. (a) Price DJ, Richardson T, Bruce DW (1995) J Chem Soc Chem Commun 1911; (b) Price DJ, Willis K, Richardson T, Ungar G, Bruce DWJ (1997) J Mater Chem 7: 883
58. Bruce DW, Dunmur DA, Lalinde E, Maitlis PM, Styring P (1988) Liq Cryst 3: 385
59. Koga T, Ohba H, Takase A, Sakagami S (1994) Chem Lett 2071
60. For example: (a) Hirst SC, Hamilton AD (1991) J Am Chem Soc 113: 382; (b) Vincent C, Hirst SC, Garcia-Tellado F, Hamilton AD (1991) J Am Chem Soc 113: 5466; (c) Garcia-Tellado F, Geib SJ, Goswami S, Hamilton AD (1991) J Am Chem Soc 113: 9265
61. (a) Kato T, Kubota Y, Moteki T, Uryu T (1995) Chem Lett 1127; (b) Kato T, Ogasawara M, Goodby JW, unpublished results
62. Ihata O, Yokota H, Kanie K, Ujiie S, Kato T (2000) Liq Cryst 27: 69
63. (a) Kato T, Ogasawara M, Ujiie S (1999) Kobunshi Ronbunshu 56: 410; (b) Kato T, Ogasawara M, unpublished results

64. Goldmann D, Dietel R, Janietz D, Schmidt C, Wendorff JH (1998) Liq Cryst 24: 407
65. Paleos CM, Tsiourvas D, Fillipakis S, Fillipaki L (1994) Mol Cryst Liq Cryst 242: 9
66. Xu B, Swager TM (1995) J Am Chem Soc 117: 5011
67. Fukumasa M, Takeuchi K, Kato T (1998) Liq Cryst 24: 325
68. Johnson SL, Rumon KA (1965) J Phys Chem 69: 74
69. Odinokov SE, Mashkovsky AA, Glazunov VP, Iogansen AV (1976) Spectrochim Acta 32A: 1355
70. Odinokov SE, Iogansen AV (1972) Spectrochim Acta 28A: 2343
71. Kato T, Jin C, Kaneuchi F, Uryu T (1993) Bull Chem Soc Jpn 66: 3581
72. Veytsman BA (1995) Liq Cryst 18: 595
73. Sear RP, Jackson G (1995) Phys Rev Lett 74: 4261
74. Bladon P, Griffin AC (1993) Macromolecules 26: 6604
75. (a) Finkelmann H, Rehage G (1984) Adv Polym Sci 60/61: 99; (b) Ober CK, Jin J-I, Lenz RW (1985) Adv Polym Sci 59: 1985
76. Fouquey C, Lehn J-M, Levelut AM (1990) Adv Mater 2: 254
77. Sato A, Kato T, Uryu T (1996) J Polym Sci, Part A: Polym Chem 34: 503
78. (a) Kato T, Kihara H, Ujiie S, Uryu T, Fréchet JMJ (1996) Macromolecules 29: 8734; (b) Kato T, Adachi H, Hirota N, Fujishima A, Fréchet JMJ (1992) Contem Top Polym Sci 7: 299
79. Kumar U, Kato T, Fréchet JMJ (1992) J Am Chem Soc 114: 6630
80. Kumar U, Fréchet JMJ, Kato T, Ujiie S, Iimura K (1992) Angew Chem Int Ed Engl 31: 1531
81. Araki K, Kato T, Kumar U, Fréchet JMJ (1995) Macromol Rapid Commun 16: 733
82. Kumar U, Fréchet JMJ (1992) Adv Mater 4: 665
83. Kato T, Nakano M, Moteki T, Uryu T, Ujiie S (1995) Macromolecules 28: 8875
84. Bazuin CG, Brandys FA (1992) Chem Mater 4: 970
85. Bazuin CG, Brandys FA, Eve TM, Plante M (1994) Macromol Symp 84: 183
86. Brandys FA, Bazuin CG (1996) Chem Mater 8: 83
87. Kato T, Hirota N, Fujishima A, Fréchet JMJ (1996) J Polym Sci, Part A: Polym Chem 34: 57
88. Stewart D, Imrie CT (1995) J Mater Chem 5: 223
89. Kawakami T, Kato T (1998) Macromolecules 31: 4475
90. Tal'roze RV, Kuptsov SA, Sycheva TI, Bezborodov VS, Platé NA (1995) Macromolecules 28: 8689
91. Tal'roze RV, Kuptsov SA, Lebedeva TL, Shandryuk GA, Stepina ND (1997) Macromol Symp 117: 219
92. Malik S, Dahl PK, Mashelkar RA (1995) Macromolecules 28: 2159
93. Wu X, Zhang G, Zhang H (1998) Macromol Chem Phys 199: 2101
94. Ruokolainen J, Tanner J, Ikkala O, ten Brinke G, Thomas EL (1998) Macromolecules 31: 3532
95. Ruokolainen J, Torkkeli M, Serimaa R, Komanschek BE, ten Brinke G, Ikkala O (1997) Macromolecules 30: 2002
96. Luyten MC, Alberda van Ekenstein GOR, Wildeman J, ten Brinke G, Ruokolainen J, Ikkala O, Torkkeli M, Serimaa R (1998) Macromolecules 31: 9160
97. Ruokolainen J, Makinen R, Torkkeli M, Makela T, Serimaa R, ten Brinke G, Ikkala O (1998) Science 280: 557
98. van Nunen JLM, Folmer BFB, Nolte RJM (1997) J Am Chem Soc 119: 283
99. Gulik-Krzywicki T, Fouquey C, Lehn J-M (1993) Proc Natl Acad Sci USA 90: 163
100. Alexander C, Jariwala CP, Lee CM, Griffin AC (1994) Macromol Symp 77: 283
101. Lee CM, Jariwala CP, Griffin AC (1994) Polymer 35: 4550
102. He C, Donald AM, Griffin AC, Waigh T, Windle AH (1998) J Polym Sci Part B: Polym Chem 36: 1617
103. Lee CM, Griffin AC (1997) Macromol Symp 117: 281
104. Kato T, Kubota Y, Uryu T, Ujiie S (1997) Angew Chem Int Ed Engl 36: 1617
105. Kato T, Ihata O, Ujiie S, Tokita M, Watanabe J (1998) Macromolecules 31: 3551

106. Ihata O, Kato T (1999) Sen'i Gakkaishi 55: 274
107. Kato T (1998) Korea Polym J 6: 153
108. Kato T, Kihara H, Kumar U, Uryu T, Fréchet JMJ (1994) Angew Chem Int Ed Engl 33: 1644
109. Kihara H, Kato T, Uryu T, Fréchet JMJ (1996) Chem Mater 8: 961
110. Kihara H, Kato T, Uryu T (1996) Trans Mater Res Soc 20: 327
111. Kihara H, Kato T, Uryu T, Fréchet JMJ (1998) Liq Cryst 24: 413
112. Pourcain CBSt, Griffin AC (1995) Macromolecules 28: 4116
113. Kurihara S, Mori T, Nonaka T (1998) Macromolecules 31: 5940
114. (a) Kelly SM (1995) J Mater Chem 5: 2047; (b) Kelly SM (1998) Liq Cryst 24: 71
115. (a) Broer DJ, Hikmet RAM, Challa G (1989) Makromol Chem 190: 3201; (b) Broer DJ, Lub J, Mol GN (1995) Nature 378: 467
116. (a) Finkelmann H (1983) Phil Trans Roy Soc 309: 105; (b) Cabrera I, Dittrich A, Ringsdorf H (1991) Angew Chem Int Ed Engl 30: 76; (c) Coles HJ (1985) Faraday Discuss Chem Soc 79: 201
117. (a) Ikeda T, Horiuchi S, Karanjit DB, Kurihara S, Tazuke S (1990) Macromolecules 23: 36; (b) Sasaki T, Ikeda T, Ichimura K (1992) Macromolecules 25: 3807
118. (a) Pearce EM, Kwei TK, Min BY (1984) J Macromol Sci Chem A21: 1181; (b) Lee JY, Painter PC, Coleman MM (1988) Macromolecules 21: 954; (c) Dai YK, Chu EY, Xu ZS, Pearce EM, Okamoto Y, Kwei TK (1994) J Polym Sci Part A: Polym Chem 32: 397; (d) de Meftahi MV, Fréchet JMJ (1988) Polymer 29: 477; (e) Lange RFM, Meijer EW (1995) Macromolecules 28: 782
119. (a) Hanabusa K, Miki T, Taguchi Y, Koyama T, Shirai H (1993) J Chem Soc Chem Commun 1382; (b) Jeong SW, Murata K, Shinkai S (1996) Supramol Sci 3: 83
120. Kimizuka N, Kawasaki T, Hirata K, Kunitake T (1998) J Am Chem Soc 120: 4094
121. Ashton PR, Brown GR, Hayes W, Menzer S, Philp D, Stoddart JF, Williams DJ (1996) Adv Mater 8: 564
122. (a) Valiyaveettil S, Enkelmann V, Müllen K (1994) J Chem Soc Chem Commun 2097; (b) Pfaadt M, Moessner G, Pressner D, Valiyaveettil S, Boeffel C, Müllen K, Spiess HW (1995) J Mater Chem 5: 2265
123. Valiyaveettil S, Müllen K (1998) New J Chem 89
124. Inoue K, Itaya T, Azuma N (1998) Supramol Sci 5: 163
125. Gray GW, Jones B (1953) J Chem Soc 4179
126. Gray GW, Jones B (1955) J Chem Soc 236
127. Gray GW, Jones B, Marson F (1957) J Chem Soc 393
128. Blumstein A, Clough SB, Patel L, Blumstein RB, Hsu EC (1976) Macromolecules 9: 243
129. Hoshino H, Jin JI, Lenz RW (1984) J Appl Polym Sci 29: 547
130. Barmatov EB, Pebalk DA, Barmatova MV, Shibaev VP (1997) Liq Cryst 23: 447
131. Asakawa M, Ashton PR, Brown GR, Hayes W, Menzer S, Stoddart JF, White AJP, Williams DJ (1996) Adv Mater 8: 37
132. Alcala R, Martinez-Carrera S (1972) Acta Crystallogr B 28: 1671
133. (a) Duchamp DJ, Marsh RE (1969) Acta Crystallogr B 25: 5; (b) Ermer O (1988) J Am Chem Soc 110: 3747
134. Zimmerman SC, Zeng F, Reichert DEC, Kolotuchin SV (1996) Science 271: 1095
135. (a) Beijer FH, Sijbesma RP, Kooijman H, Spek AL, Meijer EW (1998) J Am Chem Soc 120: 6761; (b) Sijbesma RP, Beijer FH, Brunsveld L, Folmer BJB, Hirschberg JHKK, Lange RFM, Lowe JKL, Meijer EW (1997) Science 278: 1601
136. (a) Ducharme Y, Wuest JD (1988) J Org Chem 53: 5787; (b) Simard M, Su D, Wuest JD (1991) J Am Chem Soc 113: 4696; (c) Gallant M, Viet MTP, Wuest JD (1991) J Am Chem Soc 113: 721
137. (a) Kleppinger R, Lillya CP, Yang C (1995) Angew Chem Int Ed Engl 34: 1637; (b) Kleppinger R, Lillya CP, Yang C (1997) J Am Chem Soc 119: 4097
138. Beginn U, Lattermann G (1994) Mol Cryst Liq Cryst 241: 215
139. Suárez M, Lehn J-M, Zimmerman SC, Skoulios A, Heinrich B (1998) J Am Chem Soc 120: 9526

140. Chandrasekhar S, Shadashiva BK, Suresh KA (1977) Pramana 9: 471
141. Kobayashi Y, Matsunaga Y (1987) Bull Chem Soc Jpn 60: 3515
142. Matsunaga Y, Nakayasu Y, Sakai S, Yonenaga M (1986) Mol Cryst Liq Cryst 141: 327
143. Kawada H, Matsunaga Y, Takamura T, Terada M (1988) Can J Chem 66: 1867; Harada Y, Matsunaga Y, Miyajima N, Sakamoto S (1995) J Mater Chem 5: 2305
144. Malthéte J, Levelut AM, Liébert L (1992) Adv Mater 4: 37
145. (a) Ungar G, Abramic D, Percec V, Heck JA (1996) Liq Cryst 21: 73; (b) Percec V, Ahn CH, Bera TK, Ungar G, Yeardley DJP (1999) Chem Eur J 5: 1070
146. Kato T, Kondo G, Kihara H (1997) Chem Lett 1143
147. Schubert H, Hoffmann S, Hauschild J, Marx I (1977) Z Chem 17: 415
148. (a) Mori A, Nimura R, Takeshita H (1991) Chem Lett 77; (b) Mori A, Takeshita H, Nimura R, Isobe M (1993) Liq Cryst 14: 821; (c) Mori A, Nimura R, Isobe M, Takeshita H (1992) Chem Lett 859
149. Vill V, Böcker T, Thiem J, Fisher F (1989) Liq Cryst 6: 349
150. Tschierske C, Lunow A, Zascheke H (1990) Liq Cryst 8: 885
151. (a) Pfannemüller B, Welte W, Chin E, Goodby JW (1986) Liq Cryst 1: 357; (b) Goodby JW, Marcus MA, Chin E, Finn PL, Pfannemüller B (1988) Liq Cryst 3: 1569; (c) Vill V, Kelkenberg H, Thiem J (1992) Liq Cryst 11: 459
152. (a) Goodby JW, Haley JA, Mackenzie G, Watson MJ, Plusquellec D, Ferrieres V (1995) J Mater Chem 5: 2209; (b) van Doren HA, van der Geest R, de Ruijter CF, Kellogg RM, Wynberg H (1990) Liq Cryst 8: 109
153. Praefcke K, Levelut AM, Kohne B, Eckert A (1989) Liq Cryst 6: 263
154. Auzély-Velty R, Benvegnu T, Plusquellec D, Mackenzie G, Haley JA, Goodby JW (1998) Angew Chem Int Ed 37: 2511
155. (a) Masuda M, Shimizu T (1996) Chem Commun 1057; (b) Shimizu T (1999) Trans Mater Res Soc Jpn 24: 431
156. (a) Praefcke K, Kohne B, Psaras P, Hempel J (1991) J Carbohydr Chem 10: 523; (b) Praefcke K, Marquardt P, Kohne B, Stephan W (1991) Mol Cryst Liq Cryst 203: 149; (c) Praefcke K, Blunk D (1993) Liq Cryst 14: 1191
157. (a) Lattermann G, Staufer G (1989) Liq Cryst 4: 347; (b) Lattermann G, Staufer G (1990) Mol Cryst Liq Cryst 191: 199; (c) Lattermann G, Staufer G, Brezesinski G (1991) Liq Cryst 10: 169
158. (a) Schellhorn M, Lattermann G (1995) Macromol Chem Phys 196: 211; (b) Staufer G, Lattermann G (1991) Makromol Chem 192: 2421
159. Tschierske C, Zaschke H (1990) J Chem Soc Chem Commun 1013
160. Tschierske C, Lunow A, Joachimi D, Hentrich F, Girdziunaite D, Zaschke H, Mdicke A, Brezesinski G, Kuschel F (1991) Liq Cryst 9: 821
161. Joachimi D, Tschierske C, Müller H, Wendorff JH, Schneider L, Kleppinger R (1993) Angew Chem Int Ed Engl 32: 1165
162. Hilderbrandt F, Schroter JA, Tschierske C, Festag R, Kleppinger R, Wendorff JH (1995) Angew Chem Int Ed Engl 34: 1631
163. Hildebrandt F, Schröter JA, Tschierske C, Festag R, Wittenberg M, Wendorff JH (1997) Adv Mater 9: 564
164. (a) Kölbel M, Beyersdorff T, Sletvold I, Tschierske C, Kain J, Diele S (1999) Angew Chem Int Ed 38: 1077; (b) Tschierske C (1998) J Mater Chem 8: 1485
165. Bunning JD, Lydon JE, Eaborn C, Jackson PH, Goodby JW, Gray GW (1982) J Chem Soc Faraday Trans 1: 713
166. Eaborn C (1952) J Chem Soc 2840
167. Polischchuk AP, Timofeeva TV, Makarova NN, Antipin MY, Struchkov YT (1991) Liq Cryst 9: 433
168. Mariani P, Mazabard A, Garbesi A, Spada GP (1989) J Am Chem Soc 111: 6369
169. Bonazzi S, Capobianco M, De Morais MM, Garbesi A, Gottarelli G, Mariani P, Bossi MGP, Spada GP, Tondelli L (1991) J Am Chem Soc 113: 5809
170. Bonazzi S, De Morais MM, Gotttarelli G, Mariani P, Spada GP (1993) Angew Chem Int Ed Engl 32: 248

171. Ciuchi F, Nicola GD, Franz H, Gottarelli G, Mariani P, Bossi MGP, Spada GP (1994) J Am Chem Soc 116: 7064
172. Gottarelli G, Masiero S, Mezzina E, Spada GP, Mariani P, Recanatini M (1998) Helv Chim Acta 81: 2078
173. Kotera M, Lehn J-M, Vifneron JP (1994) J Chem Soc Chem Commun 197
174. Terech P, Weiss RG (1995) Chem Rev 97: 3133
175. (a) Tachibana T, Mori T, Hori K (1980) Bull Chem Soc Jpn 53: 1714; (b) Hafkamp RJH, Kokke BPA, Danke IM, Geurts HPM, Rowan AE, Feiters MC, Nolte RJM (1997) Chem Commun 545; (c) Aggeli A, Bell M, Boden N, Keen JN, Knowles PF, McLeish TCB, Pitkeathly M, Radford SE (1997) Nature 386: 259; (d) Murata K, Aoki M, Suzuki T, Harada T, Kawabata H, Komori T, Ohseto F, Ueda K, Shinkai S (1994) J Am Chem Soc 116: 6664; (e) Yasuda Y, Takebe Y, Fukumoto M, Inada H, Shirota Y (1996) Adv Mater 8: 740; (f) van Esch J, De Feyter S, Kellogg RM, De Schryver F, Feringa BL (1997) Chem Eur J 3: 1238
176. Hanabusa K, Yamada M, Kimura M, Shirai H (1996) Angew Chem Int Ed Engl 35: 1949
177. (a) Tanaka R, Suzuki M, Kimura M, Shirai H (1997) Adv Mater 9: 1095; (b) Hanabusa K, Shirai H (1998) Kobunshi Ronbunshu 55: 585
178. (a) Kato T, Kondo G, Hanabusa K (1998) Chem Lett 193; (b) Kato T, Kutsuna T, Hanabusa K (1999) Mol Cryst Liq Cryst 332: 377
179. Yabuuchi K, Rowan AE, Nolte RJM, Kato T (2000) Chem Mater, in press
180. Mizoshita N, Hanabusa K, Kato T (1999) Adv Mater 11: 392
181. (a) Crawford GP, Zumer S (eds) (1996) Liquid crystals in complex geometries formed by polymer and porous networks. Taylor & Francis, London; (b) Hikmet RAM (1992) Adv Mater 4: 679
182. For example: (a) Ujiie S, Iimura K (1992) Macromolecules 25: 3174; (b) Bazuin CG, Guillon D, Skoulios A, Nicoud JF (1986) Liq Cryst 1: 181
183. For example: (a) Percec V, Johansson G, Heck J, Ungar G, Batty SV (1993) J Chem Soc Perkin Trans 1 1411; (b) Percec V, Tomazos D, Heck J, Blackwell H, Ungar G (1994) J Chem Soc Perkin Trans 2 31; (c) Lee M, Oh N-K, Lee H-K, Zin W-C (1996) Macromolecules 29: 5567; (d) Lee M, Cho B-K (1998) Chem Mater 10: 1894; (e) Ohtake T, Ito K, Nishina N, Kihara H, Ohno H, Kato T (1999) Polym J 31: 1155; (f) Ohtake T, Ogasawara M, Ito-Akita K, Nishina N, Ohno H, Kato T (2000) Chem Mater, in press
184. For example: (a) Bengs H, Renkel R, Ringsdorf H, Baehr C, Ebert M, Wendorff JH (1991) Makromol Chem Rapid Commun 12: 439; (b) Praefcke K, Singer D, Kohne B, Ebert M, Liebmann A, Wendorff JH (1991) Liq Cryst 10: 147. For electron-donor acceptor interactions, see for example: (a) Kosaka Y, Kato T, Uryu T (1994) Macromolecules 27: 2658; (b) Schleeh T, Imrie CT, Rice DM, Karasz FE, Attard GS (1993) J Polym Sci Part A Polym Chem 31: 1859

Part II
Inorganic Assemblies

Synergistic Effect of Serendipity and Rational Design in Supramolecular Chemistry

Rolf W. Saalfrank, Eveline Uller, Bernhard Demleitner, Ingo Bernt

Institut für Organische Chemie, Universität Erlangen-Nürnberg, Henkestrasse 42, 91054 Erlangen, Germany
E-mail: Saalfrank@organik.uni-erlangen.de

It is like playing poker, if you play long enough you begin to make decisions based on probability, and if you are fortunate you may recognize how the cards are marked.

Richard E.P. Winpenny

Many of the higher nuclearity Werner-type clusters are fortuitous discoveries. This clearly suggests that the principles which control cluster formation are still poorly understood from the point of view of rational design. However, the predictable nature of coordination chemistry has been used successfully for the specific generation of the metalla-topomers of the well-known organic-based coronates and cryptates. Furthermore, it has been shown that the metal coordination geometry and the orientation of the interacting sites in a given ligand provide the instruction, or blueprint, for the rational design of high symmetry coordination clusters. Based on these principles, and the interplay with serendipity, directed syntheses for polynuclear metalla-coronates, sandwich complexes, manifold metalla-cryptates, and metalla-cylinders have been developed. Only recently, the combination of detailed symmetry considerations with the basic protocols of coordination chemistry have made the design of rational strategies for the construction of a variety of nanoscale systems with procured shape and size feasible. Highlights among these species are cube-like clusters and multicompartmental cylindrical nanostructures.

Keywords: Supramolecular chemistry, Self-assembly, Cage compounds, Metal complexes, Host-guest chemistry

Structure and Bonding, Vol. 96
© Springer Verlag Berlin Heidelberg 2000

1
Introduction

Synthetic organic chemists are adept in fabricating large, complex molecules in a controlled, systematic and sequential manner. Many biological structures are built up by a modular approach from simple, basic subunits, thus minimizing the amount of information required for the specific ensemble. However, chemists are not limited to systems similar to those found in nature but they are free to create unknown non-natural species and to invent novel synthetic methodologies. Self-assembly has been recognized as a powerful tool for the construction of supramolecular scaffolds [1]. It is based on weak, reversible, non-covalent interactions, which facilitate error checking and self-correction. Initially, most assemblies were discovered by serendipity rather than by design. Only recently there has been significant progress towards the generation of numerous clusters with predesigned molecular architectures by choosing rigid ligands which have strong preferences for specific coordination modes. Additionally, there have been many endeavors to control structure by use of templates. We will not attempt to cover the literature exhaustively, but more recent notable advances in this area will be given preference, as outlined below.

2
General Principles for the Design of Ligand and Metal Controlled Multicomponent Self-Assembly Reactions

Since the principles which control cluster formation were neglected for a long time, most of the higher nuclearity Werner-type clusters were discovered by serendipity rather than by design. Only recently has the predictable nature of coordination chemistry been applied successfully for the specific generation of the metalla-topomers 4–6 of the well-known organic-based coronates 1, {2}-cryptates 2, and {3}-cryptates 3 (Fig. 1) [1c, 2]. The metalla-topomers 5 and 6 essentially differ only in the substitution of the nitrogen bridgehead atoms of 2 and 3 by metal ions. Furthermore, it has been demonstrated that the metal coordination geometry and the alignment of the interacting donor sites of the ligand store the information for the rational design of high symmetry coordination clusters [1b, 3].

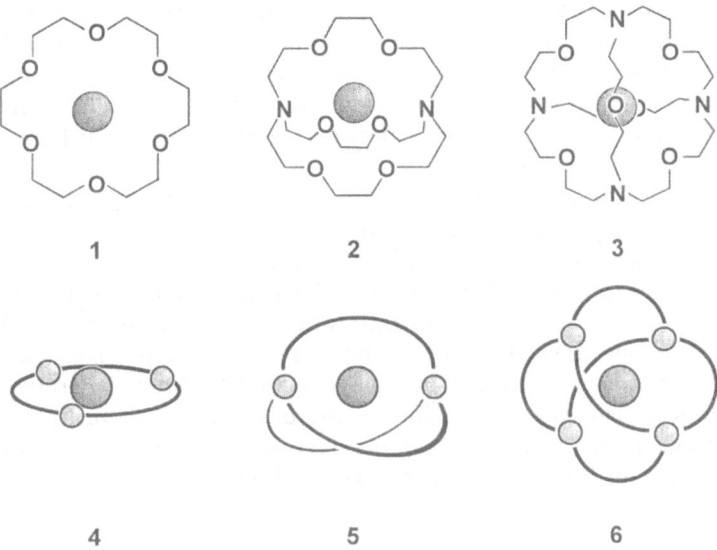

Fig. 1. Pictorial representation: Coronates **1**, {2}-cryptates **2**, {3}-cryptates **3**, and their metalla-topomers **4–6**

2.1
Metalla-Coronates

Cyclic structures generally analogous to classical organo-crown ethers are accessible by simply substituting an ethano bridge with a transition metal ion [1y]. This metalla-crown analogy has proven highly successful for the controlled preparation of homonuclear, mixed-valence, and heteronuclear assemblies. Crown ethers selectively complex alkaline and alkaline earth metal cations, and the complexation of these different sized cations leads to coronates of various structures. Besides the topological analogy between crown ethers and metalla-crown ethers, these compounds enjoy a number of advantages over their organic counterparts. In particular, they are accessible in high yields in one-pot reactions and exhibit spectroscopic, electronic, magnetic and well-defined novel host-guest properties, which are inaccessible with their organic analogues.

Since the ionic radii of alkaline and alkaline earth metal cations differ significantly in size within each group, the inclusion of small cations such as Na^+ or Ca^{2+} produces metalla-coronates with 1:1 metal to metalla-crown stoichiometry. In contrast, encapsulation of the larger K^+ cation results in the formation of metalla-crown ether sandwich complexes with 1:2 metal to metalla-crown stoichiometry.

2.1.1
Trinuclear Metalla-Coronates and Sandwich Complexes

Reaction of ketipinate H_2L^1 **7** with copper(II) acetate in the presence of calcium nitrate leads to the trinuclear metalla-coronate [Ca ⊂

$(Cu_3L_3^1)(NO_3)_2] \cdot THF \cdot H_2O$ **8**. The neutral metalla-crown scaffold encapsulates a calcium ion in the center and the resulting dipositive charge is compensated by two axially coordinated nitrate ions. However, if potassium hydroxide is used for the double deprotonation of H_2L^1 **7**, the presence of the larger potassium ion leads to the sandwich complex $[K \subset (Cu_3L_3^1)_2 (OMe)] \cdot$ 6HOMe **9** [4] (Scheme 1).

According to the X-ray crystallographic analysis of metalla-coronate **8**, the copper atoms represent the vertices of an equilateral triangle and are linked along the bis-bidentate diethyl ketipinate dianions L^1. Thereby each copper(II) ion is primarily coordinated in a square-planar fashion to four oxygen donor atoms. Additional coordination of one water molecule, one tetrahydrofuran molecule and one oxygen atom of one nitrate ion, respectively, leads to a square-pyramidal environment of each copper(II) center generated by five oxygen donors. The calcium ion in the core of **8** is axially coordinated to two oxygen atoms of one nitrate ion and one oxygen atom of the second nitrate ion which is also coordinated to a copper center (Fig. 2).

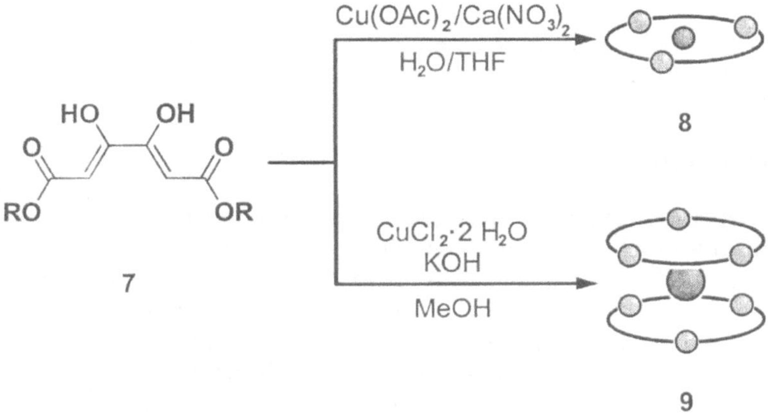

Scheme 1. Formation and schematic representation of $[Ca \subset (Cu_3L_3^1)(NO_3)_2] \cdot THF \cdot H_2O$ **8** (L^1: R = Et) and $[K \subset (Cu_3L_3^1)_2(OMe)] \cdot$ 6HOMe **9** (L^1: R = t-Bu)

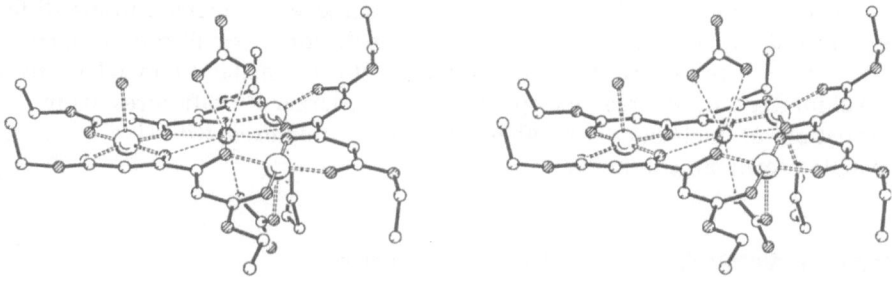

Fig. 2. Stereo representation of X-ray crystallographic structure of trinuclear metalla-coronate $[Ca \subset (Cu_3L_3^1)(NO_3)_2] \cdot THF \cdot H_2O$ **8** (L^1: R = Et)

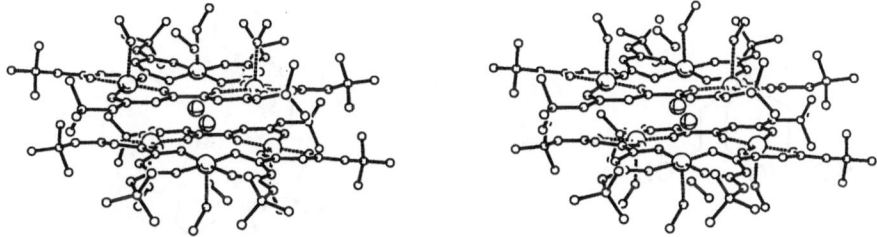

Fig. 3. Stereo representation of X-ray crystallographic structure of sandwich [K ⊂ (Cu$_3$L$_3^1$)$_2$(OMe)] · 6HOMe **9** (L^1: R = *t*-Bu). K$^+$, MeO$^-$ disordered

X-ray crystallographic analysis reveals that [K ⊂ (Cu$_3$L$_3^1$)$_2$(OMe)] · 6HOMe **9** is composed of two neutral metalla-crowns, rotated relative to each other by 60° and which are sandwich-like linked together via a potassium ion. Here the methoxide counterion is also bound to the potassium ion. Methanol molecules lead to coordinative saturation of the six copper ions. Thus, each of the six Cu^{2+} ions is surrounded by oxygen donors in a tetragonal pyramidal geometry (Fig. 3).

2.1.2
Higher Nuclearity Metalla-Coronates and Metalla-Coronands

Various polyiron-oxo species are present in aqueous solution, but the growth of polyiron complexes is difficult to control. The situation is different in the presence of additional ligands. However, defined structures are realized by serendipity rather than by rational design. A prominent member of this class is [Fe(OMe)$_2$(O$_2$CCH$_2$Cl)]$_{10}$, better known as *ferric wheel*, which is prepared in methanolic solution by the reaction of [Fe$_3$O(O$_2$CCH$_2$Cl)$_6$(H$_2$O)$_3$](NO$_3$) and Fe(NO$_3$)$_3$ · 9H$_2$O [1r, 5]. The hitherto largest cyclic iron cluster [Fe(OH)(XDK)Fe$_2$(OMe)$_4$(OAc)$_2$]$_6$ [where H$_2$XDK = *m*-xylilene diamine bis-(Kemp's triacid imide)], which contains eighteen iron(III) ions in the ring, was prepared in the presence of tetraalkylammonium carboxylate salts from slightly alkaline methanolic solutions of the diiron(III) complex [Fe$_2$O(XDK)(HOMe)$_5$(H$_2$O)](NO$_3$)$_2$ · 4H$_2$O [6].

In an attempt to generate supramolecular host-guest compounds with tripod-metal templates as building blocks, contrary to what was expected, treatment of triethanolamine **10** with sodium hydride and iron(III) chloride in THF afforded yellow crystals of {Na ⊂ (Fe$_6$[N(CH$_2$CH$_2$O)$_3$]$_6$)}Cl **11** (Scheme 2) [7].

For unequivocal characterization of **11** an X-ray crystallographic structure analysis was carried out. According to this analysis, **11** is present in the crystal as a cyclic iron(III) complex with a [12]metalla-crown-6 structure, in which each triethanolatamine acts as a tritope tetradentate ligand and links three adjacent iron(III) ions. The sodium ion is encapsulated in the center of the ring, and chloride is the counterion. The six crystallographically equivalent iron

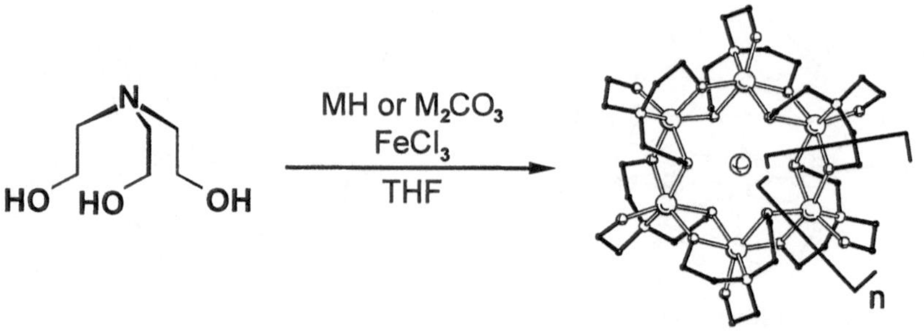

Scheme 2. Formation and schematic representation of $\{M \subset (Fe_{5+n}[N(CH_2CH_2O)_3]_{5+n})\}^+$ of 11: M = Na, $n = 1$ and 12: M = Cs, $n = 3$. Chloride is not depicted

centers of the centrosymmetric super cation $\{Na \subset (Fe_6[N(CH_2CH_2O)_3]_6)\}^+$ of 11 are located in the corners of a regular hexagon. Three sets of six oxygen donor atoms are located in the corners of three pairs of equilateral triangles that are rotated 60° relative to each other. These triangles form parallel and equidistant pairs, with one plane above and one below the hexagonal plane generated by the six iron centers. The distorted octahedral coordination sphere of the iron centers is composed of one nitrogen donor, one μ_1- and two μ_2- and μ_3-oxygen donors. The encapsulated sodium ion has a distorted octahedral coordination sphere due to the six μ_3-O donor atoms.

Template-mediated self-assembly of a supramolecular system is character-ized by the fact that, amidst a set of possibilities, the one combination of building blocks is realized that leads to the best receptor of the substrate [8]. Therefore, the six-membered cyclic structure 11 is selected from all the possible iron triethanolatamine oligomers in the presence of sodium ions. Thus, variant structures are expected when cations with ionic radii that differ from that of sodium are present. In accordance to what was expected, treatment of triethanolamine 10 with cesium carbonate and iron(III) chloride in THF afforded yellow crystals of $\{Cs \subset (Fe_8[N(CH_2CH_2O)_3]_8)\}Cl$ 12 (Scheme 2) [7]. An X-ray crystallographic analysis of this product revealed a cyclic iron(III) complex with a [16]metalla-crown-8 structure. The cesium ion is located exactly above the center of the ring. It is shifted by about 0.5 Å towards the chloride counterion. The eight iron centers of the nearly centrosymmetric cation $\{Cs \subset (Fe_8[N(CH_2CH_2O)_3]_8)\}^+$ of 12 are located in the corners of a regular octagon. Three sets of eight, of a total of twenty-four oxygen atoms, are located in the corners of three pairs of squares that are rotated 45° relative to each other. The square faces are located parallel and almost equidistant in pairs, with one plane above and one below the octagon generated by the eight iron centers (Fig. 4).

The serendipitous discovery of these six- and eight-membered rings demonstrates once again that the rules which control their formation are still poorly understood from the point of view of predesign. However, given the working plan, one realizes that only two ethanolate groups out of the

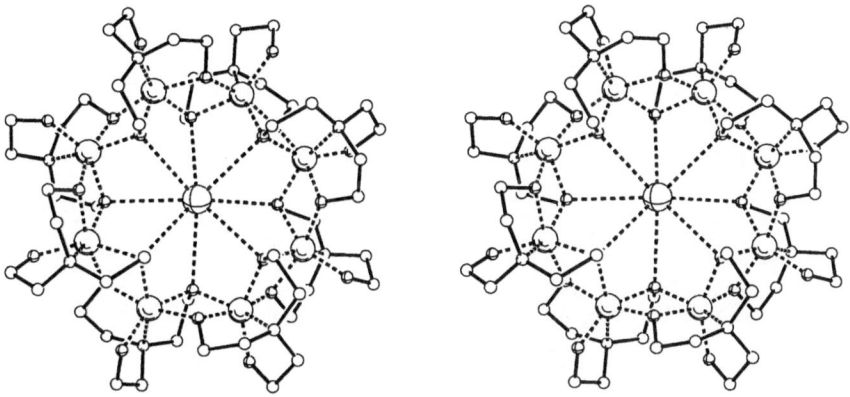

Fig. 4. Stereo representation of X-ray crystallographic structure of the cation $\{Cs \subset (Fe_8[N(CH_2CH_2O)_3]_8)\}^+$ of **12**

triethanolatamine trianion are cross-linking, while the third simply acts as a ligand to coordinatively saturate iron center. A logical consequence for the rational design of further metalla-cycles is the introduction of alkyl dietha-nolamines instead of triethanolamine **10**. For example, N-butyldiethanolamine **13** was allowed to react with calcium hydride and iron(III) chloride (Scheme 3) and an X-ray crystallographic analysis was carried out of the resulting precipitate. According to this analysis, the product is present in the crystal as the unoccupied neutral iron(III) complex $\{Fe_6[H_3C(CH_2)_3N(CH_2CH_2O)_2]_6Cl_6\}$ **14** with a [12]metalla-crown-6 structure. The six iron atoms of the centro-symmetric neutral molecule are located in the corners of an almost regular hexagon. The distorted octahedral coordination sphere of the iron atoms comprise one chloride ion, one nitrogen donor, and four μ_2-O donors [9].

The iron wheel $\{Fe_6[H_3C(CH_2)_3N(CH_2CH_2O)_2]_6Cl_6\}$ **14** crystallizes in space group $R\bar{3}$ with three molecules in the unit cell. The disk-like clusters

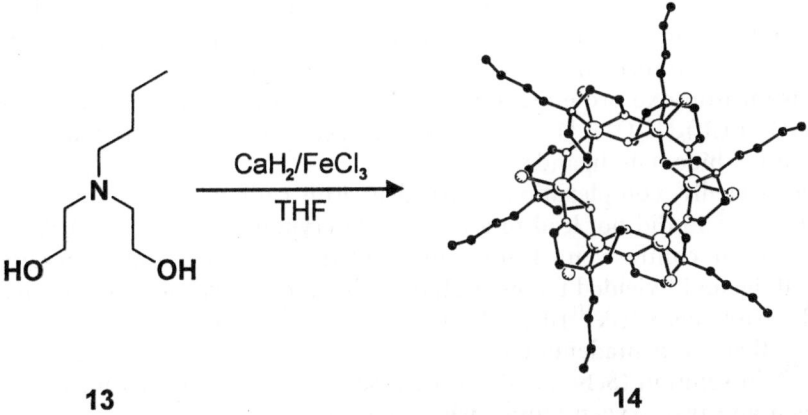

13   **14**

Scheme 3. Formation and X-ray representation of $\{Fe_6[H_3C(CH_2)_3N(CH_2CH_2O)_2]_6Cl_6\}$ **14**

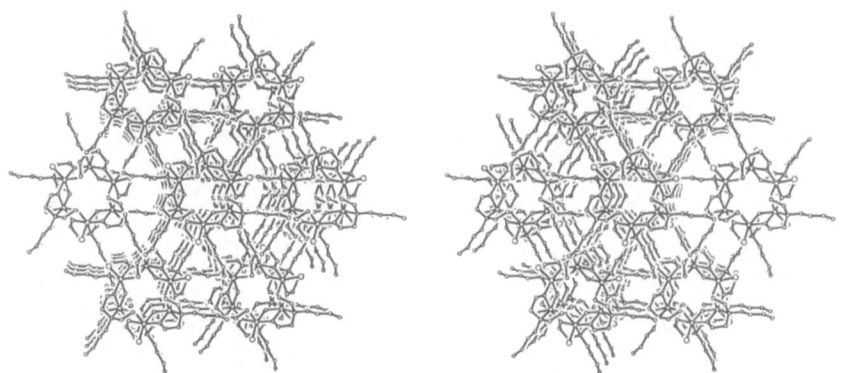

Fig. 5. Stereo representation of the crystal packing of $\{Fe_6[H_3C(CH_2)_3N(CH_2CH_2O)_2]_6Cl_6\}$ **14**

of **14** are assembled in cylindrical columns with mean intra-columnar midpoint distances d of 12.6 Å, with all the iron centers superimposed. Each column is surrounded by six parallel columns which are alternately dislocated by 1/3 d and 2/3 d against the central one (Fig. 5). There is no evidence that the iron hexagons of **14** contain cations as guests as observed for $\{Na \subset (Fe_6[N(CH_2CH_2O)_3]_6)\}Cl$ **11**.

2.1.3
Oxo-Centered Circular Helicates

Helicates consist of linear arrays of metal atoms held together by bridging polytopic ligands, helically arranged around the central axis through the metal ions [1r, 1s, 1u]. The majority of these structures consist of di- and trinuclear complexes, while more recently the research effort has been further extended to include tetra- and pentanuclear alignments. Depending on whether the metal ions prefer tetrahedral or octahedral coordination, they form double or triple helices. Among cyclic transition metal complexes, circular helicates have specific features and may be considered as toroidal helices [8, 10].

The interest in polynuclear supramolecular iron complexes mainly arises from the importance of oxo-centered polyiron aggregates as model compounds for iron-oxo proteins. In principle, μ_3-oxo-centered complexes related to $[Fe_3O(O_2CR)_6(H_2O)_3]$ [11] should also be accessible with doubly negatively charged bis-bidentate ligands.

Unexpectedly, complexation of trivalent iron cations with doubly deprotonated H_2L^2 **15** did not lead to a {2}-metalla-cryptate $[K \subset (Fe_2L_3^{2A})]PF_6$ (see Sect. 2.2). In contrast, nucleation of iron(III) with the doubly negatively charged ligand L^2 yielded the neutral, trinuclear, mixed-valence, oxo-centered, circular triple-helix $[(Fe^{III})_2Fe^{II}OL_3^{2B}]^0$ **16a** [12]. This result can be explained by the fact that the pentadentate ligand L^2 exists as two different rotamers, L^{2A} and L^{2B}, in solution (Scheme 4). In the case of L^{2A} the donor atoms are three nitrogen and two oxygen atoms, whereas in the case of L^{2B} the coordinating donors are five nitrogen atoms.

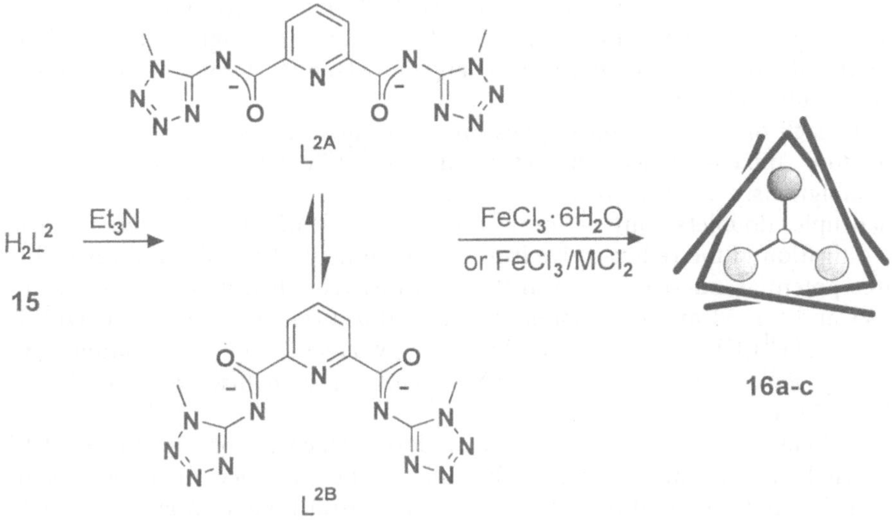

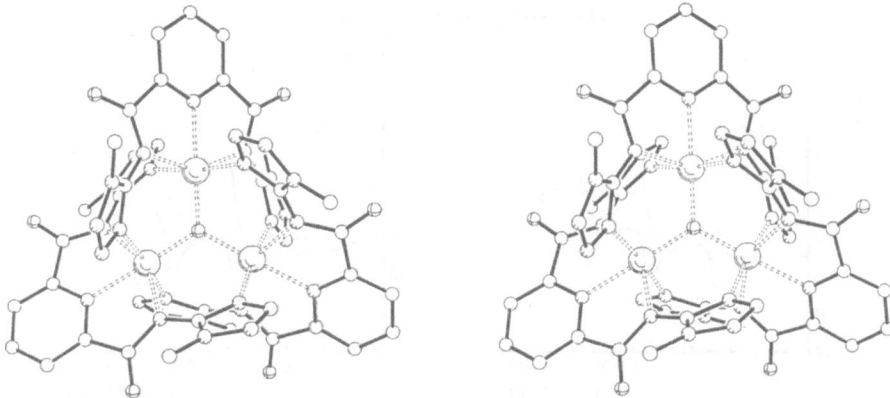

Scheme 4. Formation and schematic representation of $[(Fe^{III})_2Fe^{II}OL_3^{2B}]^0$ **16a**, $[(Fe^{III})_2Co^{II}OL_3^{2B}]^0$ **16b**, and $[(Fe^{III})_2Cu^{II}OL_3^{2B}]^0$ **16c**

For inambiguous structural determination a single crystal X-ray structure analysis was carried out on the circular triple-helix **16a**. The core of this compound is generated by an equilateral triangle with a μ_3-O^{2-} ion in the center and three iron cations at the apices. The O^{2-} ion stems from deprotonated water of crystallization of $FeCl_3 \cdot 6H_2O$. Consequently, an excess of base leads to higher yields. Each of the three doubly negatively charged pentadentate ligands L^{2B} links three metal centers. As a result, a distorted octahedral coordination geometry is observed for each iron center to five N-donors and to the bridging μ_3-O^{2-} ligand shared by all three iron ions. The

Fig. 6. Stereo representation of X-ray crystallographic structure of circular triple-helix $[(Fe^{III})_2Fe^{II}OL_3^{2B}]^0$ **16a**

iron centers of the racemic complex $[(Fe^{III})_2Fe^{II}OL_3^{2B}]^0$ **16a** are homochiral. Since the pentadentate tritopic ligands L^{2B} coordinate to neighboring iron ions at opposite sides of the (Fe,Fe,Fe)-triangular plane, a D_3-symmetric circular helix results (Fig. 6).

The lack of a counterion implies intramolecular charge compensation and therefore mixed-valence character for $[(Fe^{III})_2Fe^{II}OL_3^{2B}]^0$ **16a**. This was unambiguously confirmed by a Mössbauer spectrum which exhibits two quadruple doublets with a peak area ratio of 1:2. Cyclic voltammetric investigation of the redox-active iron centers of neutral **16a** shows a reversible three-potential one-electron transfer process. The half-wave potentials of −635 and −1230 mV correspond to the redox processes $[(Fe^{III})_2Fe^{II}OL_3^{2B}]^0$ $\leftrightarrow [Fe^{III}(Fe^{II})_2OL_3^{2B}]^- \leftrightarrow [(Fe^{II})_3OL_3^{2B}]^{2-}$, whereas the oxidation of $[(Fe^{III})_2Fe^{II}OL_3^{2B}]^0 \leftrightarrow [(Fe^{III})_3OL_3^{2B}]^+$ occurs at a half-wave potential of +90 mV (Fig. 7a).

The heteronuclear mixed-valence complexes **16b,c** were obtained from H_2L^2 **15**, triethylamine, and iron(III) chloride in the presence of an excess of cobalt(II) chloride or copper(II) chloride. In neutral **16b,c** iron is present only in the oxidation state 3^+, as unambiguously confirmed by Mössbauer spectroscopy, which reveals only one quadruple doublet. Cyclic voltammetric investigation of the redox-active metal centers of neutral **16b** shows a reversible three-potential one-electron transfer process. The half-wave potentials of −660 and −1310 mV correspond to the redox processes $[(Fe^{III})_2Co^{II}OL_3^{2B}]^0 \leftrightarrow [Fe^{III}Fe^{II}Co^{II}OL_3^{2B}]^- \leftrightarrow [(Fe^{II})_2Co^{II}OL_3^{2B}]^{2-}$, whereas

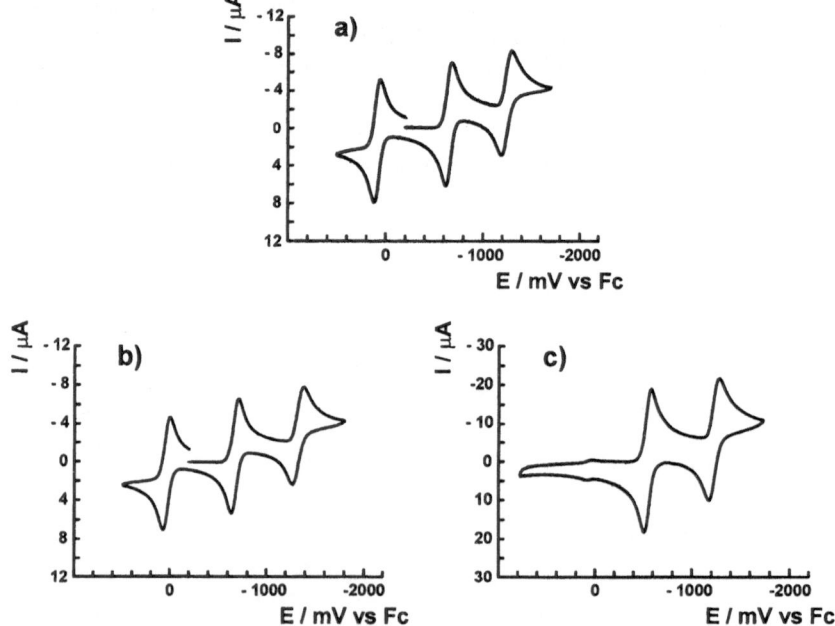

Fig. 7a–c. Cyclic voltammograms (methylene chloride) of **a** $[(Fe^{III})_2Fe^{II}OL_3^{2B}]^0$ **16a**; **b** $[(Fe^{III})_2Co^{II}OL_3^{2B}]^0$ **16b**, and **c** $[(Fe^{III})_2Cu^{II}OL_3^{2B}]^0$ **16c**

the oxidation of $[(Fe^{III})_2Co^{II} OL_3^{2B}]^0 \leftrightarrow [(Fe^{III})_2Co^{III}OL_3^{2B}]^+$ occurs at a half-wave potential of +35 mV (Fig. 7b). The cyclic voltammetric investigations of the redox-active metal centers of neutral **16c**, however, show only two reversible two-potential one-electron transfer processes. The half-wave potentials of −550 and −1220 mV correspond to the redox processes $[(Fe^{III})_2Cu^{II}OL_3^{2B}]^0 \leftrightarrow [Fe^{III}Fe^{II} Cu^{II}OL_3^{2B}]^- \leftrightarrow [(Fe^{II})_2Cu^{II}OL_3^{2B}]^{2-}$ (Fig. 7c).

2.2
{2}-Metalla-Cryptates

The predictable nature of coordination chemistry has been used successfully for the specific generation of metalla-coronates such as [Ca ⊂ $(Cu_3L_3^1)(NO_3)_2$] · THF · H_2O **8**, which may be considered as metalla-topomers **4** of the well-known organic-based coronates **1** (see Sects. 2 and 2.1.1). However, the use of suitable tailor-made ligands made available not only metalla-coronates, but also {2}-metalla-cryptates **18** [13], which may be considered as metalla-topomers **5** of the corresponding organic-based {2}-cryptates **2**. Deprotonation of *m*-pyridylene spacer ligand H_2L^3 **17** with potassium hydride in THF, addition of iron(III) chloride and subsequent workup with aqueous potassium hexafluorophosphate, yielded rust-colored microcrystals of [K ⊂ $(Fe_2L_3^3)$](PF$_6$) **18** (Scheme 5) [2a, 13a].

According to an X-ray crystallographic structure analysis, **18** is a C_{3h}-symmetric {2}-metalla-cryptate. In the center of the $[Fe_2L_3^3]$ cavity of **18** is a potassium ion with nine-fold coordination to six ligand oxygens and

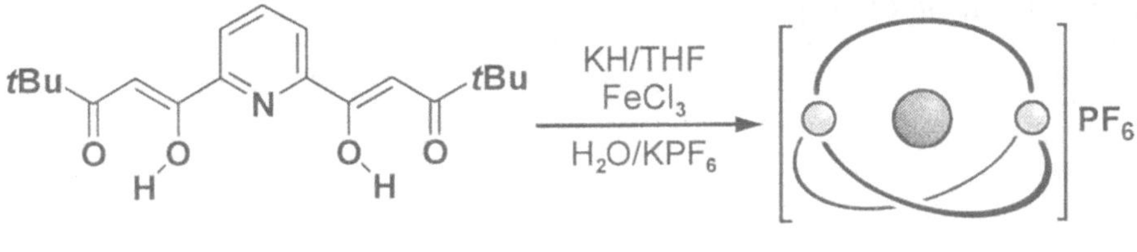

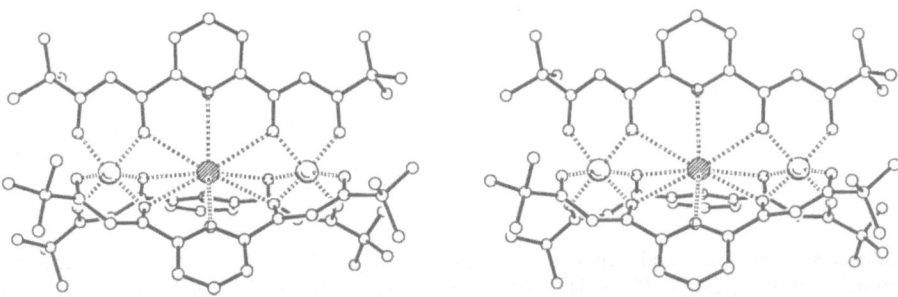

Scheme 5. Formation and schematic representation of [K ⊂ $(Fe_2L_3^3)$](PF$_6$) **18**

Fig. 8. Stereo representation of X-ray crystallographic structure of cation [K ⊂ $(Fe_2L_3^3)$]$^+$ of **18**

to three pyridine nitrogen atoms (Fig. 8). A hexafluorophosphate anion acts as counterion. The two iron centers in the cluster have opposite chirality [(Δ,Λ-*fac*] and therefore **18** represents a *meso*-complex (mesocate) [1b].

2.3
{3}-Metalla-Cryptates of [M₄L₆] Stoichiometry

The first tetranuclear metal chelate complex [(NH$_4$)$_4$ ⌒ (Mg$_4$L$_6^4$)] (H$_2$L^4 = **19**, Sect. 2.3.1) was discovered by serendipity [2b, 14], but this discovery had a considerable impact on the development of the chemistry of further tetrahemi-spheraplexes, related {3}-metalla-cryptands and also {3}-metalla-cryptates. These compounds may simply be considered as metalla-topomers **6** of the corresponding organic-based {3}-cryptates **3** (see Sects. 2 and 2.1.1). This concept has been improved considerably since it was shown that the metal coordination geometry and the orientation of the interacting sites in a given ligand provide the instruction, or blueprint, for the rational design of high symmetry coordination clusters [1b, 3].

2.3.1
Uni-Valence {3}-Metalla-Cryptates

In a reaction of dialkyl malonate with methyllithium and a stoichiometric amount of iron(II) chloride and oxalyl chloride at −78 °C in tetrahydrofuran, followed by rapid workup with 10% aqueous sodium chloride solution, uni-valence iron(III) metalla-cryptate {H$_2$O ⊂ [(Fe^{III})$_4$L$_6^4$]} · 4MeCN **20** was formed (Scheme 6) [15].

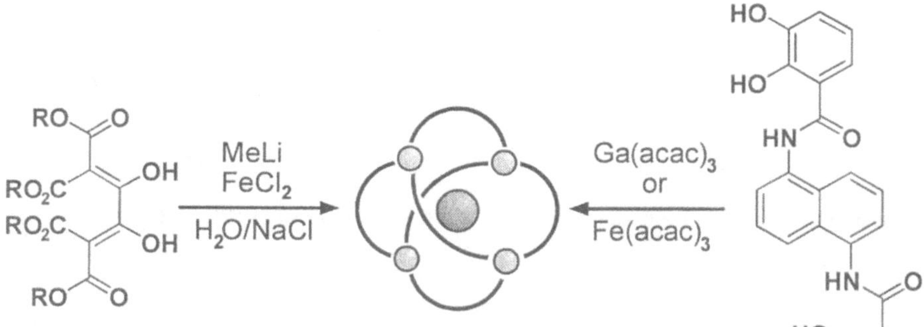

                19   20, 22, 23   21

Scheme 6. Formation and schematic representation of {H$_2$O ⊂ [(Fe^{III})$_4$L$_6^4$]} **20**, and the undeca anions [Et$_4$N ⊂ (Ga$_4$L$_6^5$)]$^{11-}$ of **22**, and [Et$_4$N ⊂ (Fe$_4$L$_6^5$)]$^{11-}$ of **23**. For reasons of clarity, the cartoon represents only the core of the clusters without solvent molecules and counterions

The bis-bidentate bridging ligand $(L^4)^{2-}$ is formally obtained by coupling two dialkyl malonate monoanions with oxalyl chloride to form H_2L^4 19, followed by double deprotonation. For a definitive characterization of 20 an X-ray crystallographic analysis was carried out. The result of this investigation showed that 20 is present in the crystal as a neutral, tetranuclear, uni-valence iron(III) chelate complex with an endohedrally encapsulated water molecule. The core of the host molecule consists of a regular tetrahedron with corners defined by the iron(III) ions. The six edges of the tetrahedron are each formed by the bis-bidentate, doubly negatively charged bridging ligands $(L^4)^{2-}$. Each of the four iron(III) ions is identically coordinated by six oxygen atoms. Hence the chiral racemic complex exists as either a $(\Delta,\Delta,\Delta,\Delta)$-*fac* or a $(\Lambda,\Lambda,\Lambda,\Lambda)$-*fac* stereoisomer and has nearly T molecular symmetry. The water molecule in the center of the tetrahedron of 20 presumably becomes enclosed during the construction of the system. The four acetonitrile molecules are not solvent molecules in the common sense as they are found directly centered above the equilateral triangular faces of the tetrahedral core (Fig. 9).

Overnight stirring of a methanolic solution containing four equivalents of $[Ga(acac)_3]$, six equivalents of H_4L^5 21, twelve equivalents of KOH, and twelve equivalents of Et_4NCl produced a yellow microcrystalline precipitate of $\{K_5[Et_4N]_6[Et_4N \subset (Ga_4L^5_6)]\} \cdot 8H_2O$ 22 (Scheme 6) [3a]. The ^1H NMR spectrum of the isolated complex exhibited all the signals with chemical shifts expected for the formation of the predesigned cluster. The extreme upfield chemical shift of the protons attributed to one Et_4N^+ group in solution can be interpreted as a direct indication of the encapsulation of one Et_4N^+ ion by the tetrahedral core $(Ga_4L^5_6)^{12-}$ of 22.

Ultimate proof for the formation of the racemic, homochiral target cluster and the presence of a host-guest complex is provided by single-crystal X-ray diffraction (Fig. 10). Suitable crystals were grown from the similar iron(III) complex $\{K_5[Et_4N]_6[Et_4N \subset (Fe_4L^5_6)]\} \cdot 8H_2O \cdot 3MeOH$ 23. The tetranuclear core $[Fe_4L^5_6]^{12-}$ of the cluster has nearly T molecular symmetry. Hence the

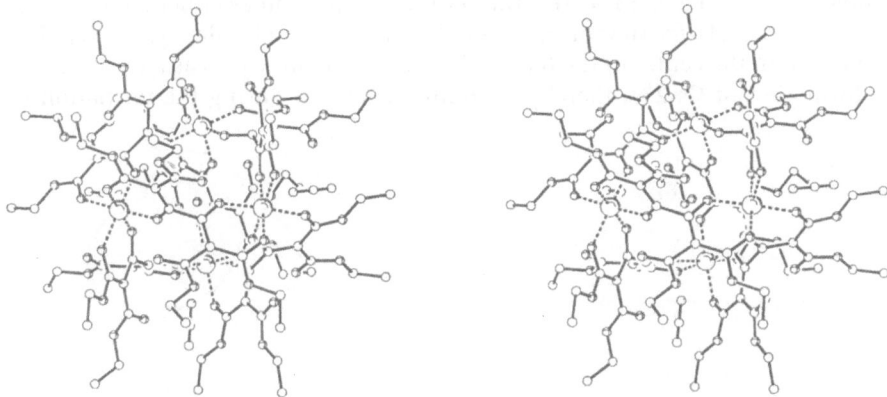

Fig. 9. Stereo representation of X-ray crystallographic structure of $\{H_2O \subset [(Fe^{III})_4L^4_6]\} \cdot$ 4MeCN 20 (L^4: R = Et)

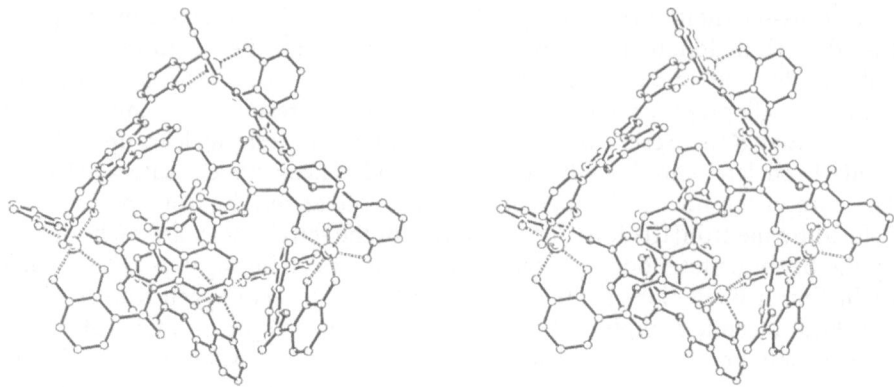

Fig. 10. Stereo representation of X-ray crystallographic structure of undecan anion $[Et_4N \subset (Fe_4L_6^5)]^{11-}$ of **23**

crystal is a racemic mixture of tetrahedra that have homochiral $(\Delta,\Delta,\Delta,\Delta)$-*fac* or $(\Lambda,\Lambda,\Lambda,\Lambda)$-*fac* iron centers. One of the Et_4N^+ counterions is located inside the cluster.

2.3.2
Mixed-Valence {3}-Metalla-Cryptates

Contrary to the formation of uni-valence iron(III) metalla-cryptate $\{H_2O \subset [(Fe^{III})_4L_6^4]\} \cdot 4MeCN$ **20** (see Sect. 2.3.1), the mixed-valence {3}-metalla-cryptates $\{M \subset [(Fe^{III})_3Fe^{II}L_6^4]\}$ **23a** (M = NH_4, K, Cs), $\{NH_4 \subset [(Fe^{III})_4L_6^4]\}Cl$ **23b**, $NH_4\{NH_4 \subset [(Fe^{III})_2(Fe^{II})_2L_6^4]\}$ **23c**, are available in one-pot reactions from dialkyl malonate, oxalyl chloride, iron(II) chloride, methyllithium and the corresponding MCl via intermediate H_2L^4 **19** (Scheme 7) [14c, 15].

According to an X-ray crystallographic analysis, the neutral, mixed-valence complex $\{Cs \subset [(Fe^{III})_3Fe^{II}L_6^4]\} \cdot 4Et_2O$ **Cs-23a** is present as either a $(\Delta,\Delta,\Delta,\Delta)$-*fac* or a $(\Lambda,\Lambda,\Lambda,\Lambda)$-*fac* stereoisomer and has nearly T molecular symmetry. The cesium ion in the center of the tetrahedron presumably serves as a template for the formation of **Cs-23a**, thereby becoming enclosed during the formation of

$$
\begin{array}{ccc}
\text{19} & \xrightarrow[\substack{\text{MeLi} \\ \text{FeCl}_2 \\ \text{H}_2\text{O / CsCl}}]{} & \text{Cs - 23a}
\end{array}
$$

Scheme 7. Formation and schematic representation of $\{Cs \subset [(Fe^{III})_3Fe^{II}L_6^4]\}$ **Cs-23a**

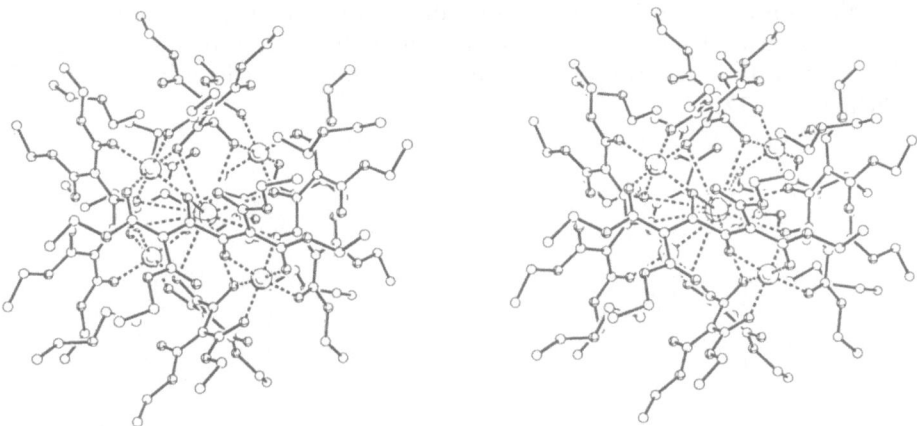

Fig. 11. X-ray crystallographic structure of neutral, mixed valence $\{Cs \subset [(Fe^{III})_3Fe^{II}L_6^4]\}$ · $4Et_2O$ **Cs-23a**

the system. The four diethyl ether molecules are found to be directly centered above the equilateral triangular faces of the tetrahedral core (Fig. 11).

2.4
{3}-Metalla-Cryptands

The synthesis of $[M_4L_6]$ {3}-metallacryptates (see Sect. 2.3) was realized by the combination of bis-bidentate ligands of two-fold symmetry (tetrahedron edges) and octahedral metal centers with three-fold symmetry (tetrahedron vertices) [1b, 2b, 3, 14]. An alternative approach to the formation of symmetric tetrahedral clusters has been described in which the octahedral metal centers (three-fold axis) occupy the vertices of the tetrahedron and trigonally symmetric tris-bidentate ligands (three-fold axis) occupy the faces of the equilateral triangles [16].

2.4.1
{3}-Metalla-Cryptands of [M₄L₄] Stoichiometry

After reaction of a solution of $Ti(OBu)_4$ in methanol with tris-catecholamide H_6L^6 **24** and triethylamine and subsequent elaborate workup the racemic, homochiral complex $(HNEt_3)_8[Ti_4L_4^6]$ **25** was isolated. Main features of the octa-anionic core $[Ti_4L_4^6]^{8-}$ are the hexa negatively charged tris-bidentate catecholamide ligands $(L^6)^{6-}$, which are centered above the triangular faces and coordinate to the four Ti^{IV} ions in the apices of the tetrahedron (Scheme 8) [16]. Likewise, reaction of potassium tris[3-(2′-pyridyl)pyrazol-1-yl]hydroborate KL^7 **26** with one equivalent of $Mn(OAc)_2 \cdot H_2O$ in methanol followed by treatment with KPF_6 yielded $[Mn_4L_4^7](PF_6)_4$ **27**. The four Mn^{II} ions are identical and constitute approximately the apices of a tetrahedron. Each tris-bidentate ligand $(L^7)^-$ is placed above one triangular face of the tetrahedron and each Mn^{II} ion at the corners of that triangle is linked to one bidentate arm. Therefore, each metal ion in the apices of the tetrahedron interacts with one arm of each of three different tris-bidentate ligands (Scheme 8) [17].

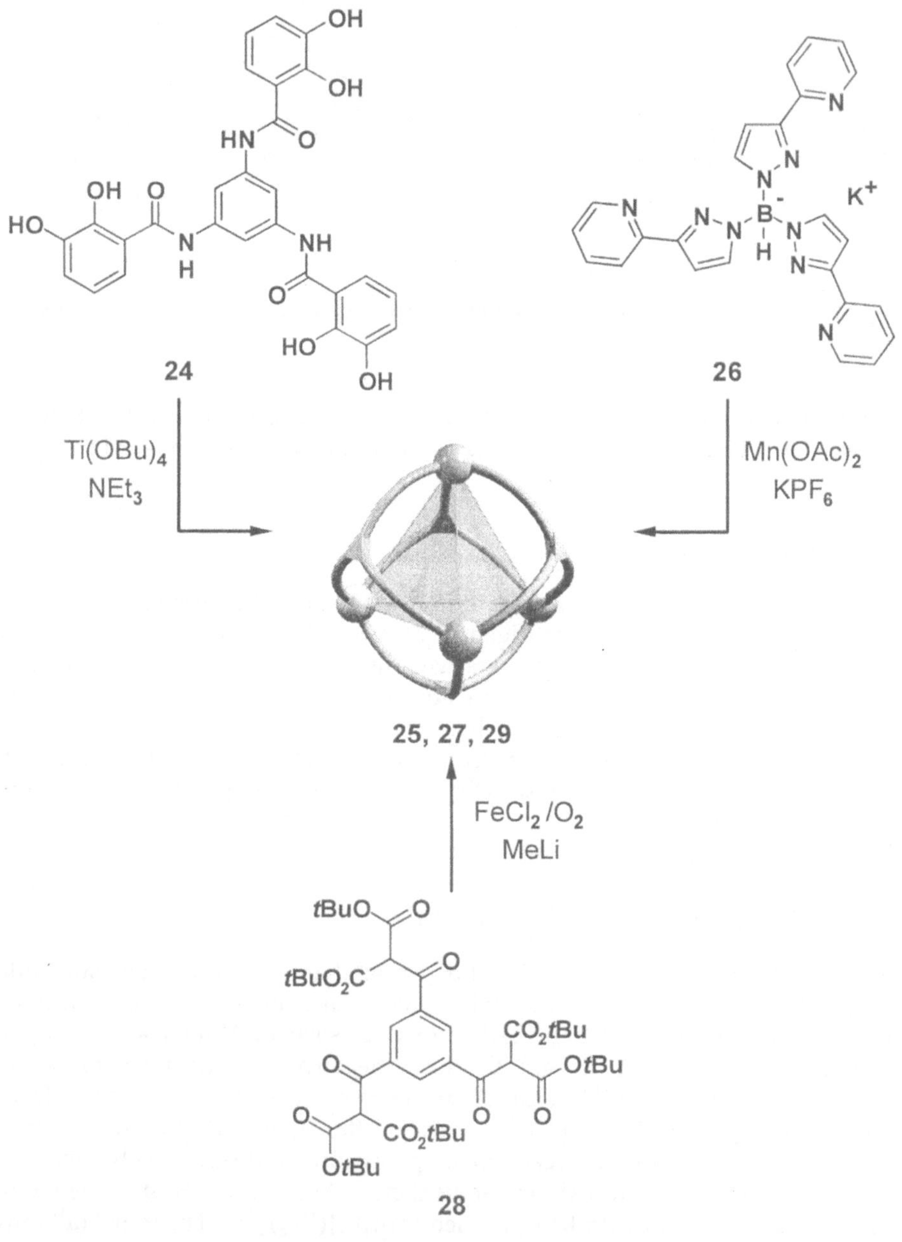

Scheme 8. Formation and schematic representation of $(HNEt_3)_8[Ti_4L_4^6]$ **25**, $[Mn_4L_4^7](PF_6)_4$ **27**, and $[Fe_4L_4^8]$ **29**

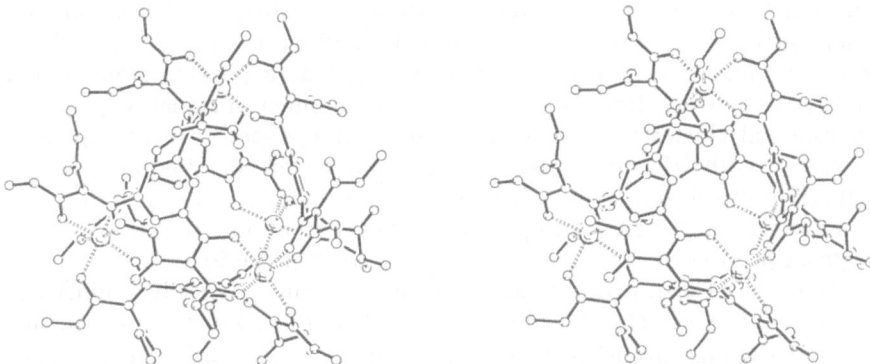

Fig. 12. Stereo representation of $[Fe_4L_4^8]$ **29**; Me groups of the t-BuO substituents have been omitted for clarity

Initiated by the progress towards developing a rational, symmetry-based synthetic approach to the formation of predesigned metal complex clusters, an attempt to generalize this methodology resulted in the design of the three-fold symmetric tris-bidentate ligand $(L^8)^{3-}$. On treatment of di-*tert*-butyl malonate with methyllithium, iron(II) chloride and benzene-1,3,5-tricarboxylic acid trichloride in THF, subsequent workup with saturated aqueous potassium chloride in the presence of atmospheric oxygen, followed by thin-layer chromatography, led to the isolation of bordeaux crystals from acetone. According to the elemental analysis and fast-atom bombardment (FAB) mass spectral data, the compound was of composition $[FeL^8]_n$, with $n = 4$. The ligand $(L^8)^{3-}$ is formally accessible starting from di-*tert*-butyl malonate and methyllithium by the coupling of three di-*t*-Bu-malonate monoanions with benzene-1,3,5-tricarboxylic acid trichloride and spontaneous triple deprotonation from H_3L^8 **28**. The NMR spectra were inconclusive due to strong signal broadening by paramagnetic iron, and could not unambiguously establish the structure of the tetranuclear chelate complex $[Fe_4L_4^8]$ **29**. Consequently an X-ray crystallographic structure analysis of this complex was carried out. Hence, the complex has nearly T molecular symmetry (characterized by three C_2 and four C_3 axes) and the crystal is comprised of a racemic mixture of homochiral $(\varDelta,\varDelta,\varDelta,\varDelta)$-*fac* or $(\varLambda,\varLambda,\varLambda,\varLambda)$-*fac* stereoisomers (Scheme 8, Fig. 12) [18]. There is no evidence that the small cavity in the tetrahedron contains a guest, as observed in some mixed-valence systems [14c, 14d] or in larger M_4L_6 tetrahedra [3a].

2.4.2
Metalla-Cylinder of [M₆L₆] Stoichiometry

As a consequence of the success of rationally designed tris-bidentate ligands for the straightforward synthesis of $[M_4L_4]$ tetrahedra (see Sect. 2.4.1), the three-fold symmetric, tris-bidentate pyrazolone-based ligand H_3L^9 **30** was prepared. Tris-diketone ligand **30** is generated from 3-methyl-1-phenyl-2-

pyrazolin-5-one with 1,3,5-benzenetricarbonyl trichloride and calcium oxide under nitrogen. In particular, on treatment of **30** with Ga(acac)$_3$ in DMSO, a microcrystalline precipitate **31**, which analyzed as a [GaL^9]$_n$ stoichiometry cluster, was isolated. However, contrary to the number of signals expected for a tetrahedral cluster ($n = 4$) with T molecular symmetry, a tripling of the number of both ^{1}H and ^{13}C NMR signals was observed. This indicates that each arm of the ligand is unequivalent and thus the C_3 symmetry of the ligand is disrupted upon metal complexation. The parent ion in the FAB mass spectrum of **31** corresponds to [Ga$_6$L$_6^9$] ($n = 6$) (Scheme 9) [19].

Single-crystal X-ray diffraction was used to unequivocally confirm the structure of [Ga$_6$L$_6^9$] **31**. According to these data, the gallium centers define a distorted trigonal antiprism in which six ligands (L^9)$^{3-}$ are centered above the six equatorial alternating up and down triangles. Cylindrical **31** has idealized D_3 molecular symmetry and the Ga$_6$ scaffold is axially compressed. However, there is no C_3 axis through the tris-bidentate ligands. This matches what is observed in the NMR spectra of **31**. Hexanuclear cluster **31** exists as a racemic mixture of homochiral ($\Delta,\Delta,\Delta,\Delta$)-*fac* and ($\Lambda,\Lambda,\Lambda,\Lambda$)-*fac* stereoisomers. Due to strong chiral coupling, there is no evidence of temperature-dependent interconversion of the various stereoisomers on the NMR time scale (Fig. 13).

2.5
{3}-Metalla-Cryptands of [M$_6$L$_4$] and [M$_{12}$L$_8$] Stoichiometry

From the point of view of rational design, there is still much to do in order to understand the principles which control the formation of supramolecular assemblies. However, in this field much progress has been made. Symmetry considerations in combination with the basic standards of coordination chemistry have made perceptible the construction of a variety of nanoscale systems which procure shape, size and, ultimately, function. It has been demonstrated that two types of building blocks are required. Rigid complexes

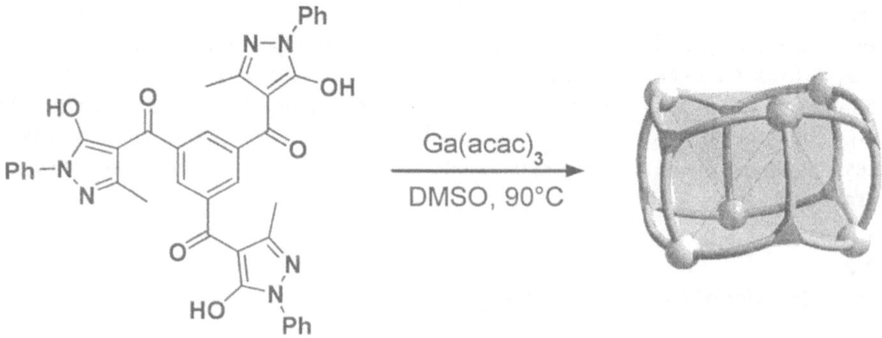

30   31

Scheme 9. Formation and schematic representation of [Ga$_6$L$_6^9$] **31**

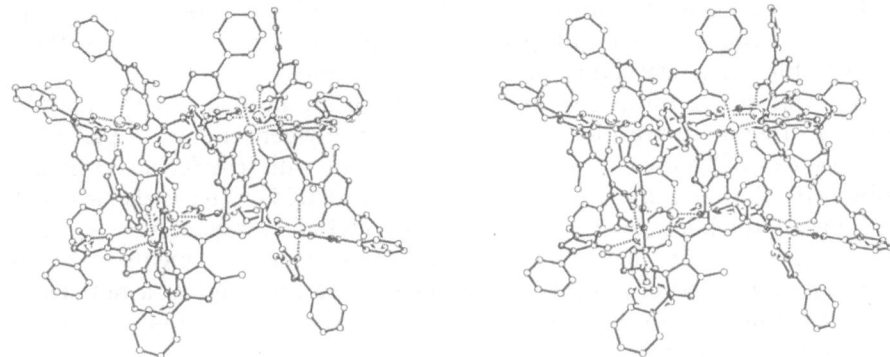

Fig. 13. Stereo representation of $[Ga_6L_6^9]$ **31**

can be synthesized, in which the overall structure is governed by the binding constraints of the ligands and the coordination geometry of the metal ions. Thus, the design of a planar triangle requires the combination of three bidentate linear linkers (L_3^2) with four bidentate angular components (A_4^2) and three 60° bidentate angular components (A_3^2) and the combination of four bidentate linear linkers (L_4^2) will result in a square, etc.

The design of three-dimensional polyhedra requires at least tridentate linkers. For example, an adamantanoid scaffold can be formed with six angular bidentate linkers (A_6^2) and four angular tridentate linkers (A_4^3). The six angular bidentate ligands comprise the apices of an octahedron and the four tridentate ligands occupy alternating faces. Likewise twelve linear bidentate units (L_{12}^2) and eight tridentate angular components may result in a cube (Fig. 14) [1j, 1z, 20].

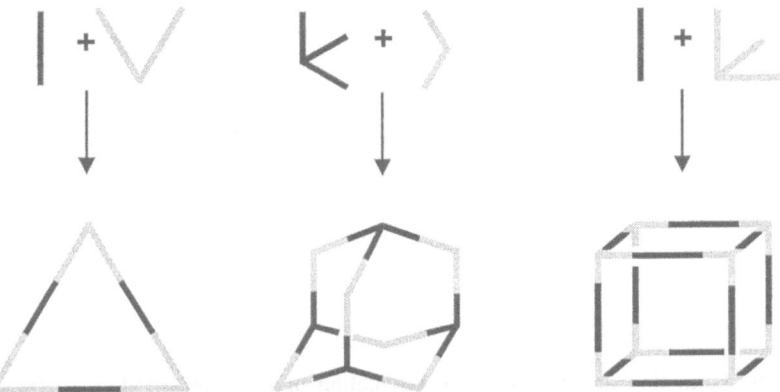

Fig. 14. Schematic representation of the rational design of two- and three-dimensional supramolecular clusters using linear and angular linking units

The use of tritopic ligands in which three metal binding sites radiate from a central aryl ring together with palladium or platinum salts offers a synthetic strategy for the construction of container type molecules of M_6L_4 stoichiometry. This structural arrangement is found in the products of the reaction between 1,3,5-tris(pyrazol-1-ylmethyl)-2,4,6-triethylbenzene with palladium dichloride or of 2,4,6-tris(4′-pyridyl)-1,3,5-triazene with (ethylenediamine)palladium(II) dinitrate. In a similar way, an adamantanoid cluster is also available with the rigid tritopic ligand tris[1,3,5-(4′-pyridyl)ethynyl]benzene. Typically, in all these clusters, an octahedral array of six metal fragments is bound together by four tritopic ligands. The ligands are arranged in a tetrahedral array, each binding to three metal fragments and occupying four alternating faces of the octahedron [1e, 1j, 1o, 1z, 21].

In the presence of templating anions, 2:3 molar mixtures of triphos[MeC(CH$_2$PPh$_2$)$_3$] 32 and silver(I) cations give dicationic hexanuclear adamantanoid cages {Ag$_6$[MeC(CH$_2$PPh$_2$)$_3$]$_4$(SO$_3$CF$_3$)$_4$}(SO$_3$CF$_3$)$_2$ 33 (Scheme 10) [22]. The structure of 33 was unequivocally determined by single-crystal X-ray diffraction techniques. The molecular structure of the dicationic core {Ag$_6$[MeC(CH$_2$PPh$_2$)$_3$]$_4$(SO$_3$CF$_3$)$_4$}$^{2+}$ of 33 has approximately T molecular symmetry. It may be simply presented as a truncated tetrahedron, in which the four triphos ligands define the truncated apices and the six silver ions occupy the edges (Fig. 15).

A triflate ion is located above each of the tetrahedral faces with the oxygen atoms directed towards the three silver ions. A novel feature is the "endo-methyl" conformation of the triphos ligands, leaving only a small cavity at the center.

Symmetry considerations in combination with the basic principles of coordination chemistry proved to be a powerful strategy for the construction of polynuclear clusters with predesigned size and shape. Consequently, twelve

32   33

Scheme 10. Formation and schematic representation of {Ag$_6$[MeC(CH$_2$PPh$_2$)$_3$]$_4$(SO$_3$CF$_3$)$_4$} (SO$_3$CF$_3$)$_2$ 33. For reasons of clarity, the cartoon represents only the dicationic core {Ag$_6$[MeC(CH$_2$PPh$_2$)$_3$]$_4$(SO$_3$CF$_3$)$_4$}$^{2+}$ of 33 without the four (F$_3$CSO$_3$)$^-$ ions located above alternating faces of the octahedron

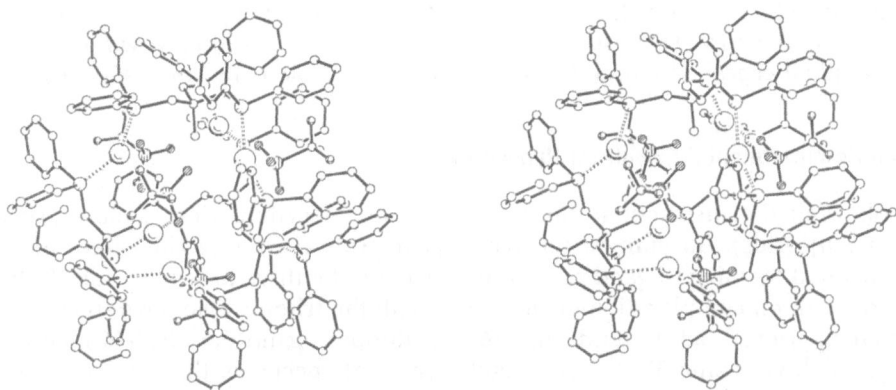

Fig. 15. Stereo representation of the dication $\{Ag_6[MeC(CH_2PPh_2)_3]_4(SO_3CF_3)_4\}^{2+}$ of **33**

linear bidentate units (L_{12}^2; L = linker; L = ligand) and eight tridentate angular components (A_8^3) might result in a cube (Fig. 14). NMR evidence has been presented for the generation of a cube-like arrangement with eight ruthenium corners and twelve bipy edges [23]. Very recently, reaction of 2,4,6-triphenylazo-1,3,5-trihydroxybenzene H_3L^{10} **34**, $Cu(NO_3)_2$, and tetramethylammonium hydroxide in 2:3:6 proportions in DMF yielded dark brown crystals of $[Cu_{12}L_8^{10}]$ **35** (Scheme 11) [24].

According to single-crystal X-ray diffraction studies each ligand ($L^{10})^{3-}$ is attached to three Cu(II) centers by virtue of a bidentate N,O chelating system. Thus, each Cu^{2+} ion is in turn linked to two ligands acquiring a distorted square planar coordination environment. The centroids of the eight ligands are located very close to the corners of a cube. The solid diagonals are very close to perpendicular to the phloroglucinate rings of the two opposite ligands. The twelve Cu ions comprise approximately the corners of a cuboctahedron.

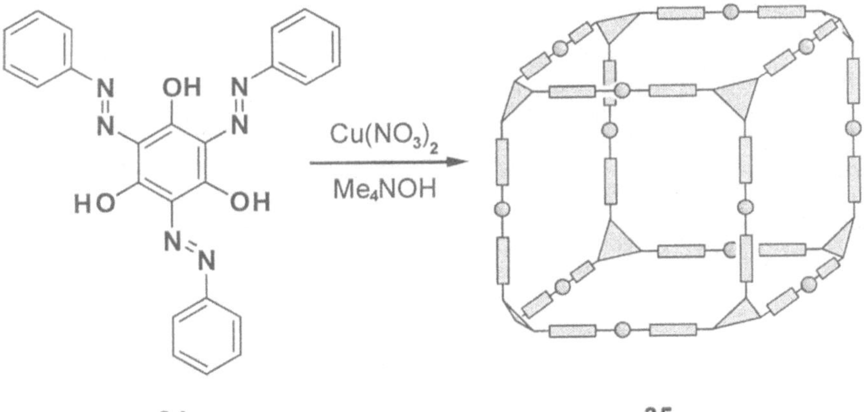

34   **35**

Scheme 11. Formation and schematic representation of $[Cu_{12}L_8^{10}]$ **35**

The six cube faces are effectively blocked off by four phenyl groups from four separate ligands. Thus the shell of the cluster is essentially windowless. Individual cages are chiral, but the crystal as a whole is racemic (Fig. 16).

2.6
Multicompartmental Cylindrical Nanostructures

Following the instructions for the generation of symmetrical, square grids ([n × n]-grids), unsymmetrical, rectangular grids ([n × m]-grids) have been generated by mixed-ligand recognition. Among the three possible grids 38–40 formed upon complexation of silver ions with the tritopic ligand 6,6′-bis[2-(6-methylpyridyl)]-3,3′-bipyridazine 36 and ditopic ligand 3,6-bis[2-(6-methyl-pyridyl)]pyridazine 37, the rectangular [n × m]-species is the main product (Scheme 12) [1d, 1r, 25].

The designed self-assembly of multicomponent and multicompartmental cylindrical nanoarchitectures represents an abiological analogue of numerous biological processes mediated by collective interactions and recognition events. Thus, monocompartmental, bicompartmental and tricompartmental complex cations containing anions in their cavities were proposed and finally realized. The compartmental cylindrical architectures 41–43 are generated in a single operation from the corresponding stoichiometric mixtures of Cu^+ ions with the linear oligobipyridine 44 and disk-like hexaazatriphenylene 45 ligands (Scheme 13) [26].

In the case of $[(Cu_{12}L_3^{11}L_4^{12}) \subset (PF_6)_6](PF_6)_6$ 43 $[L^{11} = 44 \ (n = 2); L^{12} = 45]$, the highest level of structural complexity is attained by the correct association of twenty-five particles. The cation $[(Cu_{12}L_3^{11}L_4^{12}) \subset (PF_6)_6]^{6+}$ of 43 is triple-helical and comprises three L^{11} ligands and four L^{12} ligands. Two L^{12} ligands define the ends of the cylinder, and the two inner ones divide the cylindrical cage into three compartments. The cavities in the cation of 43 are occupied by guest molecules, that are two PF_6^- ions in each of the three compartments (Fig. 17).

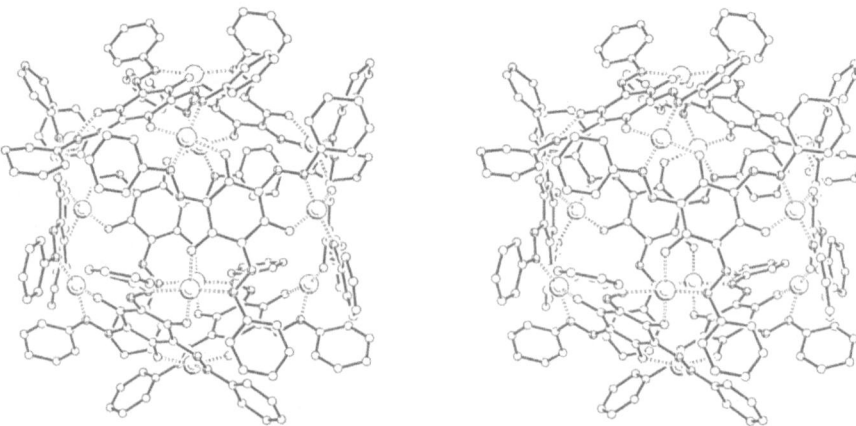

Fig. 16. Stereo representation of $[Cu_{12}L_8^{10}]$ 35

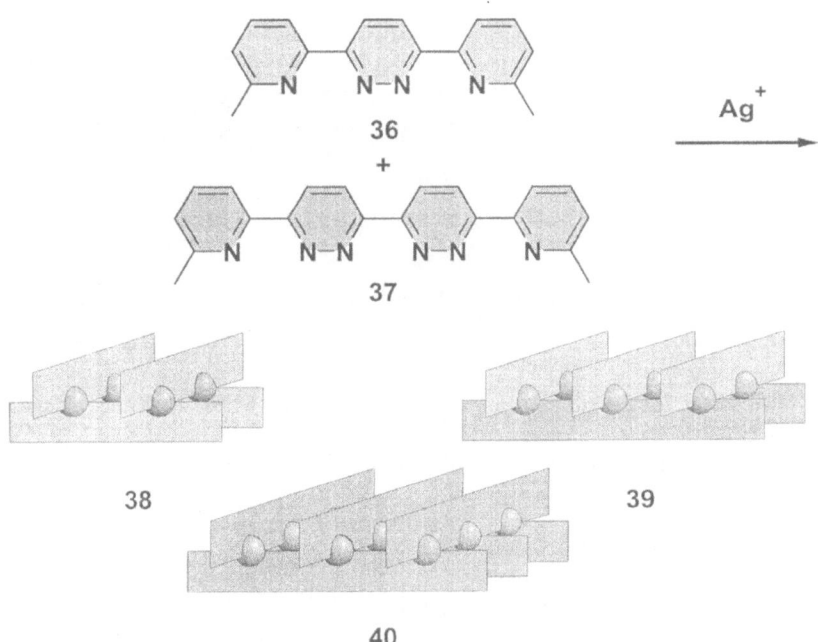

Scheme 12. Formation and schematic representation of [n × n]-grid **38**, [n × m]-grid **39**, and [m × m]-grid **40**

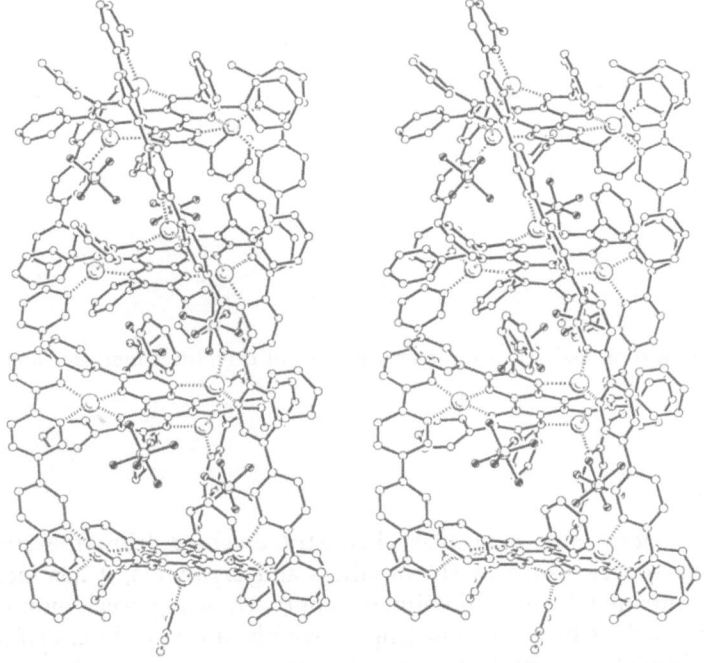

Fig. 17. Stereo representation of cation $[(Cu_{12}L_3^{11}L_4^{12}) \subset (PF_6)_6]^{6+}$ of **43**

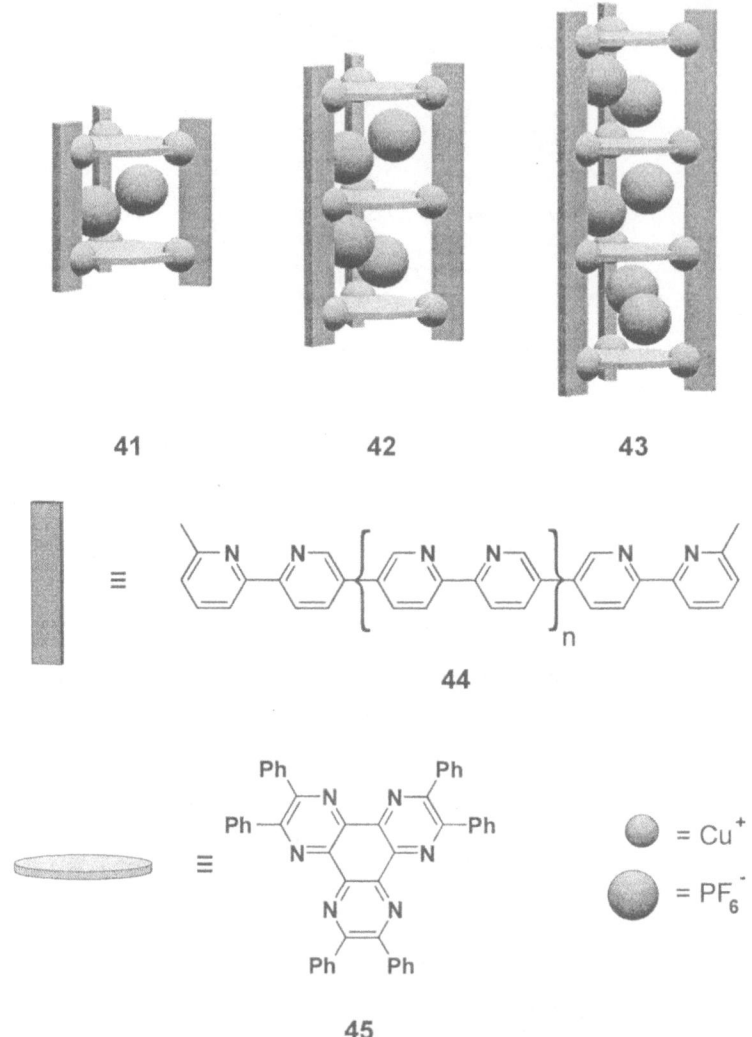

Scheme 13. Formation and schematic representation of cylindrical inorganic architectures 41–43

3
Perspectives

Although modern supramolecular chemistry emerged from the studies of covalent macrocycles such as crown ethers and cryptands, it has meanwhile become dominated by the biomimetic approach using weak non-covalent interactions such as hydrogen bonding. Currently, a very different strategy for the formation of supramolecular species via spontaneous self-assembly of precursor building blocks is the use of metals and dative bonding via coordination. This methodology has been successfully employed in the formation of infinite networks, but it has been much less explored in the self-assembly of discrete nanoscopic supramolecular species with predetermined shapes, geometries, and symmetries. In this review we have illustrated the synergistic effect of serendipity and the utility and generality of the approach to the designed synthesis of supramolecular clusters. However, the potential of fortuitously discovered supramolecular clusters, once decoded and translated into rationally designed syntheses, remains to be further explored. The pictorial representations of the complex supramolecular clusters facilitate the recognition of relationships between structures which might not otherwise be obvious. Finally, the scope of this review is not intended to be a complete collection of all the known metallo-assembled supra-structures, but rather to illustrate the strategies developed over the years represented in the most recent literature.

Acknowledgements. The financial support of the Deutsche Forschungsgemeinschaft and the Fonds der Chemischen Industrie is greatly appreciated. Particular thanks are due to the enthusiastic co-workers mentioned in the references, who have actively taken part in our own research.

4
References

1. For recent reviews, see (a) MacGillivray LR, Atwood JL (1999) Angew Chem 111: 1080; (1999) Angew Chem Int Ed Engl 38: 1018; (b) Caulder DL, Raymond KN (1999) J Chem Soc Dalton Trans 1185; (c) Saalfrank RW, Demleitner B (1999) Ligand and metal control of self-assembly in supramolecular chemistry. In: Transition metals in supramolecular chemistry. Wiley, Weinheim, vol 5, p 1; (d) Saalfrank RW, Bernt I (1998) Curr Opin Solid State Mater Sci 3: 407; (e) Jones CJ (1998) Chem Soc Rev 27: 289; (f) Lehn J-M (1990) Angew Chem 102: 1347; (1990) Angew Chem Int Ed Engl 29: 1304; (g) Lehn J-M (1995) Supramolecular chemistry. VCH, Weinheim, p 17; (h) Philp D, Stoddart JF (1996) Angew Chem 108: 1243; (1996) Angew Chem Int Ed Engl 35: 1154; (i) Hayes W, Stoddart JF (1996) Cyclic molecules formed by self-assembly. In: Semlyen JA (ed) Large ring molecules. Wiley, New York, p 433; (j) Stang PJ, Olenyuk B (1997) Acc Chem Res 30: 502; (k) Dietrich B, Viout P, Lehn J-M (1993) Macrocyclic chemistry. VCH, Weinheim, p 171; (l) Vögtle F (1992) Supramolekulare Chemie. Teubner, Stuttgart, p 45; (m) Fujita M, Oguro D, Miyazawa M, Oka H, Yamaguchi K, Ogura K (1995) Nature 378: 469; (n) Stang PJ (1998) Chem Eur J 4: 19; (o) Hartshorn CM, Steel PJ (1997) Chem Commun 541; (p) Fredericks JR, Hamilton AD (1996) Metal template control of self-assembly in supremolecular chemistry In: Hamilton AD (ed) Supramolecular control of structure and reactivity. Wiley, New York, p 1; (q) Fyfe MCT, Stoddart JF (1997) Acc Chem Res 30: 393; (r) Baxter PNW (1996) Metal ion directed assembly of complex molecular architectures and nanostructures. In: Atwood JL, Davies JED, Macnicol DD, Vögtle F, Lehn J-M, Sauvage J-P, Hosseini MW (eds) Comprehensive supramolecular chemistry, vol 9. Pergamon, Oxford, p 165; (s) Constable EC (1996) Polynuclear transition metal helicates. In: Atwood JL, Davies JED, Macnicol DD, Vögtle F, Lehn J-M, Sauvage J-P, Hosseini MW (eds) Comprehensive supramolecular chemistry, vol 9. Pergamon, Oxford, p 213; (t) Constable EC (1992) Tetrahedron 48: 10013; (u) Piguet C, Bernardinelli G, Hopfgartner G (1997) Chem Rev 97: 2005; (v) Fujita M (1996) Self-assembled macrocycles, cages, and catenanes containing transition metals in their backbones. In: Atwood JL, Davies JED, Macnicol DD, Vögtle F, Lehn J-M, Sauvage J-P, Hosseini MW (eds) Comprehensive supramolecular chemistry, vol 9. Pergamon, Oxford p 253; (w) Vögtle F, Hoss R, Händel M (1996) In: Cornils B, Hermann W (eds) Applied

homogeneous catalysis with organometallic compounds, vol 2. VCH, Weinheim, p 801; (x) Chambron J-C, Dietrich-Buchecker C, Sauvage J-P (1996) Transition metals as assembling and templating species: synthesis of catenanes and molecular knots. In: Atwood JL, Davies JED, Macnicol DD, Vögtle F, Lehn J-M, Sauvage J-P, Hosseini MW (eds) Comprehensive supramolecular chemistry, vol 9. Pergamon, Oxford, p 43; (y) Pecoraro VL, Stemmler AJ, Gibney BR, Bodwin JJ, Wang H, Kampf JW, Barwinski A (1997) Prog Inorg Chem 45: 83; (z) Fujita M (1998) Chem Soc Rev 27: 417; (α) Linton B, Hamilton AD (1997) Chem Rev 97: 1669; (β) Conn MM, Rebek J Jr (1997) Chem Rev 97: 1647

2. (a) Saalfrank RW, Dresel A, Seitz V, Trummer S, Hampel F, Teichert M, Stalke D, Stadler C, Daub J, Schünemann V, Trautwein AX (1997) Chem Eur J 3: 2058; (b) Saalfrank RW, Stark A, Peters K, von Schnering HG (1988) Angew Chem 100: 878; (1988) Angew Chem Int Ed Engl 27: 851

3. (a) Caulder DL, Powers RE, Parac TN, Raymond KN (1998) Angew Chem 110: 1940; (1998) Angew Chem Int Ed Engl 37: 1840; (b) Beissel T, Powers RE, Raymond KN (1996) Angew Chem 108: 1166; (1996) Angew Chem Int Ed Engl 35: 1084

4. (a) Saalfrank RW, Löw N, Kareth S, Seitz V, Hampel F, Stalke D, Teichert M (1998) Angew Chem 110: 182; (1998) Angew Chem Int Ed Engl 37: 172; (b) Saalfrank RW, Löw N, Demleitner B, Stalke D, Teichert M (1998) Chem Eur J 4: 1305; (c) Saalfrank RW, Löw N, Hampel F, Stachel H-D (1996) Angew Chem 108: 2353; (1996) Angew Chem Int Ed Engl 35: 2209

5. (a) Taft KL, Delfs CD, Papaefthymiou GC, Foner S, Gatteschi D, Lippard SJ (1994) J Am Chem Soc 116: 823; (b) Gatteschi D, Caneschi A, Sessoli R, Cornia A (1996) Chem Soc Rev 101

6. Watton SP, Fuhrmann P, Pence LE, Caneschi A, Cornia A, Abbati GL, Lippard SJ (1997) Angew Chem 109: 2917; (1997) Angew Chem Int Ed Engl 36: 2774

7. Saalfrank RW, Bernt I, Uller E, Hampel F (1997) Angew Chem 109: 2596; (1997) Angew Chem Int Ed Engl 36: 2482

8. (a) Hasenknopf B, Lehn J-M, Kneisel BO, Braun G, Fenske D (1996) Angew Chem 108: 1987; (1996) Angew Chem Int Ed Engl 35: 1838; (b) Lehn J-M (1999) Chem Eur J 5: 2455

9. Uller E, Heinemann F, Saalfrank RW, unpublished results

10. Hasenknopf B, Lehn J-M, Boumediene N, Dupont-Gervais A, Van Dosselaer A, Kneisel B, Fenske D (1997) J Am Chem Soc 119: 10965

11. (a) Gorun SM, Papaefthymiou GC, Frankel RB, Lippard SJ (1987) J Am Chem Soc 109: 4244; (b) Lippard SJ (1988) Angew Chem 100: 353; (1988) Angew Chem Int Ed Engl 27: 344; (c) Shweky I, Pence LE, Papaefthymiou GC, Sessoli R, Yun JW, Bino A, Lippard SJ (1997) J Am Chem Soc 119: 1037

12. (a) Saalfrank RW, Trummer S, Krautscheid H, Schünemann V, Trautwein AX, Hien S, Stadler C, Daub J (1996) Angew Chem 108: 2350; (1996) Angew Chem Int Ed Engl 35: 2206; (b) Saalfrank RW, Löw N, Trummer S, Sheldrick GM, Teichert M, Stalke D (1998) Eur J Inorg Chem 559

13. (a) Saalfrank RW, Seitz V, Caulder DL, Raymond KN, Teichert M, Stalke D (1998) Eur J Inorg Chem 1313; (b) Grillo VA, Seddon EJ, Grant CM, Aromi G, Bollinger JC, Folting K, Christou G (1997) Chem Commun 1561; (c) Hannon MJ, Painting CL, Jackson A, Hamblin J, Errington W (1997) Chem Commun 1807; (d) Albrecht M, Blau O (1997) Chem Commun 345; (e) Albrecht M, Fröhlich R (1997) J Am Chem Soc 119: 1656; (f) Albrecht M, Riether C (1996) Chem Ber 129: 829

14. (a) Saalfrank RW, Stark A, Bremer M, Hummel H-U (1990) Angew Chem 102: 292; (1990) Angew Chem Int Ed Engl 29: 311; (b) Saalfrank RW, Hörner B, Stalke D, Salbeck J (1993) Angew Chem 105: 1223; (1993) Angew Chem Int Ed Engl 32: 1179; (c) Saalfrank RW, Burak R, Breit A, Stalke D, Herbst-Irmer R, Daub J, Porsch M, Bill E, Müther M, Trautwein AX (1994) Angew Chem 106: 1697; (1994) Angew Chem Int Ed Engl 33: 1621; (d) Saalfrank RW, Burak R, Reihs S, Löw N, Hampel F, Stachel H-D, Lentmaier J, Peters K, Peters EM, von Schnering HG (1995) Angew Chem 107: 1085; (1995) Angew Chem Int Ed Engl 34: 993

15. Breit A, Glaser H, Saalfrank RW, unpublished results

16. Brückner C, Powers RE, Raymond KN (1998) Angew Chem 110: 1937; (1998) Angew Chem Int Ed Engl 37: 1837
17. Amoroso AJ, Jeffery JC, Jones PL, McCleverty JA, Thornton P, Ward MD (1995) Angew Chem 107: 1577; (1995) Angew Chem Int Ed Engl 34: 1443
18. Saalfrank RW, Glaser H, Hampel F, Raymond KN, unpublished results
19. Johnson DW, Xu J, Saalfrank RW, Raymond KN (1999) Angew Chem 111: 3058; (1999) Angew Chem Int Ed Engl 38: 2882
20. Fujita M, Ogura K (1996) Coord Chem Rev 148: 249
21. (a) Fujita M, Ogura K (1996) Bull Soc Chem Soc Jpn 69: 1471; (b) Stang PJ, Olenyuk B, Muddiman DC, Smith RD (1997) Organometallics 16: 3094
22. James SL, Mingos DMP, White AJP, Williams DJ (1998) Chem Commun 2323
23. Roche S, Hasläm C, Adams H, Heath SL, Thomas JA (1998) Chem Commun 1681
24. Abrahams BF, Egan SJ, Robson R (1999) J Am Chem Soc 121: 3535
25. (a) Hanan GS, Volkmer D, Schubert US, Lehn J-M, Baum G, Fenske D (1997) Angew Chem 109: 1929; (1997) Angew Chem Int Ed Engl 36: 1842; (b) Baxter PNW, Lehn J-M, Kneisel BO, Fenske D (1997) Angew Chem 109: 2067; (1997) Angew Chem Int Ed Engl 36: 1978; (c) Duan C, Liu Z, You X, Xue F, Mak TCW (1997) Chem Commun 381; (d) Jeffery JC, Jones PL, Mann KLV, Psillakis E, McCleverty JA, Ward MD, White CM (1997) Chem Commun 175
26. (a) Baxter PNW, Lehn J-M, Baum G, Fenske D (1999) Chem Eur J 5: 102; (b) Baxter PNW, Lehn J-M, Kneisel BO, Baum G, Fenske D (1999) Chem Eur J 5: 113

Molecular Paneling Through Metal-Directed Self-Assembly

Makato Fujita

Graduate School of Engineering, Nagoya University, Chikusaku,
Nagoya 464-8603, Japan
E-mail: *mfujita@apchem.nagoya-u.ac.jp*

Metal-directed self-assembly is widely recognized as an efficient method for constructing well-defined discrete structures. Among the many fascinating structures that have been prepared by this method, recent interest is in large three-dimensional compounds with a large cavity in the framework. Following on from the preceding chapter, which dealt with various three-dimensional coordination assemblies, this chapter focuses on a unique approach to three-dimensional assemblies via "molecular paneling". Families of planer exo-multidentate organic ligands (molecular panels) are found to assemble into large three-dimensional assemblies through metal coordination. In particular, *cis*-protected square planer metals [e.g. (en)Pd^{2+}, en = ethylenediamine] are shown to be very useful to panel molecules. Metal-assembled cages, bowl, tube, capsule, and polyhedra, are efficiently constructed by this approach.

Keywords: Self-assembly, Molecular recognition, Nanotechnology, Cage, Bowl, Tube, Capsule, Polyhedra

1
Introduction

Metal-directed self-assembly has recently been the subject of explosive development which has made possible the synthesis of many fascinating and complex structures [1]. By combining building blocks derived from various families of molecules and metal ions, the structure of the resulting multicomponent systems can, in principle, be varied infinitely at will. Early examples involve simple linear bridging ligands with two linking sites, such as 4,4'-bipyridine and aromatic dicyanide or dicarboxylic acids, as summarized in Table 1 [2–7].

These ligands are one-dimensional (1D) components and hence they are assembled in most cases into two-dimensional (2D) structures (i.e., macrocycles) upon treatment with transition metals. On the other hand, 2D components with more than two linking sites are expected to assemble into 3D coordination assemblies, if the ligand and metal components are appropriately designed. These 3D assemblies are currently attracting considerable interest. Among many possible 2D components, panel-like ligands are particularly interesting because of their simplicity and facile design and the ability to construct 3D assemblies by paneling these planer components. Table 2 summarizes typical examples of flat panel-like ligands (or *molecular panels*) which will be discussed in the following sections [8–13]. Accordingly, this review discloses the self-assembly of 3D structures by *"molecular paneling"* which may be a new concept for molecule-by-molecule construction of large, well-defined structures.

To obtain discrete 3D coordination assemblies, molecular panels should be linked with an appropriate structural turn (e.g., 90°) and, thus, the use of *cis* coordination cites of transition metals are effective to provide such turns in the assembled structures. In this regard, *cis*-protected Pd(II) block 1, which was developed in our laboratories ten years ago, is a useful metal component which makes possible the design of simple self-assembly from non-sophisticated ligands (e.g., 4-pyridyl-functionalized molecular panels).

1

In contrast to naked metal ions such as tetrahedral Cu(I), octahedral Fe(II), Co(II), Ni(II), and square planar Pd(II), the number of coordination sites is reduced and hence the direction of the coordination of the ligands can be easily controlled. In fact, this and related coordination blocks with an appropriate

Table 1. Examples of typical 1D bridging ligands with two linking sites and their 2D coordination assemblies

Table 1. (*Contd.*)

Coordination
Assemblies

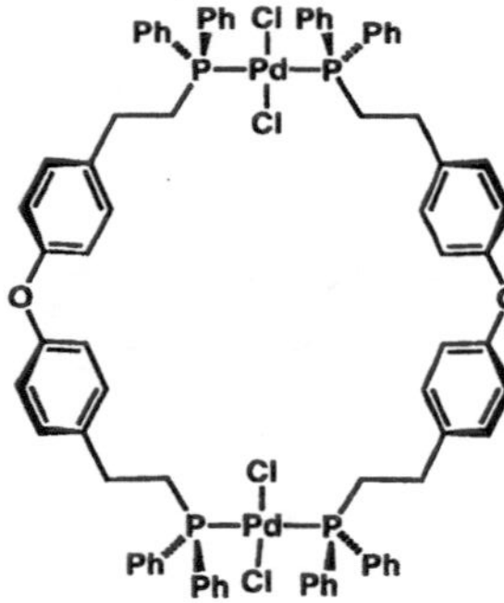

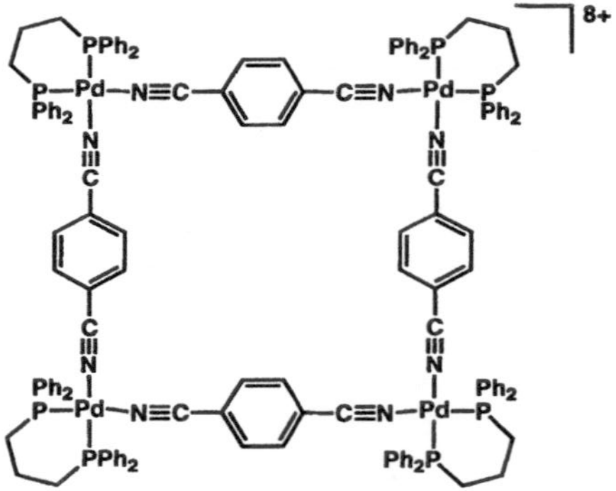

ligands

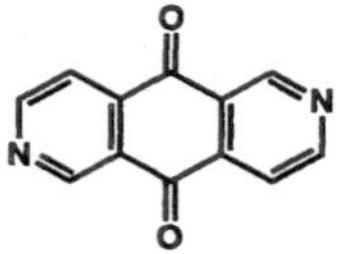

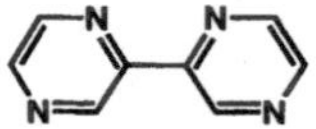

Coordination
Assemblies

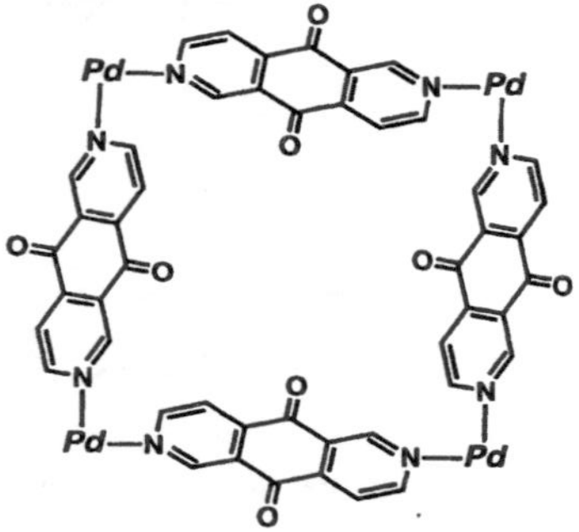

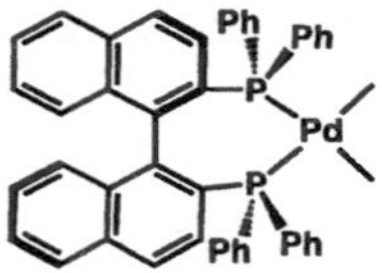

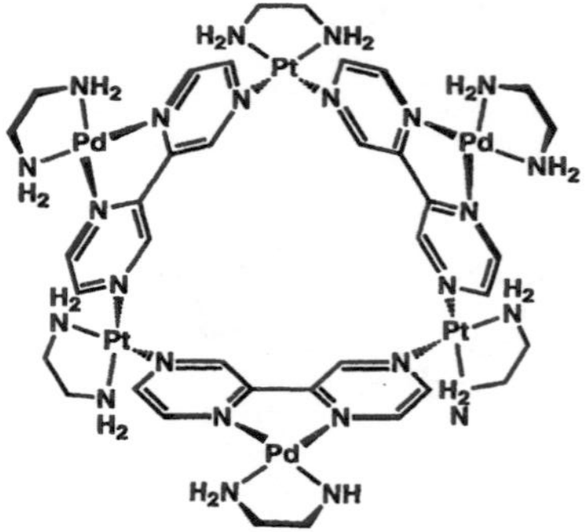

Table 2. Typical examples of panel-like ligands with more than two coordination sites

protecting group on the metal have been employed to construct a variety of discrete structures such as macrocycles, cages, and catenanes [14]. Accordingly, this review focuses on the self-assembly of 3D structures by protected-metal coordinating "molecular paneling" which may be a new concept for molecule-by-molecule construction of large, well-defined structures.

Three-dimensional coordination assemblies using precisely designed bridging ligands with more than two linking sites are discussed by Saalfrank in the preceding chapter of this volume [15].

2
Assemblies from Triangular Molecular Panels

2.1
2,4,6-Tris(4-pyridyl)-1,3,5-triazine

Triangles are the simplest 2D units and are components of a family of polyhedrons. Thus we first discuss the assembly of triangles. One of the simplest triangular molecular panels is 2,4,6-tris(4-pyridyl)-1,3,5-triazine (2), which has been employed frequently for infinite coordination assemblies by Robson [9, 16].

2

2.1.1
Self-Assembly of Octahedral M_6L_4 Cages

The first example of the use of this molecular panel to obtain a 3D discrete structure is the M_6L_4 octahedral assembly 3 reported in 1995 [9c]. By treating 1 with 2 in a 3:2 ratio, the octahedral complex 3 is quantitatively assembled (Scheme 1). In this complex, the four triangular panels are linked together at the corners of the triangles in such a way that the triangles locate alternately on the four faces of an octahedron. The basic molecular design of 3 can be regarded as an extension of the previously reported square compound 4 in structural dimensions. Namely, by linking with a 90° coordination block, 1D molecular rods assemble into a 2D square whereas 2D molecular panels form a 3D octahedron (Fig. 1). Complex 3 is a thermodynamically stable product because the formation of the product is not affected by the presence of excess 1. Due to the simple synthetic procedure, 10-g scale preparation of 3 can be carried out in the laboratory.

It has been shown that the cage complex 3 effectively binds various organic guest molecules in its cavity. The structure of the clathrate complex of 3 with the adamantane carboxylate ion has been determined by X-ray crystallography

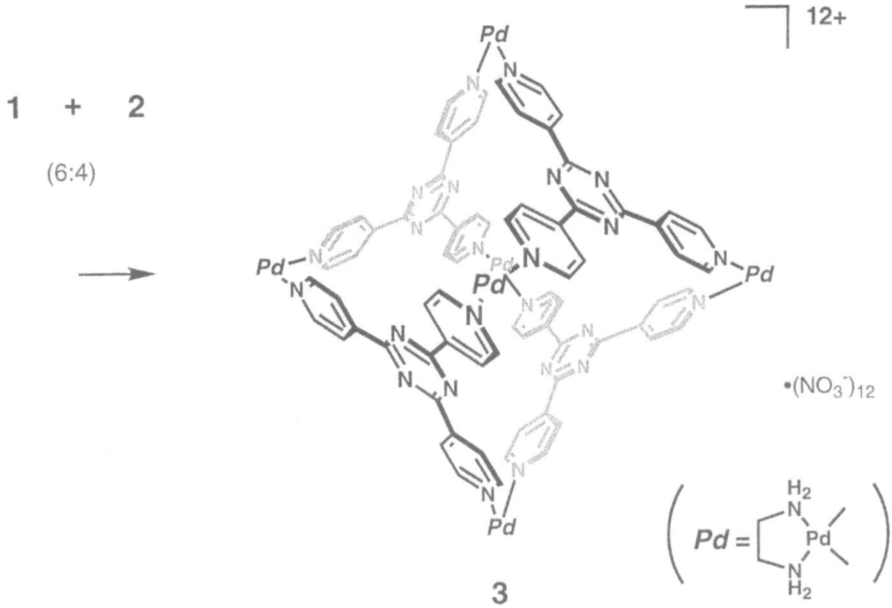

1 + 2

(6:4)

$\longrightarrow$

12+

•$(NO_3^-)_{12}$

$$Pd = \left(\begin{array}{c} H_2 \\ N \\ \vert \\ Pd \\ \vert \\ N \\ H_2 \end{array} \right)$$

3

Scheme 1. Self-assembly of coordination nanocage 3

(Fig. 2), which showed that four guest molecules are tightly encapsulated inside the nano-sized cavity of **3**. The inclusion geometry of the guest in the cavity is interesting as the hydrophobic and hydrophilic groups (COO⁻) are located inside and outside of the cavity, respectively. The space filling presentation of **3** shows that the dimension of the portal of the cage is comparable to that of adamantane, whereas the interior space can hold as many as four guest molecules. A ^1H NMR study showed that the same host-guest aggregates were also organized even in aqueous media.

2.1.2
Molecular Recognition of Large Guests by the M_6L_4 Cage

Generally, cage compounds prepared by covalent synthesis can encapsulate only one or two small molecules because the construction of the larger frameworks by conventional covalent synthesis is very difficult. Recent advances in noncovalent synthesis have made it easy to create such larger frameworks with very large cavities. The above described cage compound **3** has a very large cavity with a diameter of 1 nm and exhibits a remarkable ability to encapsulate large, neutral molecules. In addition to the binding of aqueous guests (adamantane carboxylate, as discussed in Sect. 2.1.1), neutral and spherical guest molecules such as adamantane, 1- and 2-adamantanol, o-carborane, and aromatic compounds such as 1,3,5-trimethoxybenzene, anisole, and toluene, are efficiently bound in the cavity [17]. Interestingly, adamantane was found to transfer into the aqueous phase even in a solid–

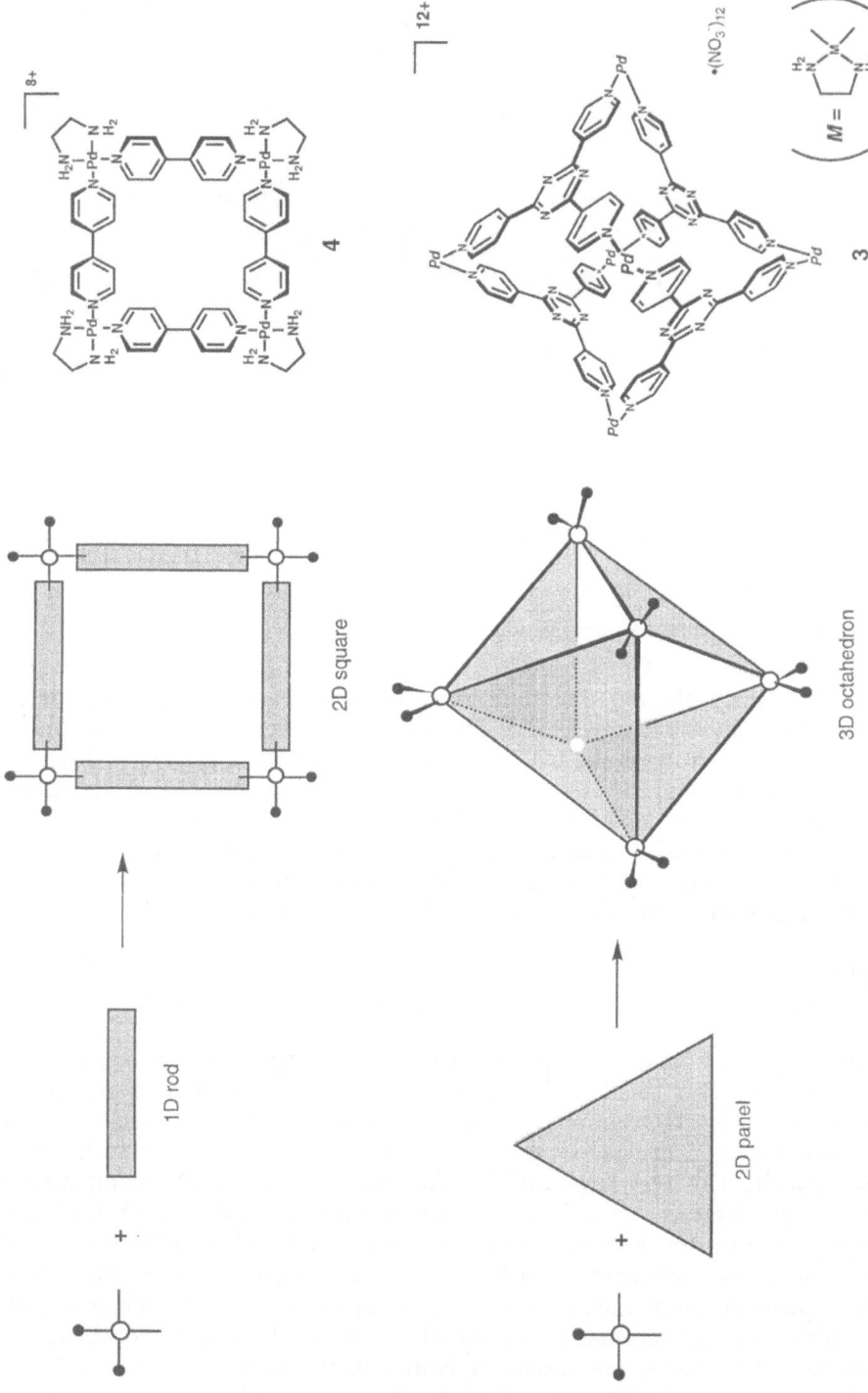

◄──

Fig. 1. Schematic presentation of molecular paneling with an extension of 2D assembling from s1D rods

liquid two-phase system. This very efficient guest binding is ascribed to the amphiphilic nature of the cage: i.e. the inside of **3** is surrounded by 16 aromatic rings and thus hydrophobic, whereas the outside surface of the cage is hydrophilic due to the exposure of six charged Pd(II) centers.

Notably, complexation is faster with smaller guest molecules and slower with larger guest molecules. For example 1,3,5-tri-*tert*-butylbenzene, which is slightly larger than the portal of **3**, was encapsulated very slowly. Tetrabenzylsilane required a few hours to be completely encapsulated by **3'** (which is a 2.2'-bpy-protected analog of **3**) and intermediates in which the guest had unsymmetrically interacted with the host were observed at the early stage of the complexation (Fig. 3). Crystallographic analysis showed the good fit of the tetrahedral symmetry of the guest in the octahedral cage (Fig. 4) [18].

2.1.3
Formation of Hydrophobic Dimers in the M_6L_4 Cage

The cage compound **3** also exhibited a remarkable ability to encapsulate C-shaped molecules such as *cis*-azobenzene (**5**) and -stilbene (**6**) derivatives [19]. These guest molecules enclathrated in the cavity via the formation of a hydrophobically interacted dimer with a topology reminiscent of a hydrogen-bonded tennis ball [20] (Fig. 5a). This hydrophobic dimer is supported by an

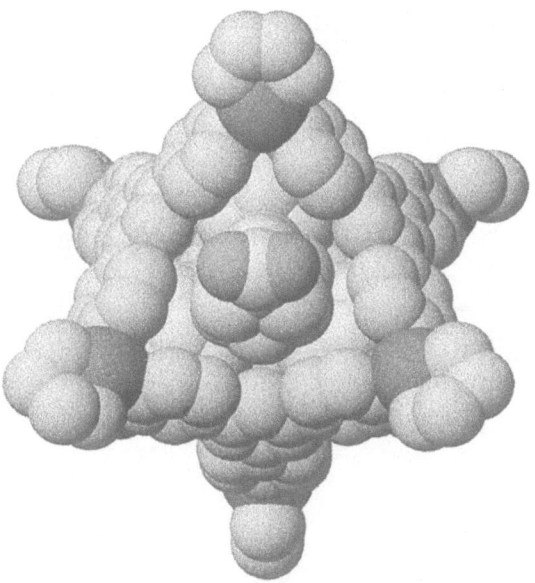

Fig. 2. X-ray crystallographic structure of **3** accommodating four molecules of adamantane carboxylate

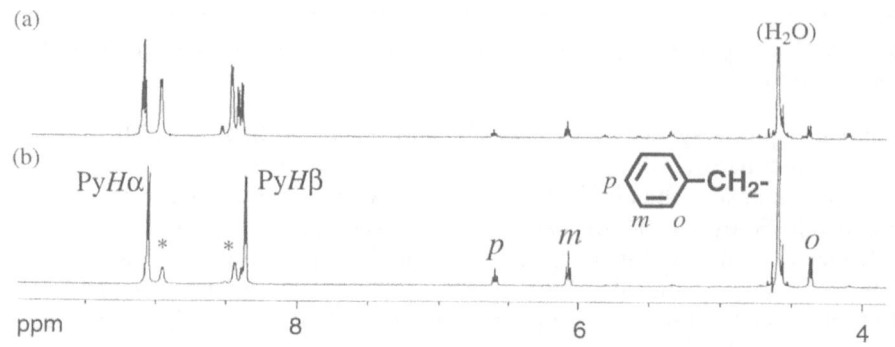

Fig. 3a,b. ¹H NMR observation of the complexation of **3** with tetrabenzylsilane. A D₂O solution of **3** was treated with an excess amount of tetrabenzylsilane and monitored by ¹H NMR. **a** After 1 h. **b** After 3 h (*: uncomplexed **3**)

NOE experiment as well as by a molecular dynamic simulation. The selective enclathration of the *cis*-isomer was observed when *cis-trans* mixtures of either **5** or **6** in hexane were stirred with a D₂O solution of **3**. The NMR spectra confirm

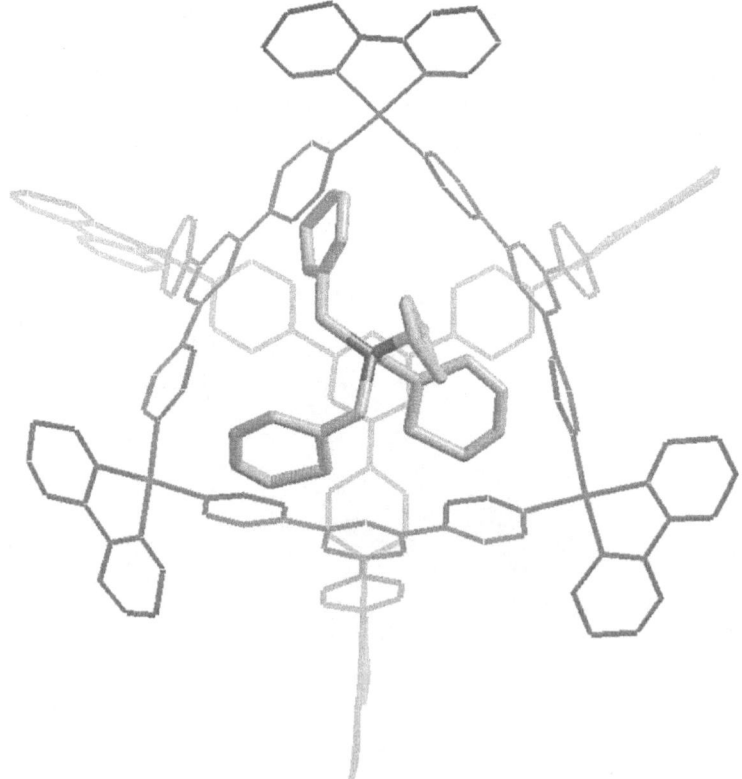

Fig. 4. X-ray crystallographic structure of **3'** enclathrating tetrabenzylsilane

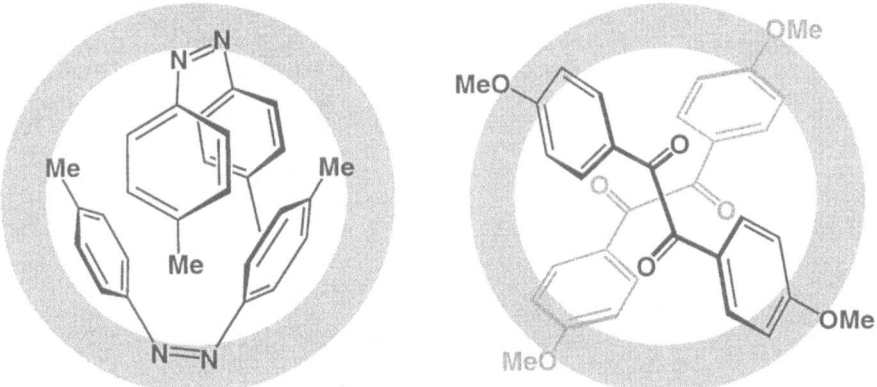

Fig. 5. Formation of a hydrophobic dimer of *cis*-stilbene and 4,4'-dimethoxydibenzoyl within the cavity of **3**

the encapsulation of dimers of *cis*-isomers in the cavity. Notably, the *cis*-isomer of **5** was significantly stabilized in the cavity and not isomerized to the *trans*-isomer even after a few weeks under visible light at room temperature. The molecular modeling calculations suggest that the hydrophobic dimers are a perfect fit for the cavity of **3**. Dimerization of the guests prior to enclathration is unlikely because the dimension of the spherical dimer (ca. 11 Å diameter) is larger than that of the window of **3** (ca. 7 Å diameter). Therefore, two guest molecules should be subsequently but not simultaneously enclathrated in the cavity and converted in situ into the stable hydrophobic dimer.

A similar dimer formation was observed when 1,2-diketone **7** was employed as a guest (Fig. 5b). In the 1:2 complex, the host ligand was dissymmetrized in NMR: four ligands are all equivalent, but 12 protons on each ligand were observed independently. This observation was clearly explained by an X-ray crystallographic analysis. As shown in Fig. 6, two guest molecules are assembled in a similar way to that of **5** and **6**. However, each guest adapts a twisted conformation: one is a *P*-form and another is an *M*-form. As a consequence, the dimer adapts a meso structure with no centrosymmetry making all protons on each ligand unequivalent [21].

2.1.4
Chemical Reactions in the M₆L₄ Cage

The reactivity and catalysis represent one of the most important features of the functional properties of self-assembled molecular systems [22]. The presence of the large cavity in **3** motivated us to test its ability to catalyze the oxidation of styrene and isomerization of allylbenzene [23]. When **1** and **2** were mixed in D₂O in a 4:2 ratio, formation of only **3** was observed and excess **1** remained in the solution. Our strategy was to use this excess of **1** as a mediator between the organic and the aqueous phase. That is to use it to cyclically and continuously transfer the substrate into the aqueous phase that contains **3** and then the

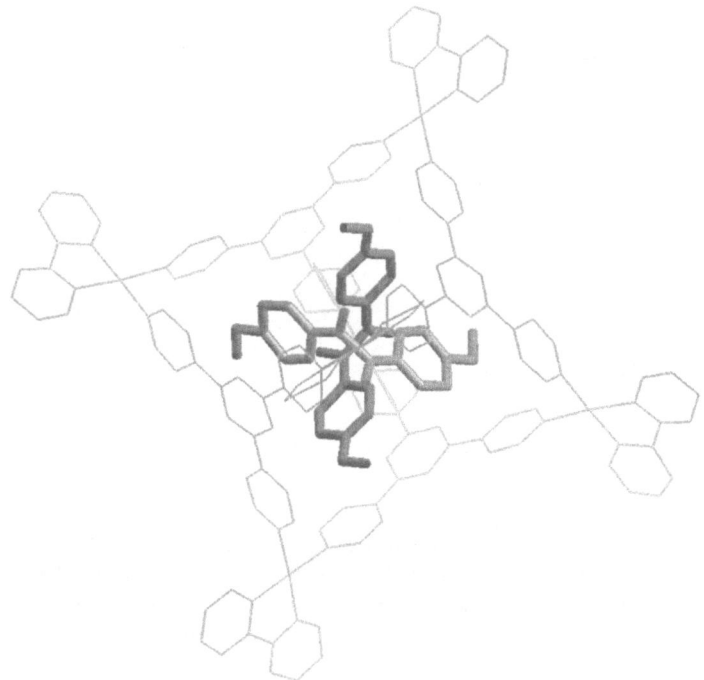

Fig. 6. X-ray crystallographic structure of **3′** enclathrating the hydrophobic dimer of 4,4′-dimethoxydibenzoyl

formed product into the organic phase (Scheme 2). It was observed that **3** can accommodate approximately three molecules of styrene in its cavity.

The oxidation of styrene at 80 °C in an aqueous solution of either **1** or **3** gave acetophenone in only 4% yield. Importantly, the presence of both **1** and **3** in the aqueous solution increased the yield of the reaction up to 86%. Similarly, the isomerization of allylbenzenes catalyzed by the presence of **1** and **3a** in aqueous solution gave β-methylstyrene **5a** in 50% yield, whereas the reaction did not occur in the absence of either **1** or **3** (Scheme 3). The presence of trimethoxybenzene in the reaction media inhibited these reactions because the cavity of **3** was strongly occupied by it. The yields of these reactions reveal that as the size and electron deficiency of the substrate increases the yield of the reaction decreases. These are good examples of organic reactions in solution but without organic solvents.

2.1.5
Guest-Templated Synthesis of a Kinetically Stable M_6L_4 Cage [M=Pt(II)]

Similar to all the Pd(II) self-assemblies described so far, **3** is also assembled as a result of thermodynamic equilibration and it is not stable enough to maintain its framework under extreme conditions (e.g., acidic, basic, or nucleophilic conditions). In order to prepare a kinetically stable M_6L_4 complex, Pt(II)

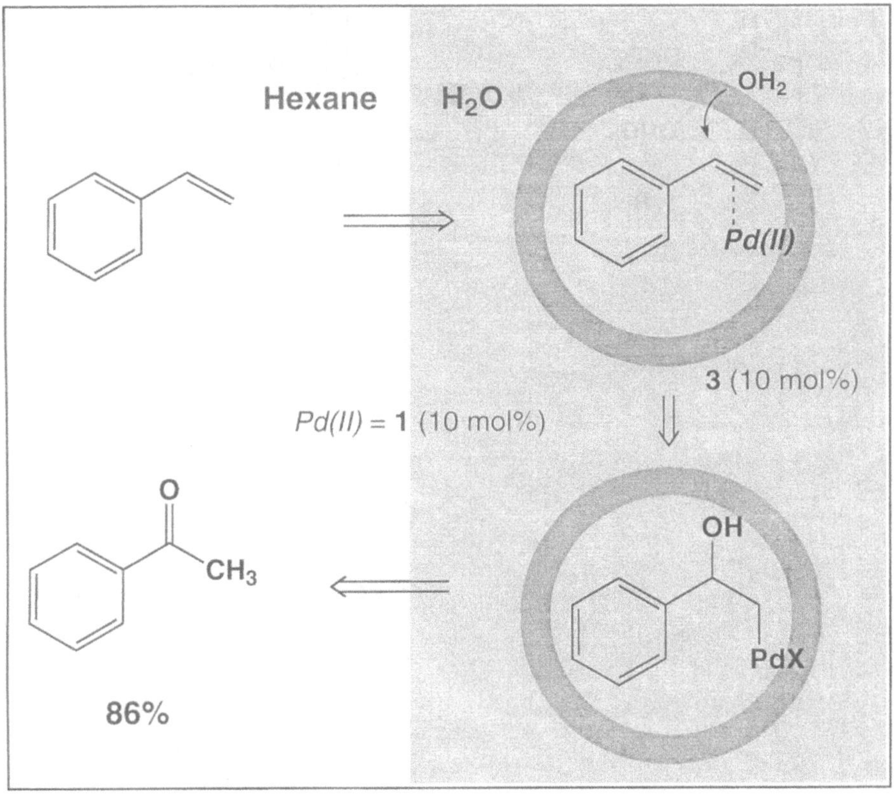

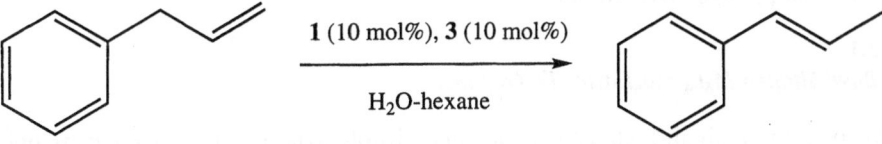

Scheme 2. Schematic presentation of reversed phase-transfer catalysis of **3**. Wacher oxidation is promoted by slight excess amount of **1**

complex **8** was used instead of **1**. In contrast to the Pd cage, the formation of Pt cage **9** was too slow to produce it in a reasonable yield. However, heating the solution and adding a guest molecule, adamantane carboxylate, dramatically improved the reaction rate as well as the yield (Scheme 4). The host-guest ratio and the guest inclusion geometry were found to be similar to those of the Pd structure **3**. Usually, a receptor framework organized by guest "induced-fit" is lost when the guest is removed. In contrast to this, Pt cage **9** did not lose its cage structure even after removal of the guest because of the locking of the Pt(II)—Py bond after the self-assembly event. As anticipated, Pt(II) complex **9** was very stable and did not decompose even in the presence of an acid (HNO₃), a base

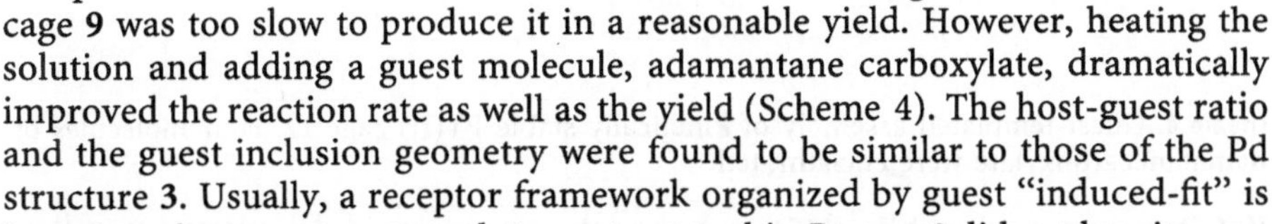

1 (10 mol%), **3** (10 mol%)

$\xrightarrow{\hspace{3cm}}$

H_2O-hexane

Scheme 3. Olefin isomerzation in an aqueous phase with the aid of **1** (10 mol%) and **3** (10 mol%)

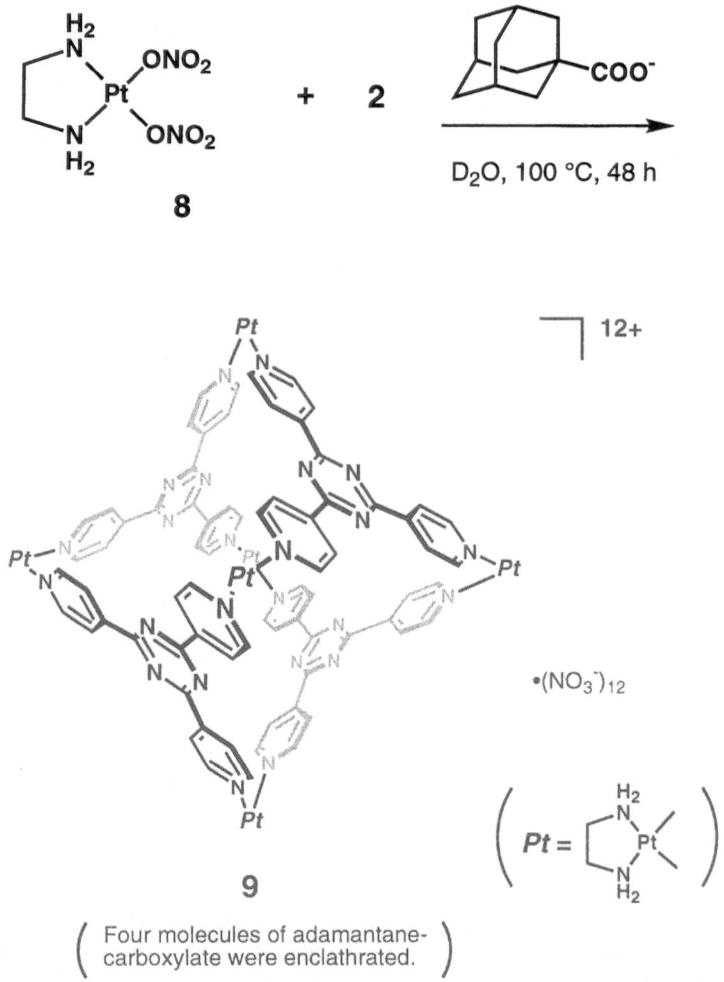

Scheme 4. Guest-templated assembly of kinetically stable PT(II) cage 12. Four molecules of adamatanecarboxylate were enclathrated

(K₂CO₃), or a nucleophile (NEt₃), due to the inertness of the Pt(II)—pyridine coordinate bond [24].

2.2
2,4,6-Tris(3-pyridyl)-1,3,5-triazine

2.2.1
A Bowl-Shaped M₆L₄ Macrotricyclic Complex

Exo-tridentate ligand 10 is also a very simple triangular molecular panel. Although 10 has a similar structure to 2, this component gives a completely different macrotricyclic complex 11 (Scheme 5) [10]. The structure of 11 was

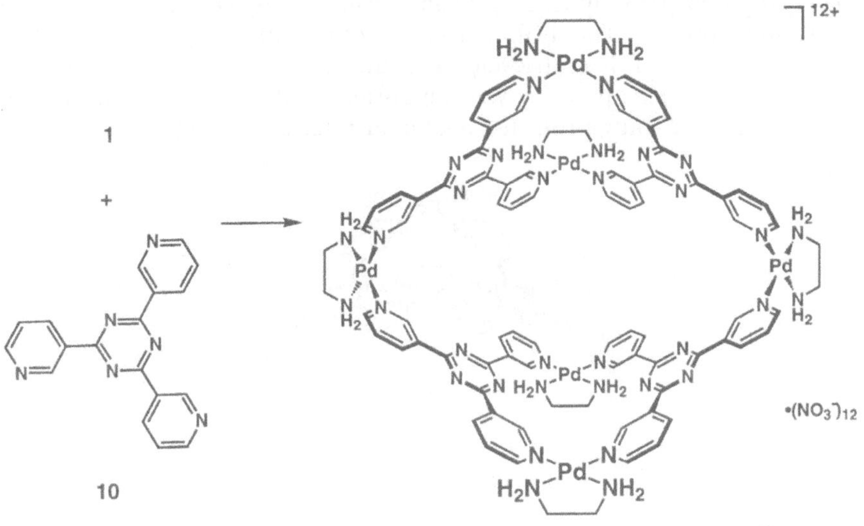

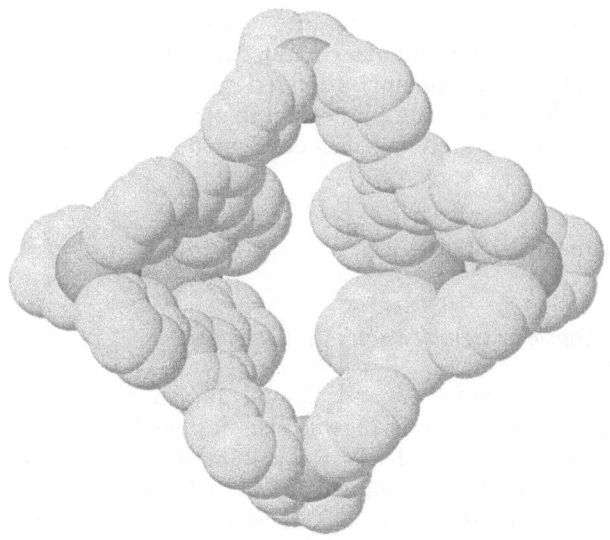

Scheme 5. Self-assembly of 11 from 1 and 8

characterized in solution by ^{1}H NMR and in the solid state by X-ray crystallography (Fig. 7): the framework is held together by 10 molecular components (six metal ions and four ligands) and it has nano-meter dimensions (ca. $3 \times 2 \times 2$ nm) in spite of the small size of the molecular components.

Fig. 7. X-ray crystallographic structure of 11

The proposed pathway leading to the formation of **11** involves a two-step self-assembly process. That is the formation of intermediate **12** in the first step and the formation of macrotricycle **11** in the second step by self-assembly of the intermediate. This pathway was supported by the isolation and the X-ray analysis of the 2,2′-bipy-protected analog of intermediate **12**.

12

2.2.2
Guest Inclusion by the M₆L₄ Bowl: Dimerization into a Capsule in the Solid State

Bowl-shaped complex **11** is expected to assemble in aqueous media into a dimeric capsule making a large hydrophobic pocket inside the framework because of its amphiphilic property: hydrophobic inside and hydrophilic outside. In fact, such a dimeric structure does assemble in the solid state. That is, X-ray structures have been obtained for some host-guest complexes with large guest molecules, all of which showed the dimeric capsule structure of the host accommodating as many as six neutral organic molecules [25]. The solid structure of the complex with *o*-terphenyl (Fig. 8a) is recognized as a dimer of 1:2 host-guest complexes because the whole structure can be divided into two identical 1:2 complexes. On the other hand, the solid structure of the complex with *m*-terphenyl (Fig. 8b) can not be divided into two halves and thus the whole structure is regarded as a 2:4 complex rather than a dimer of a 1:2 complex. With *cis*-stilbene, 1:6 complexation has been confirmed by X-ray analysis.

2.3
2,4,6-Tris(3,5-pyrimidyl)-1,3,5-triazine

2.3.1
A Coordination Capsule Assembled from 24 Components

Following the exo-tridentate ligands **2** and **10**, an exo-hexadentate ligand, 1,3,5-tris(3,5-pyrimidyl)benzene **13**, was also designed as a triangular unit. As already discussed, the triangle is a basic unit for the self-assembly of polyhedra. The ligand **13** is an almost coplanar triangle and is expected to give an edge-sharing polyhedron when it is self-assembled with **1**. Of several possibilities, the assembly of the molecular hexahedron **14** was strongly suggested by ¹H NMR (Scheme 6) [26].

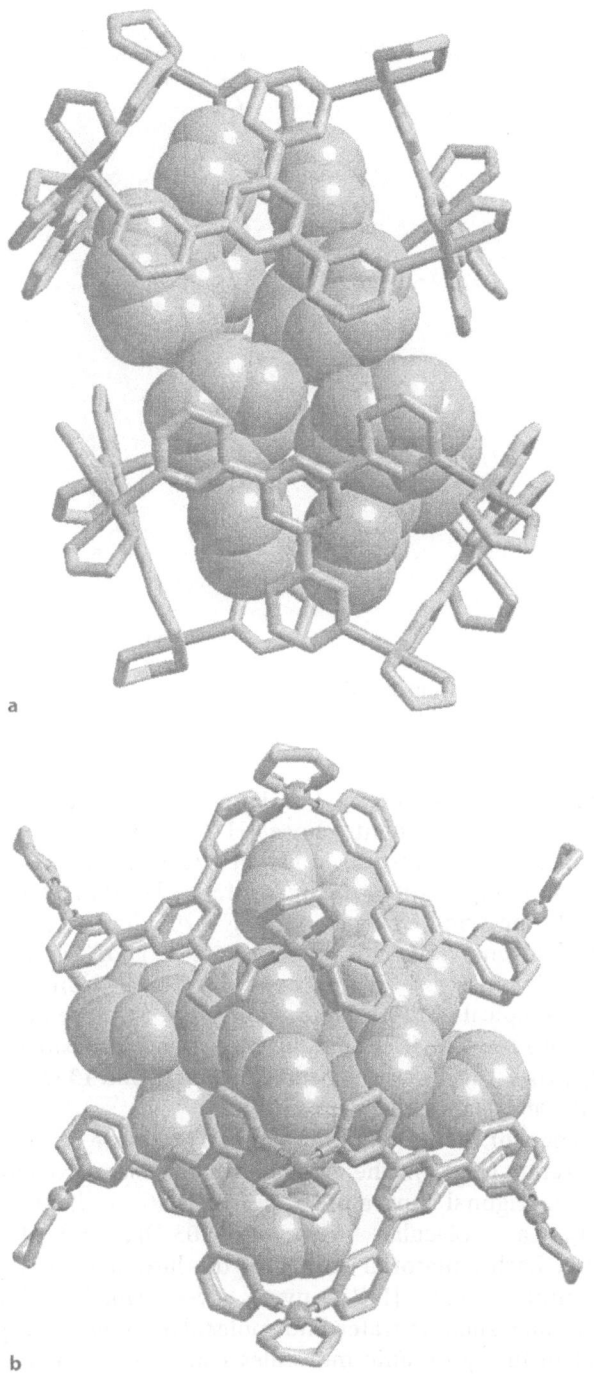

Fig. 8a,b. Dimeric capsules of **11** accommodating *o*-terphenyl (**a**) and *m*-terphenyl (**b**)

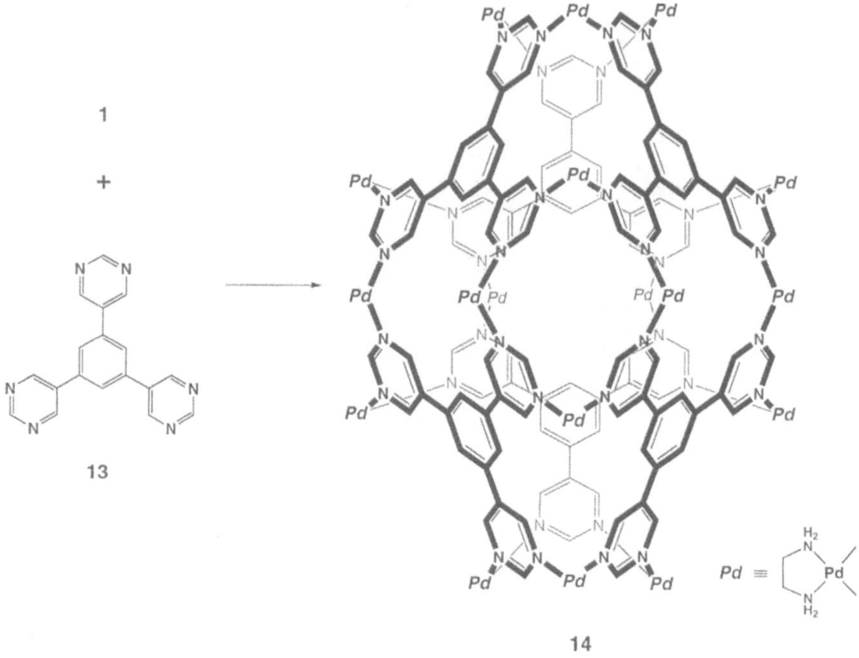

Scheme 6. Self-assembly of **14** from **1** and **13**

When ligand **13** was treated with **1** in D_2O, we observed the predominant formation of a single component whose ^1H NMR spectrum showed seven singlet-like signals in an integral ratio of 2:2:2:2:2:1:1. This observation confirms that, after complexation, the ligand **13** is placed in a less-symmetrical environment with one symmetry axis passing through a 3,5-pyramidyl (pym) ring and a core benzene ring. This symmetry is in good agreement with the trigonal bipyramidyl structure of the molecular hexahedron **14** in which the pym groups at the apical corners are unequivalent to those at the equatorial corners. The metal-linked M_2L_2 dimer and M_4L_3 trimer, possible intermediates for an assembly process of **14**, were observed when ligand **13** was treated with **1** in D_2O in a ratio of 1:1 and 3:4, respectively.

Reliable evidence for the trigonal bipyramidal structure of **14** is provided by X-ray crystallography (Fig. 9). The crystal structure clearly demonstrates that the assembly is a trigonal bipyramidal capsule with a chemical formula of $C_{144}H_{216}N_{108}Pd_{18}$, a molecular mass of 7103 Da, and dimensions of $3 \times 2.5 \times 2.5$ nm. Each equatorial corner of the hexahedron is the assembly of four triangle units, where a [Pd(II)-pym]$_4$ gives a small pin hole (2×2 Å). Only small molecules such as water and molecular oxygen can pass through these holes, but ordinary organic molecules cannot enter or escape through them. The free volume inside the capsule, in which guests can be accommodated, is ca. 900 Å^3.

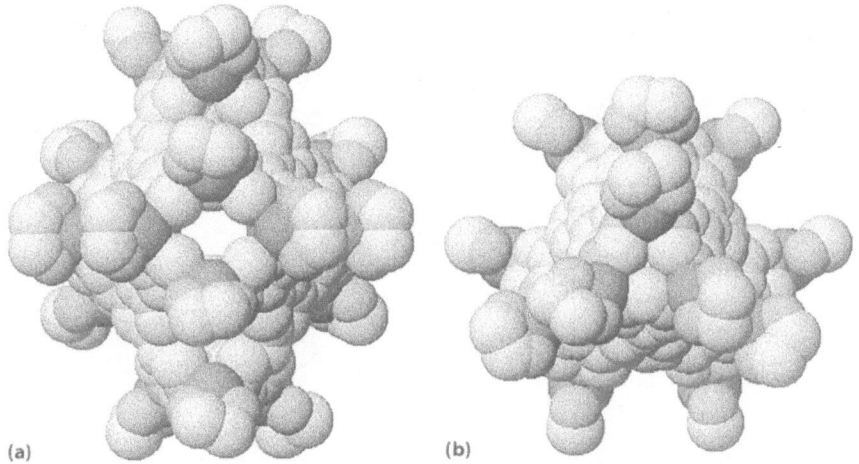

Fig. 9a,b. X-ray crystallographic structure of **14**. **a** an equatorial view; **b** an apical view

3
Assemblies from Square and Rectangular Molecular Panels

Following the triangular molecular panels, use of square and rectangular molecules as building blocks for three-dimensional coordination assemblies is discussed in this section.

3.1
Tetrapyridyl Porphyrins

Very available and promising square panels are tetrapyridyl porphyrins **15** and **16**. In fact, 4-pyridyl-substituted porphyrin **15** has been frequently employed for infinite assemblies. In contrast, there seems to be difficulty in constructing discrete 3D assemblies from tetrapyridyl porphyrins, though there are some examples of 2D assemblies such as porphyrin squares **17** [27] and **18** [28] from dipyridyl porphyrins.

15   16

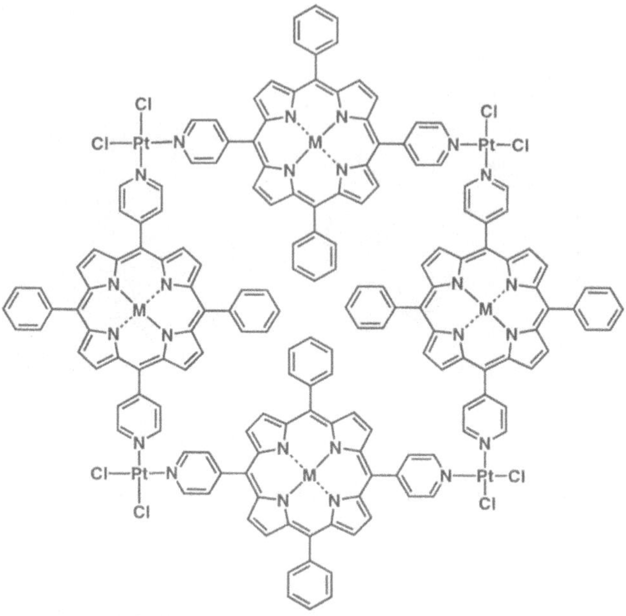

17

Whereas no 3D assemblies from 4-pyridyl-substituted porphyrins have been reported, there is a successful 3D assembly from 3-pyridyl-substituted porphyrin **19**. Upon treatment of **19** with the Pd(II) component **1**, the porphyrin prism **20** is assembled (Scheme 7) [29]. This compound can accommodate large organic molecules in the cavity surrounded by three porphyrin panels.

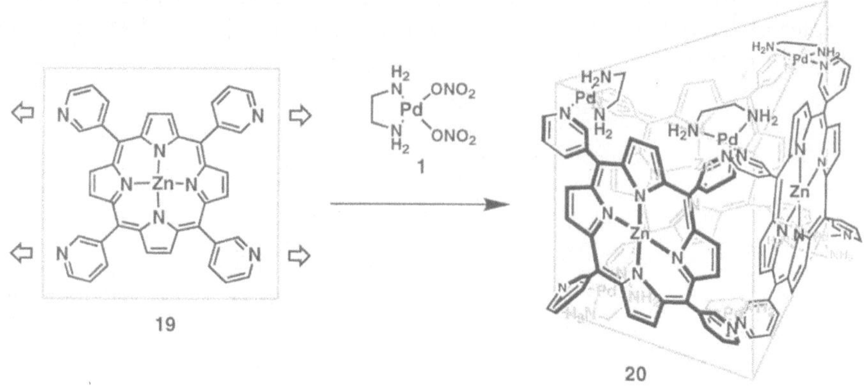

19   20

Scheme 7. Self-assembly of **20** from **1** and **19**

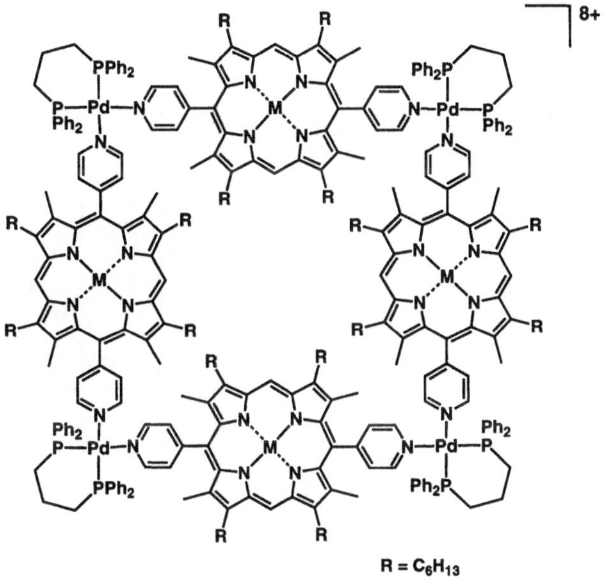

R = C6H13

18

3.2
Oligo(3,5-pyridine)s

Molecular-based tubular structures are attracting considerable current interest because of their potential ability for selective inclusion and transportation of ions and molecules and catalysis of specific chemical transformations [30]. Here we discuss the design of coordination nanotubes using oligo(3,5-pyridine)s and the *cis*-protected Pd(II) complex **1** [31]. In the case of the pentakis(3,5-pyridine) ligand **21**, a coordination nanotube **22** is expected from four molecules of **21** and 10 molecules of **1** (Scheme 8). However, the formation of coordination nanotubes was observed only in the presence of a rod-like template molecule such as sodium 4,4′-biphenylenedicarboxylate (**23**). Similarly, coordination nanotubes **24** and **25** were also obtained and characterized using NMR and electrospray ionization mass spectrometry (ESI-MS). In NMR, the protons of **23** were shifted up-field by up to 2.6 ppm indicating its encapsulation in the nanotube. A similar template effect was observed with two other rod-like molecules, biphenyl and *p*-terphenyl. Spherical and large molecules such as adamantane carboxylate failed to template the nanotubes. Interestingly, it was found that the formation of these tubes is a completely reversible process. That is the tube dissociates into its components by the removal of the guest molecule and again associates by the addition of the guest molecule.

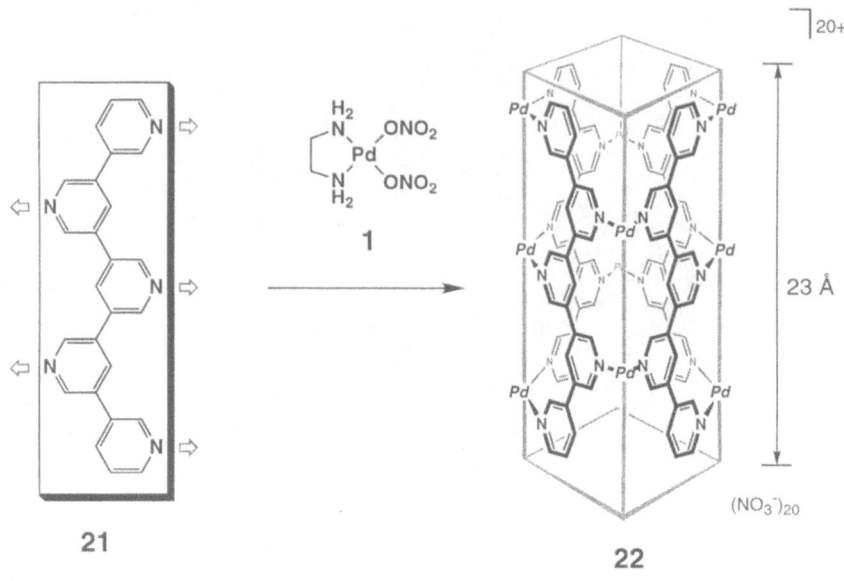

Scheme 8. Self-assembly of **22** from **1** and **22**

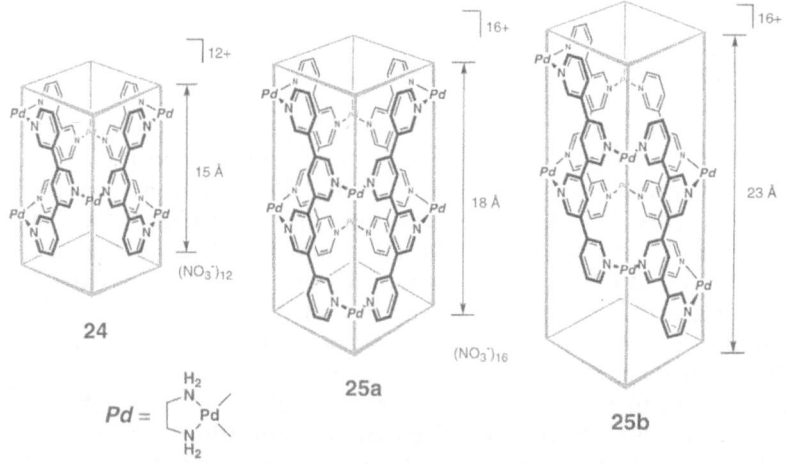

A shuttle movement of the guest molecule has been observed: the guest stays at a fixed position in the tube, shuttles on the NMR time scale at 60 °C, and rapidly moves or partially leaves above 60 °C. NMR studies of nanotube **25** revealed that it is a 1:1 mixture of structural isomers **25a** and **25b** (Scheme 9). In isomer **25b**, each ligand is placed in a C_2-symmetry and only seven protons corresponding to half of **25b** were observed. On the other hand, in isomer **25b**, all 14 protons were observed as the C_2-symmetry of the ligands was destroyed. The tubes **25b** and **24** were characterized by X-ray crystallography. The crystal structures display tubular structures for **24** and **25b** that are efficiently assembled around template **22** via strong $\pi-\pi$ and CH$-\pi$ interactions (Fig. 10). The shape of the tube, which should ideally be square, is significantly distorted in order to provide strong aromatic interactions. That is the two faces which are interacting with **23** via $\pi-\pi$ interactions are squeezed inwards, while the remaining two faces, which are interacting with **23** via CH$-\pi$

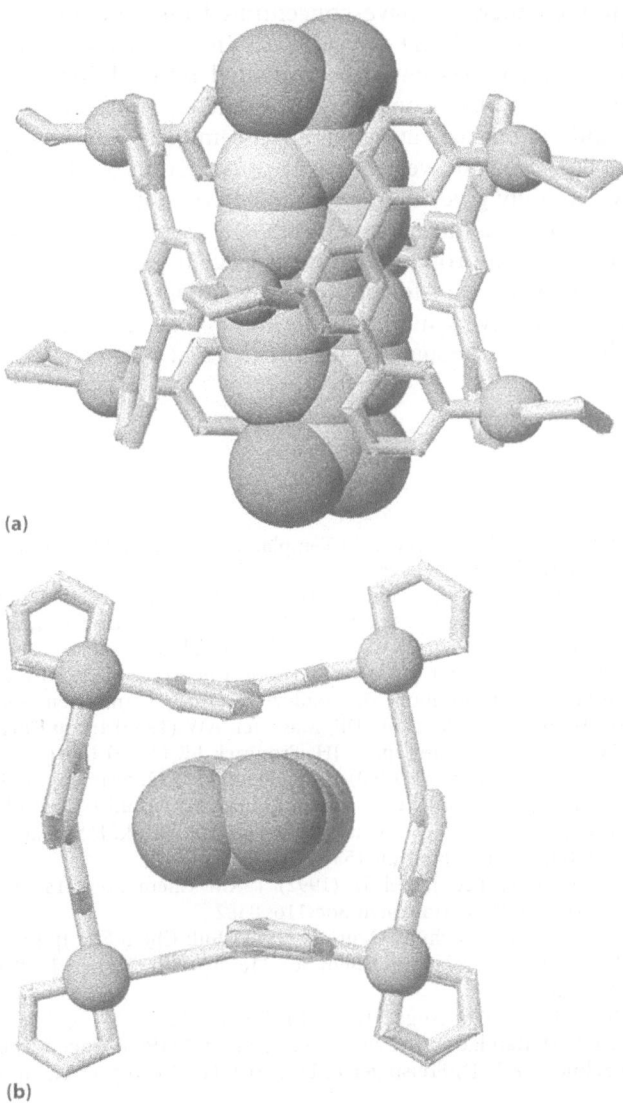

(a)

(b)

Fig. 10a,b. X-ray crystallographic structure of **24** accommodating template **23** in the tube. **a** a side view; **b** a top view

interactions, are pushed outwards. Another interesting feature of this crystal structure is the presence of a second molecule of **23** that is enclathrated between the nanotubes.

4
Conclusions

The construction of large 3D structures is attracting considerable current interest. In this chapter, we have concentrated our attention on the metal-directed self-assembly of 3D assemblies. In particular, a very efficient approach to 3D assemblies has been provided by exploiting the concept of molecular paneling. This concept has some advantages in the design, preparation and functionalization of 3D assemblies. For example, one can easily achieve the facile and very efficient construction of 3D assemblies since self-assembly generally gives the target molecules in quantitative yields under very mild conditions. Cavities created in the 3D assemblies are in general very large and novel binding properties towards large organic molecules, as well as catalysis for chemical transformations, are highly expected. In addition, transition metals involved in the frameworks may give the 3D assemblies interesting photo, redox, magnetic, and/or thermal properties.

5
References

1. Sauvage J-P, Hosseini MW (eds) (1995) Templating, self-assembly and self-organization. In: Lehn J-M (ed) Comprehensive supramolecular chemistry, vol 9. Pergamon Press, Oxford. This volume includes excellent reviews of metal-associated self-assembly. In particular, see the following: (a) Sauvage J-P, Dietrich-Buchecker C, Chambron J-C (b) Sanders JKM, chap 4 (c) Baxter PNW, chap 5 (d) Constable EC, chap 6 (e) Fujita M, chap 7
2. (a) Maverick AW, Klavetter FE (1984) Inorg Chem 23: 4129 (b) Maverick AW, Buckingham SC, Yao Q, Bradbury JR, Stanley GG (1986) J Am Chem Soc 108: 7430 (c) Bradbury JR, Hampton JL, Martone DP, Maverick AW (1989) Inorg Chem 28: 2392 (d) Maverick AW, Ivie ML, Waggenspack JH, Fronczek FR (1990) Inorg Chem 29: 2403
3. (a) Fujita M, Yazaki J, Ogura K (1990) J Am Chem Soc 112: 5645 (b) Fujita M, Yazaki J, Ogura K (1991) Tetrahedron Lett 32: 5589 (c) Fujita M, Yazaki J, Ogura K (1991) Chem Lett 1031 (d) Fujita M, Sasaki O, Mitsuhashi T, Fujita T, Yazaki J, Yamaguchi K, Ogura K (1996) J Chem Soc Chem Commun 1535
4. (a) Schwabacher AW, Lee J, Lei H (1992) J Am Chem Soc 114: 7597 (b) Lee J, Schwabacher AW (1994) J Am Chem Soc 116: 8382
5. Fujita M, Yazaki J, Kuramochi T, Ogura K (1993) Bull Chem Soc Jpn 66: 1837
6. (a) Stang PJ, Cao DH (1994) J Am Chem Soc 116: 4981 (b) Stang PJ, Olenyuk B (1997) Acc Chem Res 30: 502
7. Stang PJ, Olenyuk B (1996) Angew Chem Int Ed 35: 732
8. (a) Schneback R-D, Randaccio L, Zagrando E, Lippert B (1998) Angew Chem Int Ed Engl 37: 119 (b) Schnebeck R-D, Freisinger E, Lippert B (1999) Angew Chem Int Ed Engl 38: 168
9. (a) Batten SR, Hoskins BF, Robson R (1995) Angew Chem Int Ed Engl 34: 820 (b) Batten SR, Hoskins BF, Robson R (1995) J Am Chem Soc 117: 5385 (c) Fujita M, Oguro D, Miyazawa M, Oka H, Yamaguchi K, Ogura K (1995) Nature 378: 469

10. Fujita M, Yu S-Y, Kusukawa T, Funaki H, Ogura K, Yamaguchi K (1998) Angew Chem Int Ed Engl 37: 2082
11. (a) Baxter P, Lehn J-M, DeCian A (1993) Angew Chem Int Ed Engl 32: 69 (b) Baxter PNW, Lehn J-M, Baum G, Fenske D (1999) Chem Eur J 5: 102 (c) Baxter PNW, Lehn J-M, Kneisel BO, Baum G, Fenske D (1999) Chem Eur J 5: 113
12. (a) Gardner GB, Venkataraman D, Moore JS, Lee S (1995) Nature 374: 792 (b) Venkataraman D, Gardner GB, Lee S, Moore JS (1995) J Am Chem Soc 117: 11600
13. Drain CM, Nifiatis F, Vasenko A, Batteas JD (1998) Angew Chem Int Ed Engl 37: 2344
14. (a) Fujita M, Ogura K (1996) Bull Chem Soc Jpn 69: 1471 (b) Fujita M (1998) Chem Soc Rev 27: 417 (c) Fujita M (1999) Acc Chem Res 32: 53
15. (a) Saalfrank RW, preceding chapter (b) Saalfrank RW, Bernt I Curr Opin Solid State Mat Sci (1998) 3: 407, and references cited therein
16. Batten SR, Robson R (1998) Angew Chem Int Ed Engl 37: 1461
17. Kusukawa T, Fujita M (1998) Angew Chem Int Ed Engl 37: 3142
18. Kusukawa T, Fujita M, manuscript in preparation
19. Kusukawa T, Fujita M (1999) J Am Chem Soc 121: 1397
20. (a) Wyler R, de Mendoza J, Rebek J Jr (1993) Angew Chem Int Ed Engl 32: 1699 (b) Rebek J Jr (1996) Chem Soc Rev 255
21. Kusukawa T, Fujita M, manuscript in preparation
22. Lehn J-M (1978) Pure Appl Chem 50: 871
23. Ito H, Kusukawa T, Fujita M, submitted
24. Ibukuro F, Kusukawa T, Fujita M (1998) J Am Chem Soc 120: 8561
25. Yu S-Y, Kusukawa T, Biradha K, Fujita M (2000) J Am Chem Soc 122: in press
26. Takeda N, Umemoto K, Yamaguchi K, Fujita M (1999) Nature 398: 794
27. Drain CM, Fisher R, Nolen E, Lehn J-M (1993) J Chem Soc Chem Commun 243
28. Stang PJ, Fan J, Olenyuk B (1997) J Chem Soc Chem Commun 1453
29. Fujita N, Yamaguchi K, Fujita M, manuscript in preparation
30. (a) Iijima S (1991) Nature 354: 56 (b) Harada A (1997) In: Michl J (ed) Modular chemistry. Kluwer Academic Publishers, Dordrecht, The Netherlands, p 361 (c) Hartgerink JD, Clark TD, Ghadiri MR (1998) Chem Eur J 3: 1367
31. Aoyagi M, Biradha K, Fujita M (1999) J Am Chem Soc 121: 7457

Pythagorean Harmony in the World of Metal Oxygen Clusters of the {Mo$_{11}$} Type: Giant Wheels and Spheres both Based on a Pentagonal Type Unit

Achim Müller, Paul Kögerler, Hartmut Bögge

Faculty of Chemistry, University of Bielefeld,
Postfach 100131, D-33501 Bielefeld, Germany
E-mail: a.mueller@uni-bielefeld.de

"[...]; since, again, they (the Pythagoreans) saw that the attributes and the ratios of the musical scales were expressible in numbers; since, then, all other things seemed in their whole nature to be modeled after numbers, and numbers seemed to be the first things in the whole of nature, they supposed the elements of numbers to be the elements of all things, and the whole heaven to be a musical scale and a number." (Aristotle Metaphysics 1, 985b 31 ff).

Polyoxometalate systems provide excellent models for the study of unusual molecular growth processes based on {Mo$_{11}$}-type building motifs/blocks leading to unusual giant spherical and ring-shaped structures with stoichiometry {Mo$_{11}$}$_n$ (n = 12, 14, 16)-a Pythagorean harmony.

Keywords: Clusters, Polyoxometalates, Building Blocks

1
Introduction

The generation of complex molecular systems, even relatively large ones, can in principle be achieved by successive steps of synthesis and isolation, following the logic of retrosynthesis. In contrast, the formation of extremely large molecular species – in particular those with unusual structures and especially with very high symmetry, e.g. those comparable to simple spherical viruses, but also the wheel-shaped type – requires a different method of approach. In this respect, the use of *pentagonal-type building groups* with different symmetry plays a key role which was not realized until now. Interestingly, pentagons as such play an extraordinary role in the cultural and science history of mankind, beginning with the Pythagoreans (the pentagram was the emblem of their school), then later from Archimedes, Kepler and Dürer to Goethe's Faust and to Weyl and Penrose. In many scientific disciplines, they are of special interest: In *mathematics* (one of the five regular bodies is built up of pentagons), in *architecture* and history of arts (in the sense of Fuller's domes and in context of the Golden Section), in *solid-state physics* (in context of the fascinating quasi crystals), in *virology* (due to the structure of spherical viruses), even in *cosmology* (with respect to a recently discovered planetary nebula with fullerene structure) and last but not least in *chemistry*, allowing novel design strategies of curved species.

In polyoxometalate chemistry a large variety of compounds, clusters and solid-state structures can be formed by linking together metal-oxygen building blocks [1]. In this review we will concentrate on related (molecular growth) processes based (in a formal consideration) on such pentagonal building blocks/motifs leading to different types of the unusual giant structures mentioned above. Although details about the mechanisms of reaction pathways are not known, the herein mentioned pentagon-type motifs appear in the final reaction products. The relevant procedures lead to the high-yield syntheses of two classes of $\{Mo_{11}\}_n$ compounds, (a) *icosahedral inorganic superfullerenes* or *Keplerates* (including a cluster containing the largest number of paramagnetic centers known today for a discrete species) and (b) *giant wheel-shaped, electron-rich, mixed-valence species with nanosized cavities* (so-called "big wheel" type). Remarkably, from a structural point of view, both types of species are based on building blocks/motifs which have the same stoichiometry and a central, obviously directing pentagonal bipyramid while the number n allows to distinguish between the two rather different structural types. Remarkable in the

Pythagorean sense is that the giant cluster species – categorized on the basis of distinct, specific numbers – can themselves be used as synthons as they can be coordinatively linked together to form chains as well as layered compounds with properties relevant for materials science.

2
Polyoxometalate Cluster Synthesis Using Building Blocks: Foundations and the Special {Mo₁₁} Motif

The deliberate synthesis of discrete complex multifunctional molecular systems can be considered as an essential tool to meet future requirements in the field of materials science, e.g. in the context of developing (switchable) molecules, molecular machines as well as magnets or quantum computers. In nature, complex molecular systems with the desired functionality like proteins are formed (at least formally) in a sequence of steps under dissipative conditions, i.e. far from thermodynamic equilibrium. The challenge for the chemist is to synthesize correspondingly complex molecules by means of multicomponent one-pot reactions, and thus exempt from tedious separation and intermediate purification steps. In particular, solutions of polyoxoanions of the early transition metals (especially V, Mo and W) serve as extraordinary model systems in which novel types of molecular growth processes leading to a huge variety of compounds with unusual structures and properties can be discovered. These processes can be described as based on combinatorially linkable polyoxometalate building groups "present" in solution either as existing or virtual fragments as disposition which become available on demand from a kind of 'library' depending on the boundary conditions, i.e. on which other species are present as templates (a disposition can be interpreted as the inherent tendency of an object or system to act or react in a certain characteristic way under certain conditions, in this context molecules, ions or fragments of the system).

Accordingly, the question could be addressed: Into which single structural elements can we (formally) reduce a giant cluster systems with the possibility of generating a structural *and* functional hierarchy by starting just from these elements? This corresponds to a methodological approach for the planning and construction of a huge variety of species.

Before returning to the problem of understanding the construction pathway based on well-defined elements, it is worthwhile to have a look at conditions which favor the corresponding emergence of molecular complexity (including pathways involving several successive reaction steps).

Optimal conditions for linking of fragments leading to a huge variety of structures are:

- abundance of a variety of linkable units, i.e. building blocks and the following possibilities,
- of generation of groups (intermediates) with high free enthalpy to drive polymerization or growth processes, e.g. by formation of H_2O,
- of an easy structural change in the building units and blocks,
- of including hetero elements in the fragments,
- to form larger groups which can be linked in different ways,

- to control the structure-forming processes by templates,
- to generate structural defects in reaction intermediates, e.g. by removing building blocks from (large) intermediates due to the presence of appropriate reactants,
- to localize and delocalize electrons in different ways in order to gain versatility,
- to control and vary the charge of building parts (e.g. by protonation, electron transfer reactions, or substitution) and to limit growth by the abundance of appropriate terminal ligands,
- of generating fragments with energetically low-lying unoccupied molecular orbitals.

These conditions can be optimally fulfilled in polyoxometalate systems which possess the relevant variety of structural and electronic versatility, and not e.g. in silicate-based systems.

In this respect many comparisons can be drawn with natural processes generating an overwhelming variety of molecular species which are effected by the (directed as well as non-directed) linking of basic and well-defined pre-organized (or stepwise organized) fragments. An impressive example of this strategy, discussed in almost all textbooks on biochemistry, is the (reversible) self-aggregation-reconstitution process of the tobacco mosaic virus (TMV), which is based on preorganized units [2]. This process more or less meets our strategy in controlling the linking of fragments to form (larger) groups and linking the latter again. One possible relevant example is based on the linking of building groups containing 17 metal atoms ($\{Mo_{17}\}$ groups) to form cluster anions containing two or three of these groups. The resulting two- or three-fragment clusters are of the $\{Mo_{36}\}$ ($[\{MoO_2\}_2\{H_{12}Mo_{17}(NO)_2O_{58}(H_2O)_2\}_2]^{12-}$ $\equiv[\{Mo_1'\}_2(\{Mo_8\}_2\{Mo_1\})_2])$ or of the $\{Mo_{57}M_6\}$ type (e.g. $[\{VO(H_2O)\}_6\{Mo_2(\mu\text{-}H_2O)_2(\mu\text{-}OH)\}_3\ \{Mo_{17}(NO)_2O_{58}(H_2O)_2\}_3]^{21-}\equiv[\{VO(H_2O)\}_6\{Mo_2\}_3(\{Mo_8\}_2\{Mo_1\})_3])$ (Fig. 1) [3]. The $\{Mo_{17}\}$ groups are linked by various octahedrally coordinated transition metal centers (V^{IV}, Fe^{II}, Fe^{III}) and $\{Mo_2^V\}$ groups and can themselves be structurally reduced to smaller building groups, such as $\{Mo_8\}$, which will be discussed later (this motif-approach differs from the $\{Mo_{11}\}$-based one mainly considered here; see below).

We can now demonstrate to what exciting extent the principles of fragment linking are valid in the construction of nanosized polyoxomolybdate clusters. In particular, the significance of the directing power of pentagonal-type units for the formation of a huge variety of giant clusters should be stressed. But first of all, what makes pentagonal building groups/motifs so special? (Fig. 2). On one hand edge-sharing (condensed) pentagons cannot build (tile) a regular and infinitely repetitive (translational invariant) two-dimensional pattern. On the contrary, exactly 12 pentagons are required in connection with well-defined sets of hexagons to construct spherical systems such as that observed in the truncated icosahedron – the most spherical Archimedean solid –, in polyhedral viruses, or in the geodesic Fuller domes [4]. Remarkably, they can also be used to create curved molecular systems as in the giant wheel-type species.

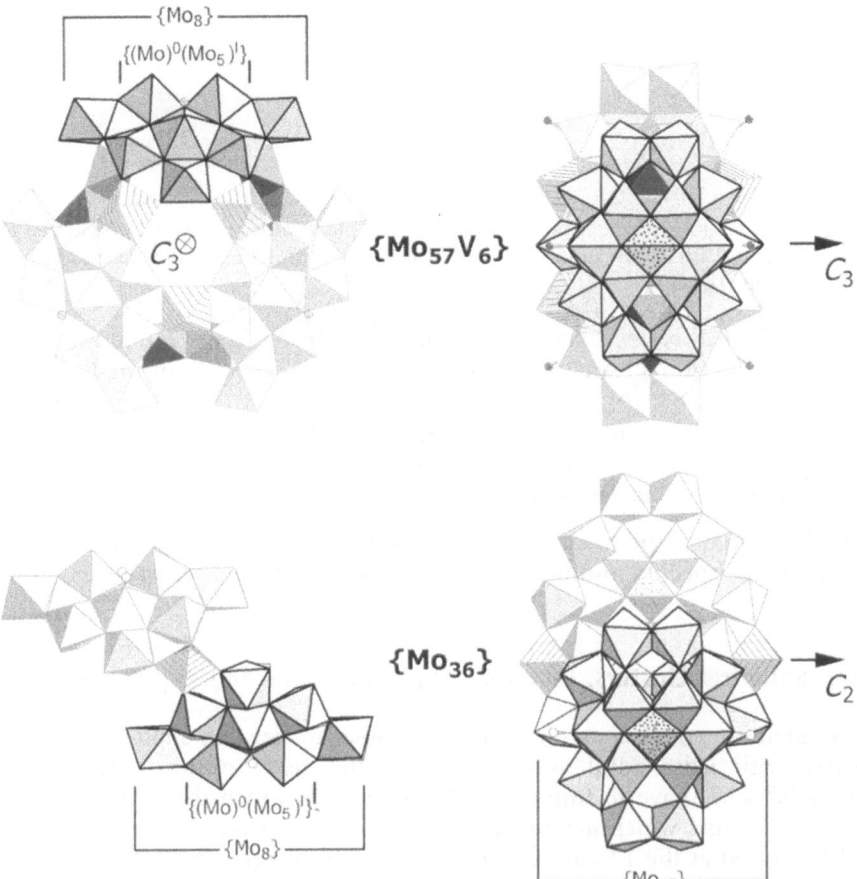

Fig. 1. Polyhedral representation of the three-fragment $\{Mo_{57}V_6\}$ cluster along the C_3 (*upper left*) and along one of the three C_2 axes (*upper right*) showing patterns of its basic building blocks and their constituents: On the *upper right*, one $\{Mo_{17}\}$ building block consisting of two $\{Mo_8\}$ groups linked via one MoO_6 octahedron, and on the *upper left*, one $\{Mo_8\}$ group is emphasized; the $\{V\}$ octahedra are *hatched*. Also highlighted are the three $\{Mo_2\}$ groups (*dark gray*, built up from two face-sharing octahedra) and $\{Mo_1\}$ units (linking the $\{Mo_8\}$ groups; *dotted*). For comparison, polyhedral representations of the two-fragment $\{Mo_{36}\}$ cluster structure, consisting of two $\{Mo_{17}\}$ building blocks linked by two $\{Mo_1'\}$ units (*hatched*), are shown in the related views, also highlighting one $\{Mo_8\}$ (*bottom left*) and one $\{Mo_{17}\}$ building block (*bottom right*). Important for our present view is that the pentagonal $\{(Mo)^0(Mo_5)^I\}$ group is a part of the $\{Mo_8\}$ group

Molybdenum-oxide based giant clusters with characteristic curvatures contain the potential $\{Mo_{11}\} \equiv \{(Mo)^0(Mo_5)^I(Mo_5)^{II}\}$ building block/motif with a central (directing) pentagonal MoO_7 bipyramid ($\equiv(Mo)^0$) in accordance with Fig. 3.

The most fascinating point is that this basic building block/motif is archetypal for different types of giant molecular constructions of the $\{Mo_{11}\}_n$ type (including also solid-state structures derived from them) (Fig. 4), e.g.

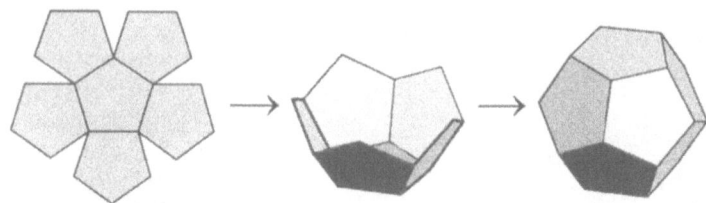

Fig. 2. Demonstration that it is not possible to tile a plane with pentagonal units (*left*) and that pentagons are ideal units for the construction of spherical patterns like the dodecahedron (*right*)

(1) for *spherical systems* such as the $\{Mo_{11}\}_{12}$ cluster if the Mo atoms of the peripheral $(Mo_5)^{II}$ shell are equivalent, i.e. if they are all of the Mo^V type, and

(2) for *lower symmetrical wheel-type systems* such as the $\{Mo_{11}\}_n$ ($n = 14, 16$) species if the relevant five Mo atoms are not equivalent (with an average oxidation number less than VI).

The differentiation between species of different types is possible by referring simply to the specific number n – perhaps in the Pythagorean sense.

3
Giant Spherical Clusters: Inorganic Superfullerenes and Keplerates

If we intend to construct a giant species similar in size and shape to spherical viruses with icosahedral symmetry – in the sense of an "Archimedean synthesis" – we have to find a reaction system in which the aforementioned pentagonal units which are present as disposition insolution can get linked, and be placed at the 12 corners of an icosahedron (see below).

3.1
The Prototype: {Mo₁₁}₁₂

In the case of polyoxomolybdates the pentagonal groups/motifs $\{(Mo)Mo_5\}\equiv\{(Mo)^0(Mo_5)^I\}$ consist of a central bipyramidal MoO_7 unit $((Mo)^0)$ sharing edges with 5 MoO_6 octahedra $((Mo_5)^I$ type). These groups are the starting point for quite a number of spherical-type clusters. When linkers are present in the reaction medium, for instance those of the classical $\{Mo_2^VO_4\}^{2+}$ type (see for instance [5]), typically formed with a high formation tendency in reduced molybdate solutions in the presence of bidentate ligands or directly used as educts, an icosahedral molecular system with 12 of the aforementioned $\{(Mo)Mo_5\}$ pentagons and 30 of the mentioned linkers is formed. When for example carboxylate-type acetate anions occur as bridging ligands for the $\{Mo_2^VO_4\}^{2+}$ groups, the spherical cluster with the stoichiometry $[Mo_{72}^{VI}Mo_{60}^VO_{372}(CH_3COO)_{30}(H_2O)_{72}]^{42-}$ ($\{Mo_{132}\}\equiv\{Mo_{11}\}_{12}$) results (Fig. 5) [6], where the central Mo positions of the $\{(Mo)^0(Mo_5)^I\}$ pentagons define the 12 corners, and the $\{Mo_2^VO_4\}^{2+}$ groups the 30 edges of an icosahedron, a situation

$$\{(Mo)^0(Mo_5)^I(Mo_5)^{II}\}$$

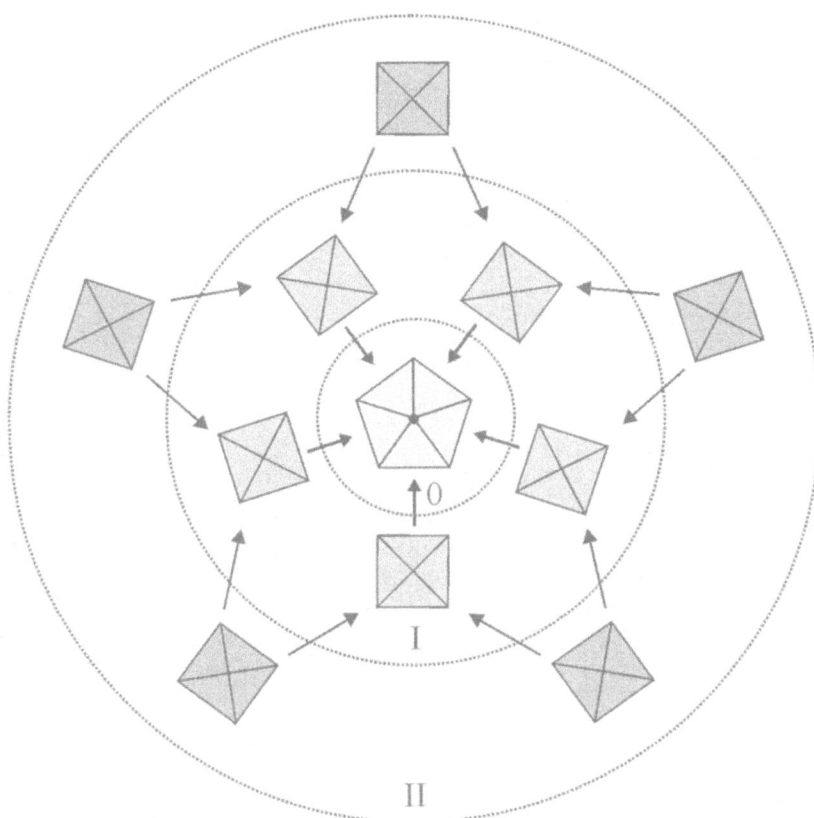

Fig. 3. Schematic representation of the $\{Mo_{11}\}\equiv\{(Mo)^0(Mo_5)^I(Mo_5)^{II}\}$ motif with its central part, a pentagonal-bipyramidal MoO_7 unit, having a directing function. Five MoO_6 octahedra of the $(Mo_5)^I$ type are condensed to the equatorial plane of this central MoO_7 unit, sharing edges, while the next five MoO_6 octahedra of the $(Mo_5)^{II}$ type share corners with the former octahedra

which is in agreement with Euler's well-known formula. (This corresponds to the formulation $[\{(Mo)Mo_5O_{21}(H_2O)_6\}_{12}\{Mo_2^VO_4(CH_3COO)\}_{30}]^{42-}$). Based on a topological or motif concept, the cluster can be formulated as $\{Mo_{11}\}_{12}\equiv\{(Mo)^0(Mo_5)^I(Mo_5^V)^{II}\}_{12}$ (Fig. 3, 6). This approach takes formally into account that the Mo atoms of each rather stable $\{Mo_2^V\}$ dumbbell belong to two adjacent $\{Mo_{11}\}$ groups. Interestingly, the $\{Mo_2^V\}$ groups themselves form a solid comparable to the C_{60} fullerene (Fig. 7).

Remarkably, we are speaking here about a molecular system having icosahedral symmetry although it is built up from more than 500 atoms.

Structural Hierarchy

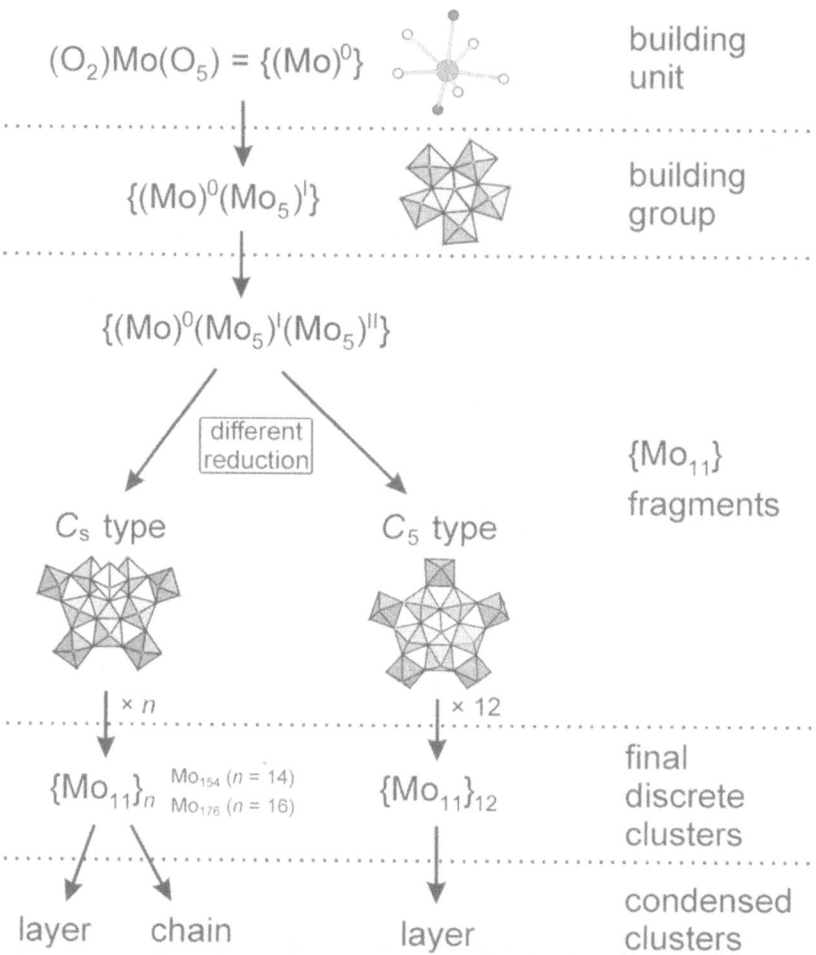

$(O_2)Mo(O_5) = \{(Mo)^0\}$ building unit

$\{(Mo)^0(Mo_5)^I\}$ building group

$\{(Mo)^0(Mo_5)^I(Mo_5)^{II}\}$

different reduction

$\{Mo_{11}\}$ fragments

C_s type C_5 type

$\times n$ $\times 12$

$\{Mo_{11}\}_n$ Mo_{154} ($n = 14$) $\{Mo_{11}\}_{12}$ final discrete clusters
          Mo_{176} ($n = 16$)

layer   chain   layer   condensed clusters

Fig. 4. Structural hierarchy in systems based on the $\{Mo_{11}\}$ motifs corresponding to Fig. 3

This giant cluster with its 60 MoO$_6$ subunits or 12 related $\{(Mo)^0(Mo_5)^I\}$ pentagons represents a topological model for spherical viruses, e.g. the most simple satellite tobacco necrosis virus (STNV) with its only 60 identical protein subunits (Fig. 6) coded by only one gene which correspondingly contains only 12 pentagonal capsomers (morphology units, each consisting of five protomers; for details see [7]). Furthermore, the wire-frame representation of the $\{Mo_{11}\}_{12}$ cluster species additionally suggests the structural/topological similarity to the more complex icosahedral virus capsids with the so-called triangulation number of $T = 3$ (Fig. 8). Note that T allows a systematic

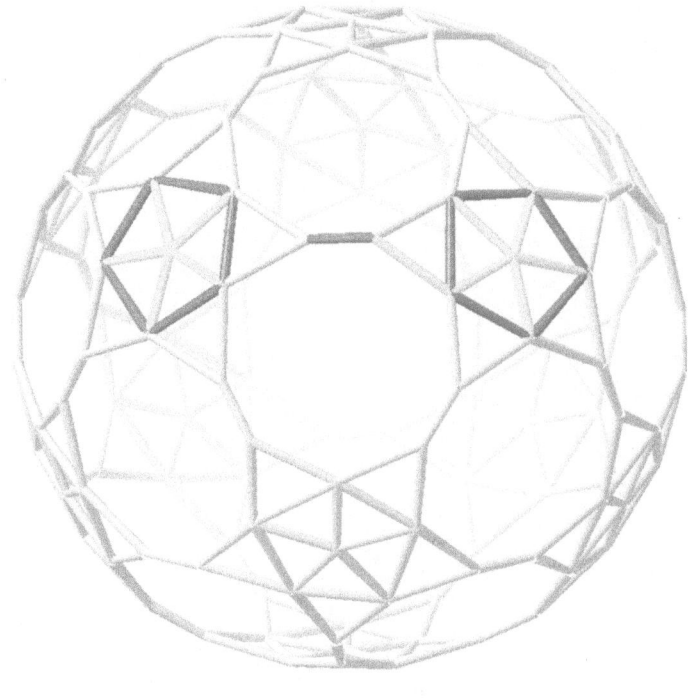

$$\{Mo_{11}\}_{12}$$

Fig. 5. Schematic representation of the 132 molybdenum atom framework of the $\{Mo_{11}\}_{12}$-type Keplerate cluster highlighting its spherical nature. Two pentagonal $\{(Mo)^0(Mo_5)^I\}$ groups linked by a $\{Mo^V\!-\!Mo^V\}$ bridge are highlighted. The related extended $\{Mo_{11}\}\equiv\{(Mo)^0(Mo_5)^I(Mo_5)^{II}\}$ motif can also be identified (see Figs. 3, 4)

structural differentiation of spherical viruses (for details on virus structures and topologies see [6–8] and literature cited therein).

Interestingly, by substituting the bridging acetate by formate ligands the smaller ligand allows a special type of organization of the encapsulated $(H_2O)_n$ guest collection resulting in a hydrogen-bonded cluster with an 'onion-type' structure, the formation of which is completely controlled by the spherical cluster shell built up by molybdenum and oxygen atoms and is not disturbed/ influenced by the ligand (Fig. 9) [9].

It has been proposed to call the $\{Mo_{11}\}_{12}$-type cluster a Keplerate corresponding to Kepler's early speculative model of the cosmos and concept of planetary motion, as described in his opus *Mysterium Cosmographicum* [10]. In accordance with this model, Kepler believed that the distances between the orbits of the planets could be explained if the ratios between the successive orbits were designed to be equivalent to the spheres successively circumscribed around and inscribed within the five Platonic solids. By analogy, the cluster correspondingly shows concentric spherical shells of 132 terminal oxygen and 132 molybdenum atoms while the centers of the 12 $\{(Mo)^0(Mo_5)^I\}$

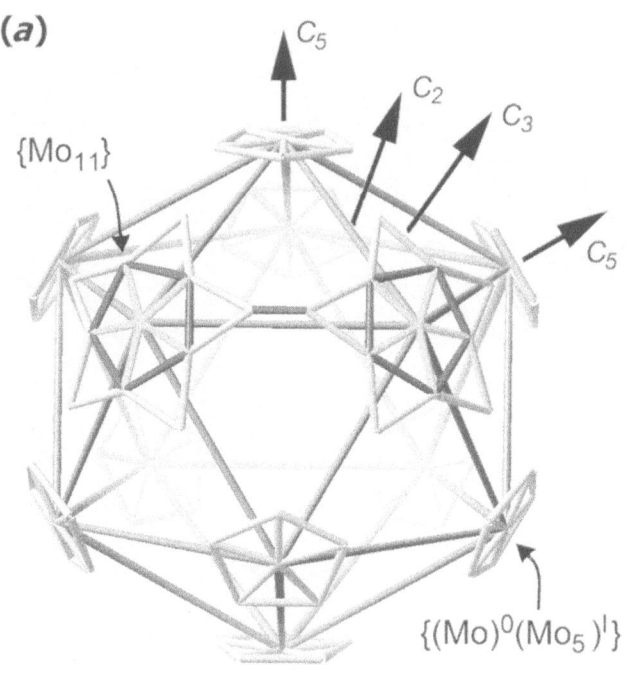

$\{Mo_{11}\}_{12}$

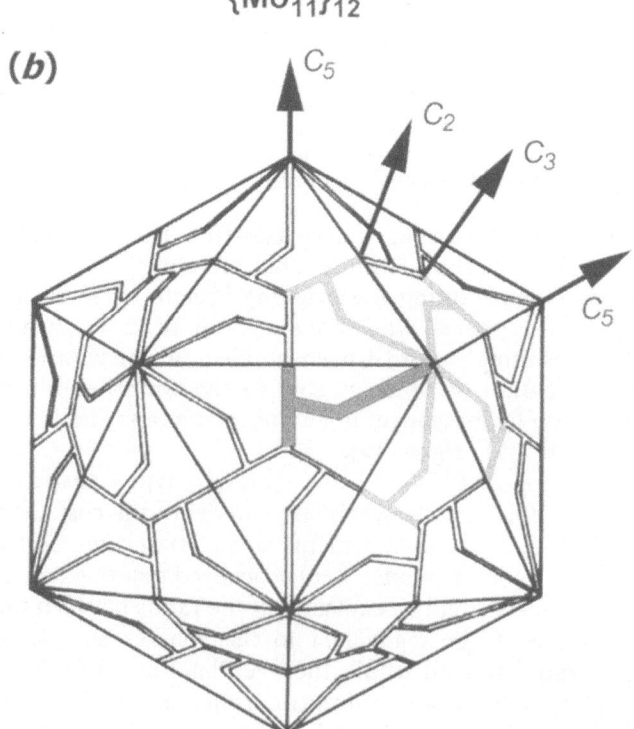

Fig. 6a,b. Schematic representation of the icosahedron spanned by the centers of the $\{Mo_{11}\}$ groups building the $\{Mo_{11}\}_{12}$-type cluster (**a**) and of the satellite tobacco necrosis virus (STNV) with triangulation number $T = 1$ highlighting five of the 60 protomers (**b**). Two $\{Mo_{11}\}$ groups formed by $\{(Mo)^0(Mo_5)^I\}$ groups and five further Mo centers (identical with Mo atoms of the five neighboring $\{Mo_2^V O_4\}^{2+}$ bridges) are emphasized

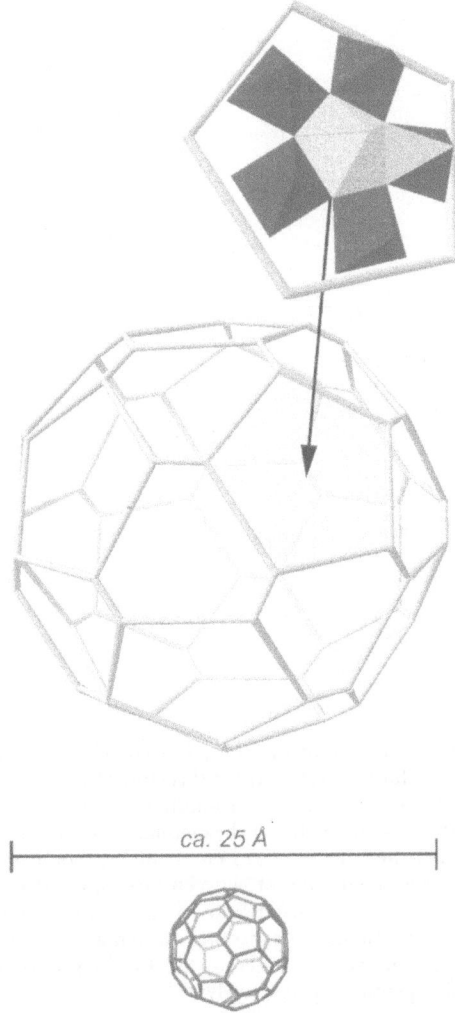

ca. 25 Å

Fig. 7. Structure of the icosahedral $\{Mo_2^V\}_{30}$ fragment with 12 regular pentagons and 20 (trigonal) hexagons as well as its coherence to the C_{60} fullerene, which is depicted on the same scale. A separate $\{(Mo)^0(Mo_5)^I\}$ pentagon in polyhedral representation is emphasized

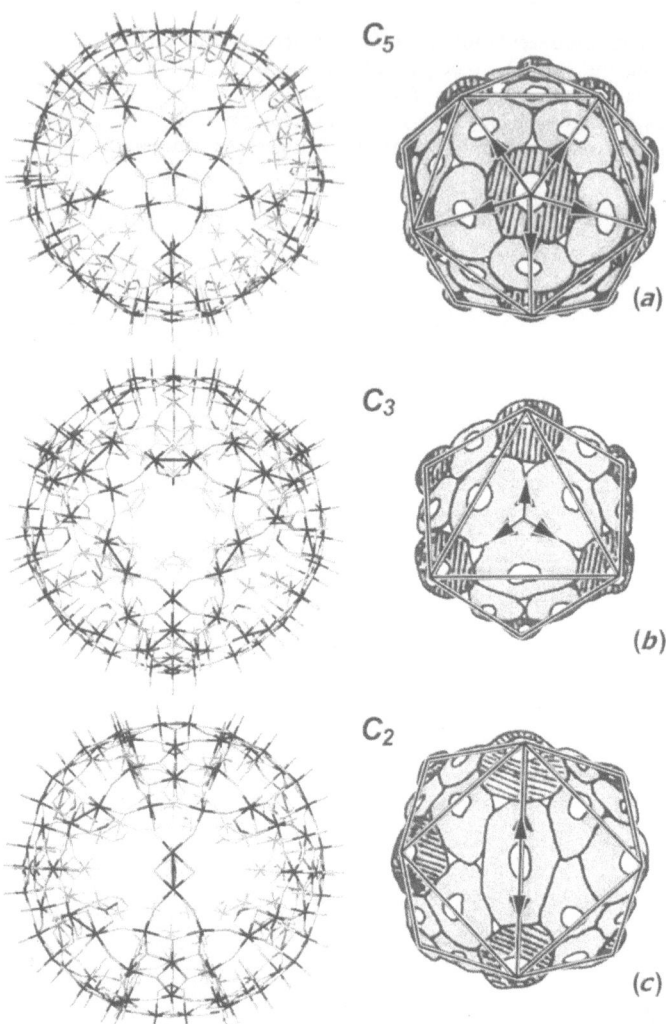

Fig. 8a–c. Illustration of the structure of the $\{Mo_{11}\}_{12}$-type cluster in wire-frame presentations with perspective views along a fivefold (**a**), a threefold (**b**), and a twofold (**c**) symmetry axis. For the purpose of comparison, corresponding schematic representations of an icosahedral virus capsid ($T = 3$) with 20 hexagonal and 12 pentagonal capsomers (morphology units) are presented. The C_5 axes cross the centers of the pentagonal groups (hatched) (**a**), the C_3 axes cross the midpoint between three units (**b**), and the C_2 axes cross the center of the units (**c**). Whereas (**a**) refers to the centers of both the pentagonal $\{(Mo)^0(Mo_5)^I\}$ groups and the pentagonal capsomers which are located at the 12 corners of an icosahedron (see Fig. 6), (**b**) and (**c**) refer to the $\{Mo_2^V\}$ groups of the $\{Mo_{11}\}_{12}$-type cluster and the hexagonal capsomers, respectively

pentagons – namely the 12 Mo atoms of the central MoO_7 bipyramids – span an icosahedron (see Fig. 6).

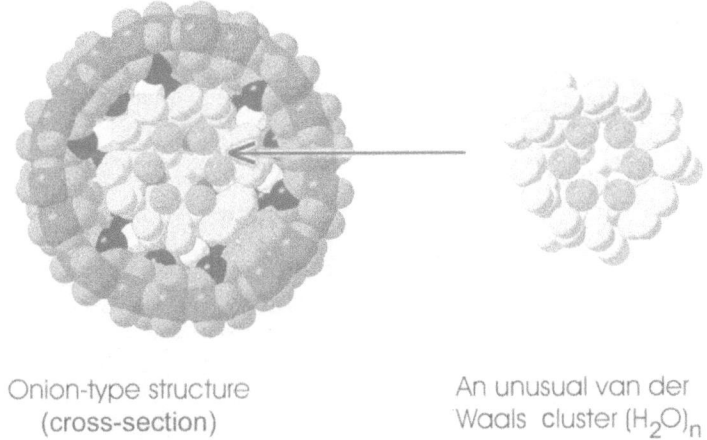

Onion-type structure
(cross-section)

An unusual van der
Waals cluster (H$_2$O)$_n$

Fig. 9. Cross section through the equator of the supramolecular species $[(H_2O)_n \subset Mo_{132}O_{372}(HCOO)_{30}(H_2O)_{72}]^{42-}$ ($n \approx 50$) allowing a view into the cavity of the cluster shell and highlighting the hydrogen-bonded cluster formed by the encapsulated H$_2$O molecules (oxygen atoms) (*left*). The different shells spanned by encapsulated H$_2$O molecules are represented as spheres with different shadings (*dark gray shell* (radius ~3.5 Å), *gray shell* (6.2–6.9 Å) and *light gray shell* (8.2–8.7 Å) (*right*)). Interestingly, the onion-like structure of the whole anion is completed by the three following outer shells consisting of (1) that of the 72 H$_2$O and 30 formate ligands coordinated to molybdenum atoms and pointing into the cavity (~10.5 Å), (2) that of the 132 molybdenum atoms (~13.1 Å) and (3) that of the terminal 132 oxygen atoms (~14.7 Å)

3.2
Deliberate Sizing of Giant Spherical Clusters

According to our concept the {Mo$_{11}$}$_{12}$-type cluster is built up from 12 {(Mo)^0(Mo$_5$)^I} pentagonal building groups interconnected by 30 {Mo$_2^V$} linkers (spacers). This concept has recently been used for the synthesis of other clusters, e.g. the neutral species of the type {Mo$_{72}$Fe$_{30}$}≡{(Mo)^0(Mo$_5$)^I (Fe$_{5/2}$)^{II}}$_{12}$ ([Mo$_{72}$Fe$_{30}^{III}$O$_{252}$(Mo$_2$O$_7$(H$_2$O)$_2$)$_2$(Mo$_2$O$_6$(OH)$_2$H$_2$O) (CH$_3$COO)$_{12}$ (H$_2$O)$_{91}$]) [11]. This is a (smaller sized) giant ball in which the 30 linking {Mo$_2^V$} groups have been substituted by 30 linking Fe^{III} centers by means of a related simple synthesis, i.e. by reaction of the (larger) {Mo$_{132}$}-type cluster anion with an aqueous solution containing Fe^{III} centers. This situation confirms unambiguously a type of unit-based construction principle. As expected, the central Mo atoms of the 12 {(Mo)^0(Mo$_5$)^I} pentagons span an icosahedron (as in the larger ball-type species), while the 30 Fe atoms as well as the 30 centers of the Mo$_2^V$ groups form an icosidodecahedron (Fig. 10). The special interest in this cluster system is also due to its magnetochemistry which is in principle governed by an only small antiferromagnetic exchange coupling between the 30 high-spin ($S = 5/2$) Fe centers (Fig. 11). Interestingly, the (smaller) {(Mo)^0(Mo$_5$)^I(Fe$_{5/2}$)^{II}}$_{12}$-type cluster represents the discrete

$$\{(Mo)^0(Mo_5)^I(Fe_{5/2})^{II}\}_{12}$$

Fig. 10. Structure of the $\{(Mo)^0(Mo_5)^I (Fe_{5/2})^{II}\}_{12}$-type cluster in polyhedral representation along a C_5 axis. Fe-based octahedra: *hatched*, $\{(Mo)^0(Mo_5)^I\}$ groups: *light gray*; $(Mo)^0$ pentagonal bipyramids: *dark gray* with emphasized edges

cluster species with the largest number of paramagnetic centers known until now.

It is even possible to condense the individual spherical clusters according to Scheme 1 to form two-dimensional layers (Fig. 12). In these, each $\{(Mo)^0(Mo_5)^I(Fe_{5/2})^{II}\}_{12}$-type entity is interconnected with four others via Fe-O-Fe bridges. In principle, it should be possible to "switch" the open shell centers of the individual units, which renders the $\{(Mo)^0(Mo_5)^I(Fe_{5/2})^{II}\}_{12}$ layer compound as possible model for information storage.

$$\cdots FeOH_2 + H_2OFe\cdots \xrightarrow{-H^+} \cdots FeOH_2 + HOFe\cdots \xrightarrow{-H_2O/-H^+} \cdots Fe-O-Fe\cdots$$

Scheme 1.

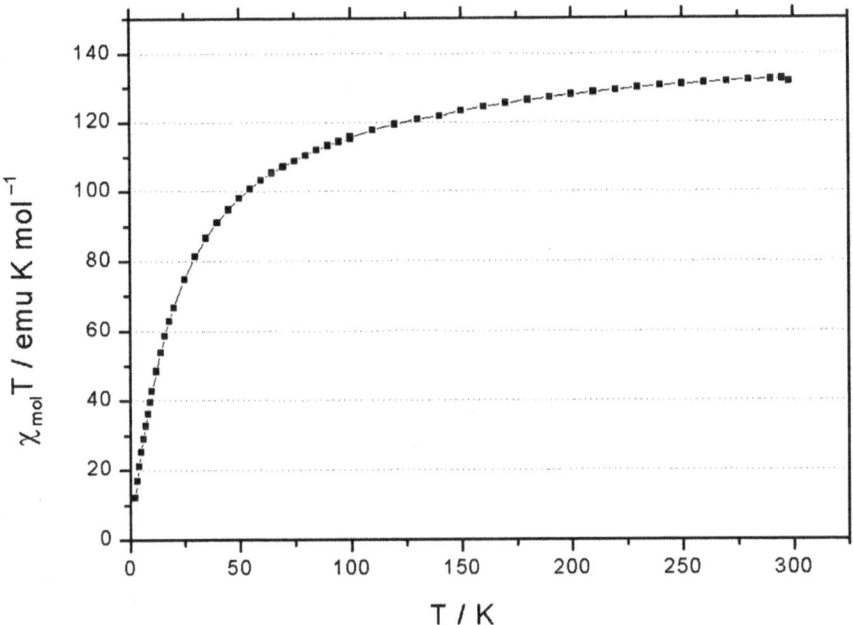

Fig. 11. $\chi_{mol}T$ vs T graph of the $\{(Mo)^0(Mo_5)^I(Fe_{5/2})^{II}\}_{12}$-type compound. The $\chi_{mol}T$ value of 130 emu K mol^{-1} at room temperature equates approximately the value expected for a system consisting of 30 uncorrelated $S = 5/2$ centers (131.25 emu K mol^{-1})

4
Construction Logic: From Giant Spherical Clusters to Giant Wheels

4.1
{Mo₁₁}₁₄- and {Mo₁₁}₁₆-type Clusters: A Topological Consideration

The crucial aspect that decides whether the $\{Mo_{11}\}$ groups will finally occur as present in the $\{Mo_{11}\}_{12}$-type cluster with fivefold local symmetry or such that the fivefold symmetry is disturbed depends mainly on the specific degree of reduction of the unit. Regular $\{Mo_{11}\}$ groups of C_5 symmetry require five equivalent $(Mo^V)^{II}$-type MoO_6 octahedra bound to the $\{(Mo)^0(Mo_5)^I\}$ pentagons (Fig. 3, 4). This is achieved only in the presence of a (comparably strong) reducing agent in the appropriate concentration which can reduce all of these five peripheral $(Mo)^{II}$-type atoms to Mo^V. If the related reduction potential is not met (and working at relatively high H^+ concentrations i.e. corresponding to pH $\approx$ 1–2 and not pH $\approx$ 4 with the formation of the $\{Mo_{11}\}_{12}$ system), the five peripheral Mo centers are only partially reduced with the consequence that they are no longer equivalent and the whole group has no longer fivefold symmetry. In the latter case other types of $\{Mo_{11}\}_n$ species such as the spherical ones are formed, for instance that with $n = 14$, which means the $\{Mo_{154}\}=\{Mo_{11}\}_{14}$-type cluster (this was highly celebrated by the scientific community; see below) (Fig. 13) [12]. Here the $\{Mo_{11}\}$ groups differ from

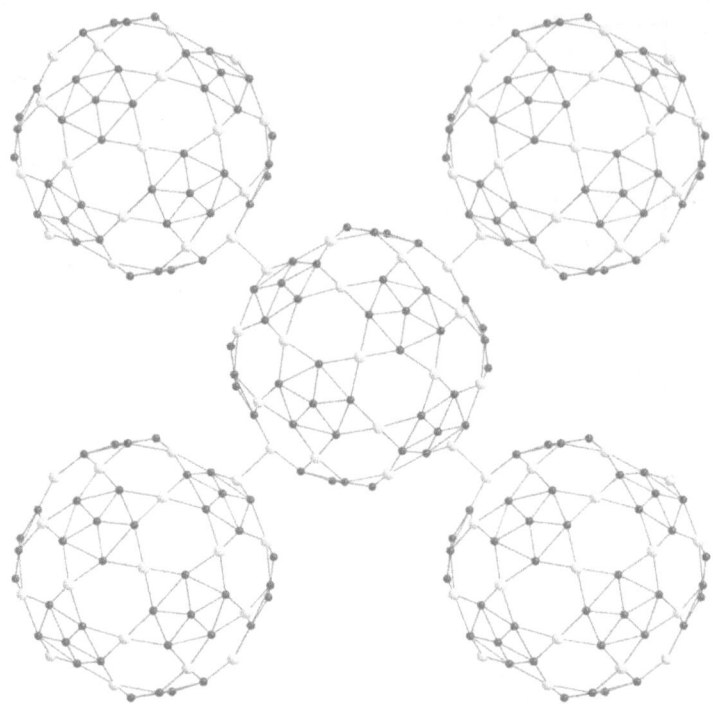

Fig. 12. Schematic representation of the layer formed by the linking of $\{(Mo)^0(Mo_5)^I$ $(Fe_{5/2})^{II}\}_{12}$-type clusters via Fe—O—Fe bridges (only the metal skeletons of the front shell parts are depicted). Each $\{(Mo)^0(Mo_5)^I(Fe_{5/2})^{II}\}_{12}$ entity is interconnected to four others

those present in the aforementioned $\{Mo_{11}\}_{12}$-type superfullerene in the sense that one of the peripheral MoO_6 octahedra is positioned in such a way that the $\{Mo_{11}\}$ group – as mentioned above – correspondingly shows only C_s symmetry.

Another topological description of the structure of the ring-type species (which was previously used by us and which refers to the relevant reactivities; see below) is based on the $\{Mo_8\}$ group as a fragment of the $\{Mo_{11}\}$ group: Discrete wheel-shaped clusters, such as $[Mo_{154}O_{462}H_{14}(H_2O)_{70}]^{14-}$, can accordingly be formulated as $[\{Mo_2\}_{14}\{Mo_8\}_{14}\{Mo_1\}_{14}]^-_{14} \equiv$ $[\{Mo_2^{VI}O_5(H_2O)_2\}_{14}^{2+}\{Mo_8^{VI/V}O_{26}(\mu_3\text{-}O)_2H(H_2O)_3Mo^{VI/V}\}^{3-14}]^{14-}$. Each of the 14 basic $\{Mo_8\}$ building blocks contains the central pentagonal-bipyramidal MoO_7 polyhedron which is symmetrically connected to five MoO_6 octahedra by edge sharing resulting in the $\{(Mo)^0(Mo_5)^I\}$ pentagon or motif mentioned above. Four of these MoO_6 octahedra are linked to two further MoO_6 octahedra via corners to form the $\{Mo_8\}$-type block. Starting from the $\{Mo_8\}$ group the complete $\{Mo_{11}\}_{14}$-type ring is built up according to the following scheme: (1) The two MoO_6 octahedra which are not directly connected to the central MoO_7 bipyramid are fused to two neighboring $\{Mo_8\}$ groups via corners. (2) Neighboring $\{Mo_8\}$ groups are additionally fused together by

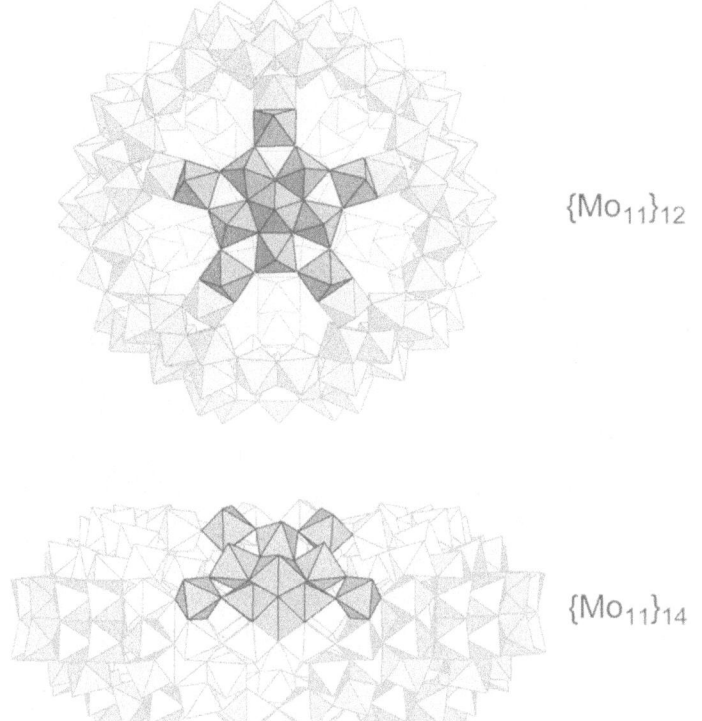

$\{Mo_{11}\}_{12}$

$\{Mo_{11}\}_{14}$

Fig. 13. Schematic comparison of the C_5- and C_S-type {Mo₁₁} motifs (*dark gray*) as present e.g. in the {Mo₁₁}₁₂- (*above*) and in the {Mo₁₁}₁₄-type cluster (*below*)

the {Mo₂} groups (which are important for interesting reactivity studies on the cluster), thus completing the inner-ring parts of the upper or lower half of the ring structure which are both built up in the same way. (3) The complete ring system is constructed when the second half is rotated around 360/14° relative to the first and fused to it through {Mo₁} units which are located at the equator of the (complete) ring (Fig. 14).

Whereas the {Mo₁₁}-based formulation is optimal to show the structural relation to the spherical clusters, a more detailed building-group view is necessary in the case of wheel-type species for the identification of sites/groups showing different reactivity. This is essential since the Mo atoms of the $(Mo_5)^{II}$ shell are not equivalent.

The first publication of the {Mo₁₁}₁₄≡{Mo₁₅₄}-type cluster [12] led to the metaphorical statement in *New Scientist*: "*Big wheel rolls back the molecular frontier*" [13].

This signifies the enormous interest of modern chemistry to 'leave' the molecular world, characterized by rather small molecules or ions, and to 'proceed' to the meso- or nanoworld with the intention of discovering and investigating the new inherent phenomena and system qualities.

$$\{Mo_{11}\}_{14}$$

Fig. 14. Polyhedral representation of the $\{Mo_{11}\}_{14}$-type cluster $[Mo_{154}O_{462}H_{14}(H_2O)_{70}]^{14-}$ in a front diagonal (*above*) and side (*below*) view with one $\{Mo_8\}$ group emphasized ($\{Mo_1\}$ units: *hatched*; $\{Mo_2\}$ groups: *dark gray*)

An even larger ring-shaped cluster than the tetradecameric $\{Mo_{11}\}_{14} \equiv \{Mo_{154}\}$ type with a hexadecameric structure ($\{Mo_{11}\}_{16} \equiv \{Mo_{176}\}$) can be obtained, too. It contains correspondingly 16 instead of 14 $\{Mo_{11}\}$ blocks or sets of each of the three mentioned building groups of the type $\{Mo_8\}$, $\{Mo_2\}$ and $\{Mo_1\}$, respectively (Fig. 15) [14].

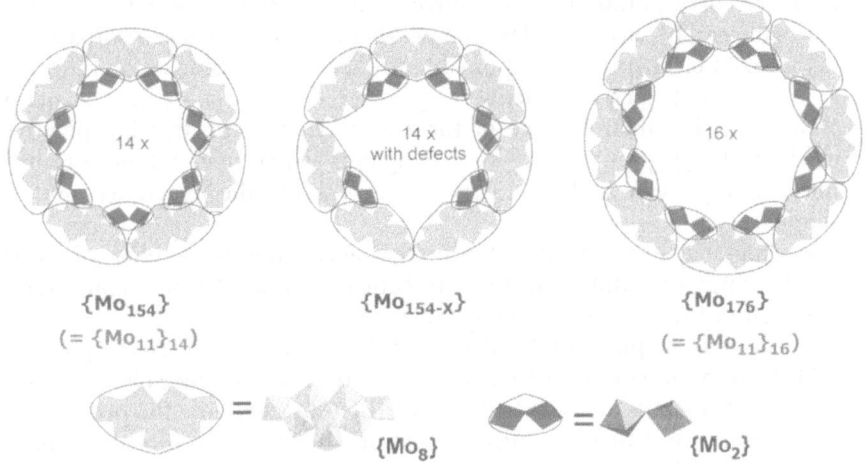

Fig. 15. Schematic comparison of the tetradecameric $\{Mo_{11}\}_{14}$ (with and without defects referring to missing $\{Mo_2\}$ groups) and hexadecameric $\{Mo_{11}\}_{16}$-type cluster with the correspondingly larger number of $\{Mo_8\}$ and $\{Mo_2\}$ groups (the equatorial $\{Mo_1\}$ units are not visible in this representation)

4.2
Multifunctionality

The $\{Mo_{11}\}_n$-type clusters ($n = 14, 16$) exhibit interesting properties and potentials:

1. They exhibit a nanometer-sized cavity, presenting new perspectives for a novel host-guest chemistry according to the different sites on the cluster surfaces.
2. They have an extended hydrophilic inner and outer surface due to the presence of $5n$ H_2O ligands ($2n$ H_2O ligands belong to the $\{Mo_2\}$ and $3n$ H_2O ligands to the $\{Mo_8\}$ groups).
3. They have a huge surface area which explains in part the high affinity towards adsorbents, such as charcoal or silk.
4. They render a molecular model for catalytically active metal-oxides of industrial importance.
5. The aqueous solution shows further aggregation tendencies: The formation of colloids of ca. 40 nm hydrodynamic radius could be detected by means of dynamic light scattering and scanning electron microscopy for the precipitated species [15].
6. The periphery of the cluster rings shows a rather high electron density with implications for the activation of small molecules coordinated to the surface sites.
7. There are n electronically practically uncoupled $\{Mo_5O_6\}$ compartments of the incomplete double-cubane type, each of which carries two delocalized 4d electrons, a situation which is comparable to a so-called electronic

necklace corresponding to an electron-storage system where the uncoupled storage elements are threaded like pearls on a string. These compartment-delocalized electrons are responsible for the intense blue color [16].

8 It is possible to generate deliberately discrete and well-defined structural defects on the inner surface of the cluster ring by abstracting positively charged {Mo$_2$} groups using special ligands which have a high affinity to these groups (see below). This changes both the reactivity and the shape.

9 The ring-shaped units can be linked according to a type of crystal engineering: Assembly due to synergetically induced functional comple-mentarity of distinct surface sites (see below).

10 It is possible to place molecules or replace ligands, e.g. H$_2$O by CH$_3$OH [17], at different sites of the surface, in particular to replace deliberately up to two H$_2$O ligands at the {Mo$_8$} groups (thereby changing the properties of the clusters) and to study direct reactions between the molecules placed inside the cavity.

4.3
En Route to Larger and More Complex Systems Based on Giant Wheel-type Clusters: Supramolecular and Network Structures

An aspect of particular interest is the fact that the large ring-shaped clusters can be used as synthons and can be linked to form chains or layers [16, 18, 19, 20]. The assembly is in principle based on the replacement of H$_2$O ligands at {Mo$_2$} groups (on rings) by oxygen atoms of terminal {Mo$_2$}-type Mo=O groups (on other rings), the nucleophilicity of which is increased for instance by introducing electron donating ligands like H$_2$PO$_2^-$ at relevant neighboring sites (Fig. 16), with the consequence that a layer compound can be obtained (Fig. 17) [19]. As this novel process is based on reactions at the same type of amphiphilic O=MoL (L = H$_2$O, H$_2$PO$_2^-$) {Mo$_2$}-type groups in different rings, the terminology 'synergetically induced functional complementarity' seems to

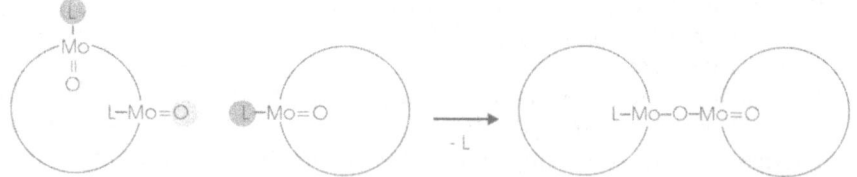

Fig. 16. Schematic representation of the basic 'assembly' principle of ring-shaped cluster units forming layers with [Mo$_{154}$O$_{462}$H$_{14}$(H$_2$O)$_{54}$(H$_2$PO$_2$)$_7$]$^{21-}$ building blocks. The linking is based on the synergetically induced functional complementarity of the {Mo$_2$}-type O=Mo(L) (L = H$_2$O, H$_2$PO$_2^-$) sites at the cluster surfaces: Each precursor ring contains corresponding functional complementary O=Mo(L) groups (L = H$_2$O, H$_2$PO$_2^-$). They become 'donors' in case of the presence of the electron-donating H$_2$PO$_2^-$ replacing H$_2$O ligands and 'acceptors' when no H$_2$PO$_2^-$ ligands are present

be justified. During the synthesis, the anion of the hypophosphoric acid acts both as reducing agent *and* exchanges with the H_2O ligands.

Linking is in principle also possible by abstracting some {Mo_2}-type fragments with bidentate ligands, such as the formate anion, from the inner surroundings of the {Mo_{11}}$_{14}$-type cluster (this also causes a higher nucleophilicity at the terminal oxygen atoms of {Mo_2} groups in the neighborhood). Compounds with chains having interesting electronic properties can be formed easily in this way [20].

Under special reducing conditions the {Mo_{11}}$_{16}$-type cluster system even starts growing again with the consequence that the two (unstable) molybdenum-oxide fragments of the type {$Mo_{36}O_{96}(H_2O)_{24}$} cover the cavity of the wheel-shaped cluster like hubcaps, resulting in an {Mo_{248}}-type cluster (Fig. 18) [21].

This is in terms of the number of metal atoms the largest known cluster anion that has been structurally characterized so far. Remarkably the structure of the {$Mo_{36}O_{96}(H_2O)_{24}$} fragment is in principle identical to a segment of the solid-state structure of the compound Mo_5O_{14} [22a,b] (Fig. 19) [22c].

This offers the possibility to model nucleation e.g. under well-defined boundary conditions, which is not understood at all. On the other hand, this type of aggregation has relevance for metal-center assembly in biological systems or special types of biomineralization in compartments.

The ring-shaped {Mo_{154}}-type clusters linked to chains can also act as hosts for smaller polyoxometalate guests, such as the stable (non-reduced) two-fragment {Mo_{36}}-type cluster mentioned above [23]. In this novel supramolecular system the interaction between host and guest, which fits exactly into the cavity of the host, is due to 16 hydrogen bonds as well as the Coulomb interaction mediated by at least four sodium cations located between the negatively charged host and negatively charged guest (Fig. 20). This compound is the only example among those mentioned here where we have difficulties with synthesis.

5
Extending the {Mo₁₁}-based Principle to More Complex Heterometallic Systems with {(Mo)⁰(Mo₅)ᴵ(Mᵢ₅)ᴵᴵ} Motifs

The directing influence of the pentagonal MoO_7 bipyramid is also valid for more complicated systems, i.e. heterometallic polyoxometalates which include other metal centers, too. In these systems, the molybdenum atoms of the peripheral $(Mo_5)^{II}$ units of the (former) {Mo_{11}} groups are in a formal consideration partially or completely substituted by heterometallic centers M^i, for instance V^{IV} or Fe^{III}, resulting in {$(Mo)^0(Mo_5)^I(M_5^i)^{II}$}-type motifs (Fig. 21). This is e.g. valid for the spherical {$(Mo)^0(Mo_5)^I(Fe_{5/2})^{II}$}$_{12}$-type system with M^i=Fe^{III} [11]. Note that the designation ($Fe_{5/2}$) in the *sum formula* corresponds to the fact that each Fe center belongs to two {$(Mo)^0(Mo_5)^I(M_5^i)^{II}$}-type building blocks.

A further example for this type of heterometallic system is presented by the cluster anion [{$Mo^{VI}O_3(H_2O)$}$_{10}${$V^{IV}O(H_2O)$}$_{20}${(Mo^{VI}/$Mo_5^{VI}O_{21}$)(H_2O)$_3$}$_{10}$

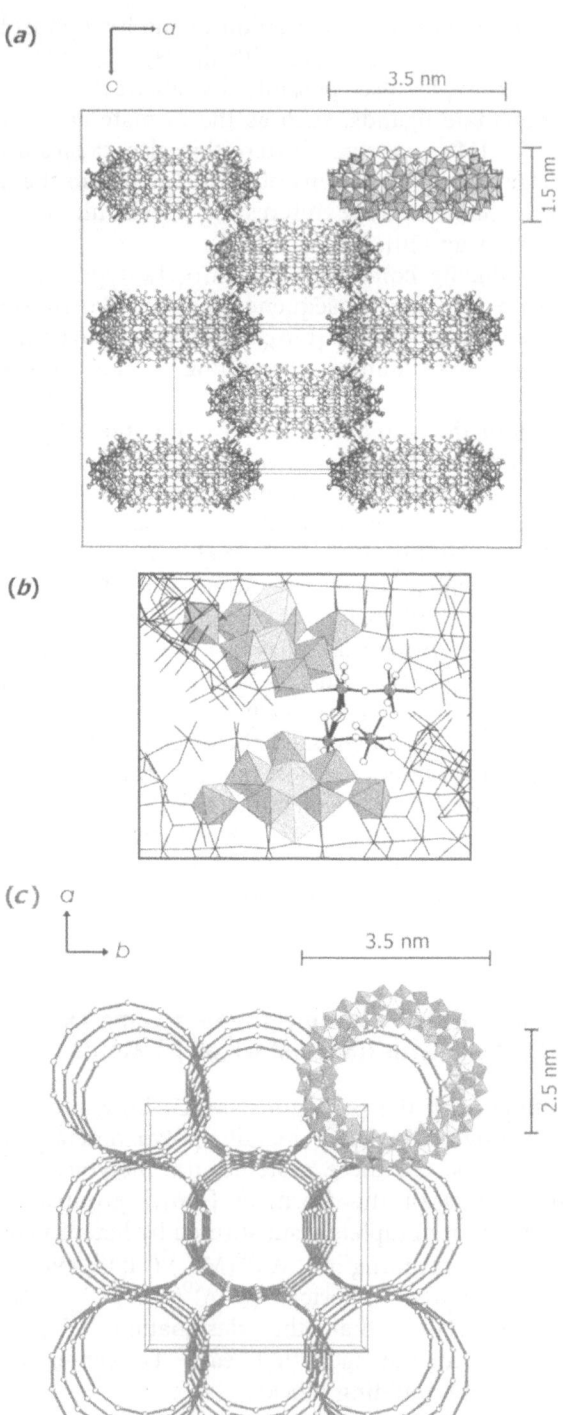

Fig. 17. a Ball-and-stick representation of the 'packing' of the covalently linked rings (in the ac plane) in crystals of $Na_{16}[Mo^{VI}_{124}Mo^{V}_{28}O_{429}(\mu_3\text{-}O)_{28}H_{14}(H_2O)_{66.5}]$ ca. 300 H_2O ({Mo₁₅₂}) viewed along the crystallographic b axis. Each ring is connected to surrounding rings via Mo—O—Mo bridges forming the O=Mo—O—Mo—OH₂ groups, thus generating layer networks parallel to the ac plane. One ring is shown as basic unit in polyhedral representation. **b** Detailed view of the bridging region between two cluster rings. One {Mo₈} group of each ring together with one {Mo₁} unit is shown in polyhedral representation and one $\{Mo_2\}^{2+}\equiv\{Mo^{VI}_2O_5(H_2O)_2\}^{2+}$ group per ring in ball-and-stick representation. The bridging (disordered) oxygen centers are depicted as *large, hatched circles* (for clarity, the disorder in the bridging positions is not shown here). **c** Perspective view along the crystallographic c axis showing the framework with nanotubes which are filled with H_2O molecules and Na^+ cations. For clarity, only one ring is shown completely in polyhedral representation. For the other rings only the centers of the {Mo₁} units are given and connected. The diameter of the central cavity inside a ring in the crystal is at least 1.9 nm

$(\{Mo^{VI}O_2(H_2O)_2\}_{5/2})_2$ $(\{NaSO_4\}_5)_2]^{20-}$ ({Mo₇₅V₂₀}) [24] which shows 10 $\{(Mo)^0(Mo_5)^I(M^i_5)^{II}\}$ ($M^i_5 = Mo^{VI} + 4V^{IV}$) motifs. Interestingly, it is now evident that the aforementioned (virtual) $\{(Mo)^0(Mo_5)^I\}$ pentagons "present" as disposition in solution can be "used" as basic building blocks under special boundary conditions, i.e. in presence of linkers/spacers in order to construct

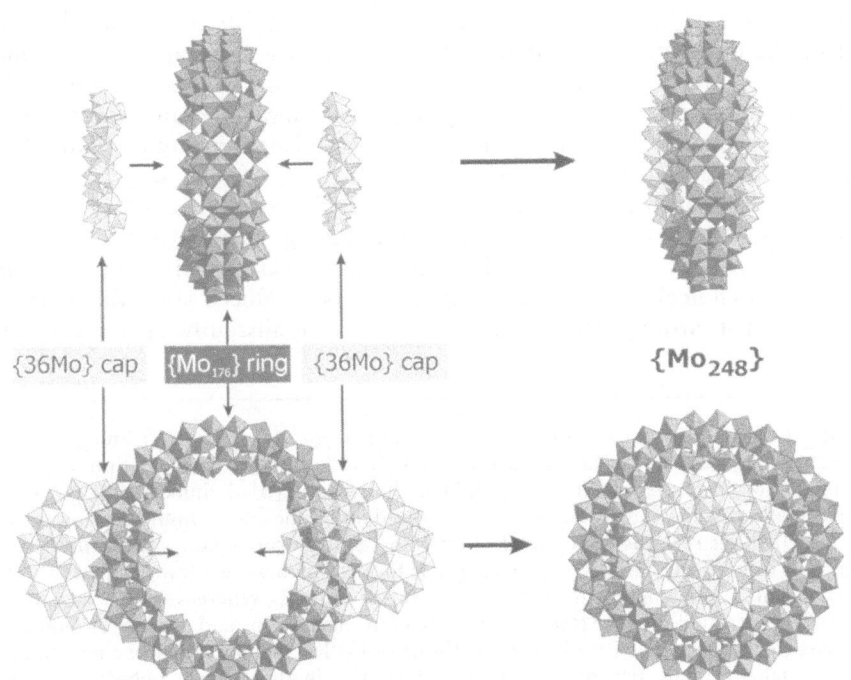

Fig. 18. Schematic representation of the growth process {Mo₁₇₆} → {Mo₂₄₈}. The structure of one {Mo₂₄₈} cluster can formally be decomposed into one {Mo₁₁}₁₆-type cluster ring (*dark gray*) and two {Mo₃₆O₉₆(H₂O)₂₄}-type hubcaps (*light gray*)

spherical species, e.g. derivatives of the $\{Mo_{11}\}_{12}$-type cluster with its 12 pentagonal $\{(Mo)^0(Mo_5)^I\}$ groups. This leads for instance to the mentioned cluster anion which can be formally described with reference to one of the Archimedean solids, namely the icosidodecahedron with 30 corners and 32 faces. Its 12 pentagonal and 20 triangular faces are built up by 20 V^{IV} and 10 Mo^{VI} centers while only ten of the 12 pentagonal faces of the icosidodeca-hedron are capped by the $\{(Mo)^0(Mo_5)^I\}$-type groups (Figs. 22, 23). Besides the remarkable aspect for magnetochemistry that the ten triangles, built up by the 20 V^{IV} centers, form an equatorial paramagnetic ring-shaped band with very strong antiferromagnetic exchange interactions, the cluster opens the doorway for new aspects of supramolecular chemistry because of the abundance of two $\{(NaSO_4)_5\}$-type rings inside the cavity. On the other hand it seems to be clear that in principle the $V^{IV}O^{2+}$ groups should be stepwise replaceable by Fe^{III} centers, which could enable a tuning of magnetic properties.

6
Facile and High-yield Synthesis of Giant Cluster Anions Including those of the Molybdenum-Blue Type – Now Possible

The precipitation and isolation of giant cluster anions, mainly those of the $\{Mo_{11}\}_n$ ($n = 14, 16$) type with extreme solubility (due to a large number of H_2O ligands, see Fig. 24), caused difficulties for chemists in the past, but also during the initial phase of our related investigations. Generations of chemists have for instance tried, without success, to isolate pure crystalline compounds (or even obtain a few crystals) from the relevant molybdenum blue solutions [25], which mainly contain the above-mentioned highly soluble giant wheel-type cluster anions $\{Mo_{11}\}_n$ ($n = 14, 16$). (This could in principle be considered as surprising as the anionic species can be easily formed in molybdate solutions using appropriate reducing agents within minutes.) Nonetheless, in the presence of electrolytes in a relatively high concentration (method of choice) the solubility is tremendously reduced since the extensive hydrate shell around the cluster anions is significantly disturbed. This

---→

Fig. 19. Structural comparison of the hubcap motif of the $\{Mo_{248}\}$ cluster and the related segment of the solid-state structure of Mo_5O_{14}. Above: Schematic representation of one half of the $\{Mo_{248}\}$ cluster with a highlighted $\{Mo_{36}O_{96}(H_2O)_{24}\}$ hubcap in polyhedral representation. Below: Structure of Mo_5O_{14} viewed along the c axis. Both the hubcaps and the Mo_5O_{14} layer sections each contain four $\{Mo_8\}$ entities surrounding two central $\{Mo_2\}$ groups. A central ring of six MoO_6 octahedra is formed by these two $\{Mo_2\}$ groups and two MoO_6 octahedra belonging to two opposite $\{Mo_8\}$ entities. Whereas in the case of the Mo_5O_{14} layer section the four $\{Mo_8\}$ entities are of the 'usual' type and occur in several of the discussed giant clusters with one central $\{(Mo)^0(Mo_5)^I\}$ pentagon which has two adjacent MoO_6 octahedra, only two of the four $\{Mo_8\}$ entities in the $\{Mo_{248}\}$ hubcaps have that structure. The other two $\{Mo_8\}$ groups consist of a central fragment in which 6 Mo atoms span an $\{Mo_6\}$ octahedron which is linked to two *trans*-positioned edge-sharing MoO_6 octahedra. The layers of the solid-state structure of Mo_5O_{14} can be formed formally by superposition of the above-mentioned segments (hup-caps) with 36 Mo centers

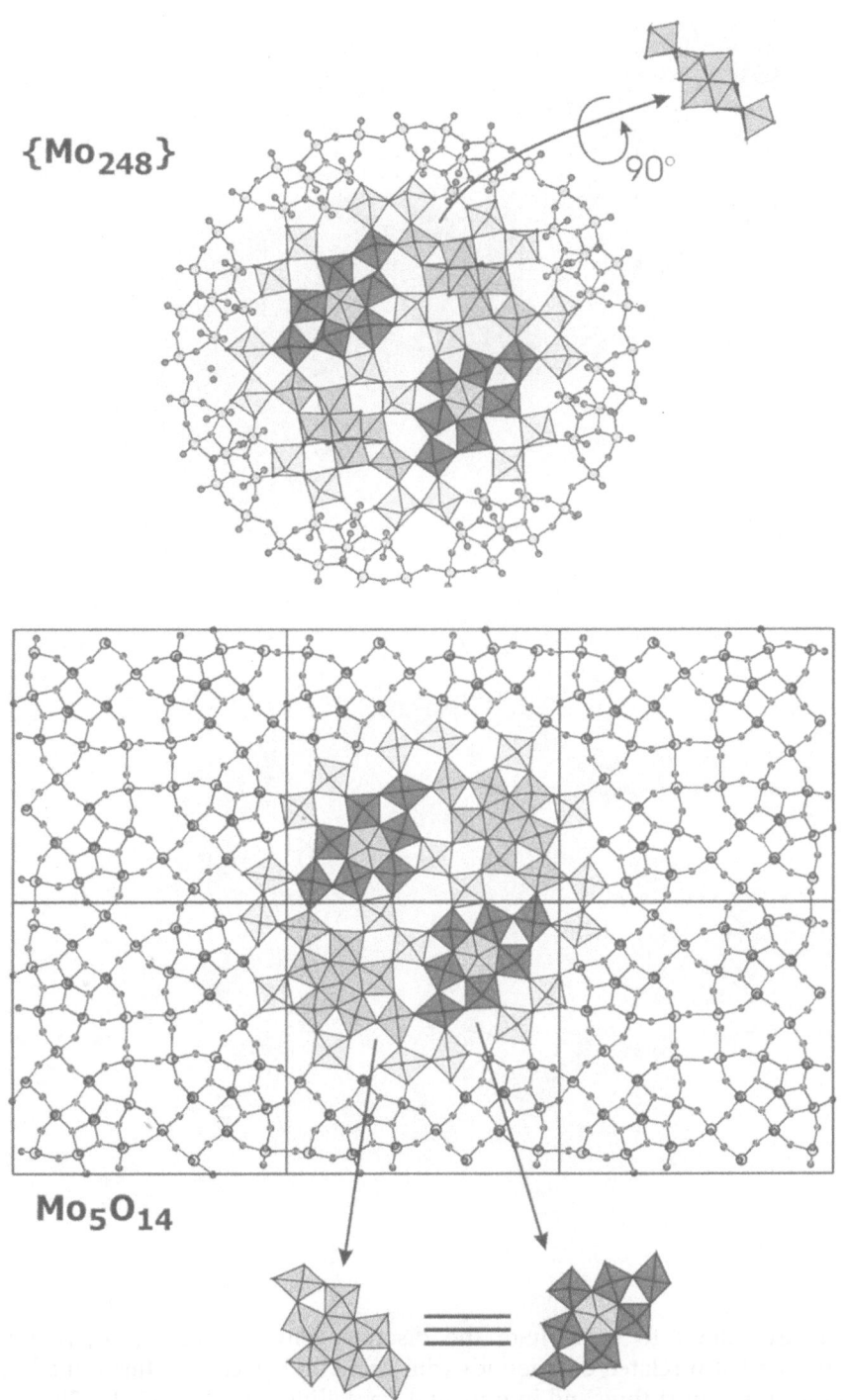

$\{Mo_{248}\}$

Mo_5O_{14}

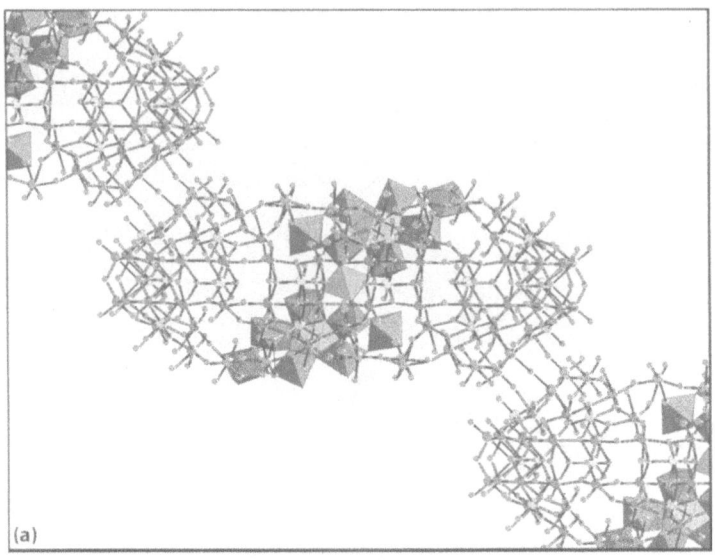

(a)

(b)

procedure allows to precipitate the discrete $\{Mo_{11}\}_{14}$- and $\{Mo_{11}\}_{16}$-type clusters and also related compounds with chain and layer structures in a high yield within a short time and in pure and crystalline form [16, 17, 19, 20]. The spherical Keplerate-type compounds are less soluble and can easily be

◀
___

Fig. 20. Some structural details of the novel supramolecular system (cluster in a cluster) $\{Mo_{36} \subset Mo_{148}\}$ (occupation of the cavities: ca. 20%). A Part of the chain structure is shown, which is built up by linking ring-shaped cluster units $\{Mo_{148}\}$ (i.e. based on the $\{Mo_{11}\}_{14}$ type with three missing $\{Mo_2\}$ groups; only the 'front' halves of the rings are shown for clarity). **b** View perpendicular to (**a**): The interaction between host and guest is due to 16 hydrogen bonds (*dotted lines*; O(host)–O(guest) = 2.744–2.965 Å) and (at least) four sodium cations (indicated by *arrow*, only two Na positions are visible in this perspective view) situated between host and guest. The sodium salt crystallizes in the space group *C2/m* ($a = 29.425$ Å, $b = 51.078$ Å, $c = 30.665$ Å, $\beta = 114.85°$)

obtained in high yield, in principle also in absence of high electrolyte concentrations.

7
Conclusions and Perspectives

The basic principles of the one-pot reactions we are particularly interested in – mainly those with feedback effects (non-linear relations) or of the induced cascade type (see below) – are of relevance not only in the context of generating complex polyoxometalate clusters, but also for general aspects of conservative self-organization and in particular of prebiotic chemistry, which is especially valid in the present case with reference to the combinatorially linkable (virtual) units as disposition. This problematic prebiotic situation is very well reflected by a citation [26] that deals with the work of a well-known member of the Santa Fe institute: "[Stuart A.] Kauffman's simulations have led him to several conclusions. One is that when a system of simple chemicals reaches a certain level of complexity [...], it undergoes a dramatic transition, or phase change. The molecules begin spontaneously combining to create larger molecules of increasing complexity and catalytic capability. Kauffman has

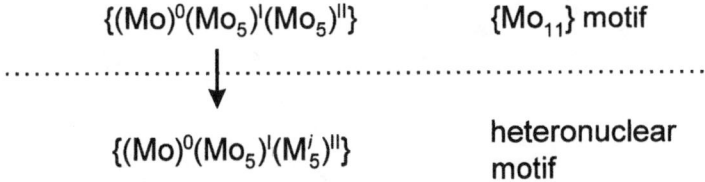

Extended Structural Hierarchy

$\{(Mo)^0(Mo_5)^I(Mo_5)^{II}\}$ $\{Mo_{11}\}$ motif

$\{(Mo)^0(Mo_5)^I(M'_5)^{II}\}$ heteronuclear motif

Examples:

$\{Mo_{72}Fe_{30}\}$: $M'_5 = 5\ Fe^{III}$ (spherical)

$\{Mo_{75}V_{20}\}$: $M'_5 = Mo^{VI} + 4\ V^{IV}$ (distorted spherical)

Fig. 21. Heterometallic extension of the structural hierarchy based on $\{(Mo)^0(Mo_5)^I(Mo_5)^{II}\}$-type pentagons

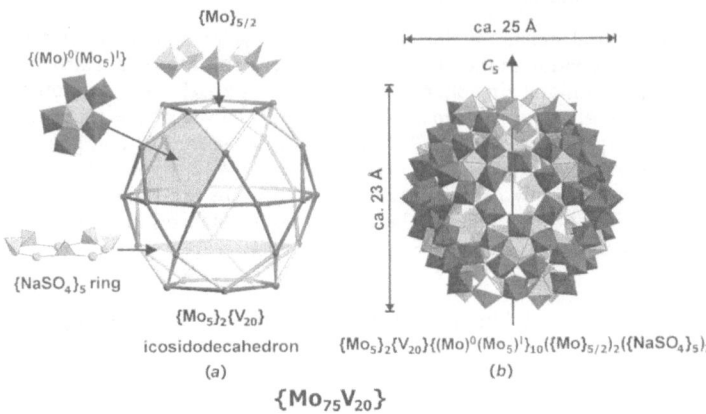

Fig. 22. a $\{V_{20}Mo_{10}\}$ framework structure of the $\{Mo_{75}V_{20}\}$-type cluster with 30 corners, 32 faces (20 triangles and 12 pentagons) spanned by 20 V^{IV} and 10 Mo^{VI} centers. Part of the framework is capped by $\{(Mo)^0(Mo_5)^I\}$ groups thus forming $\{(Mo)^0(Mo_5)^I(M_5^i)^{II}\}$ moieties (the Fe atoms of the related $\{(Mo)^0(Mo_5)^I(Fe_{5/2})^{II}\}_{12}$-type cluster form a less distorted icosidodecahedron). Additional details: (1) One of the $\{(Mo)^0(Mo_5)^I\}$ groups capping 10 pentagonal faces, (2) a pentagonal under-occupied array built up by five of the MoO_6 octahedra capping the other two pentagonal faces, and (3) one of the two encapsulated $\{NaSO_4\}_5$-type rings. **b** Polyhedral representation of the (approximately) spherical structure of the complete $\{Mo_{75}V_{20}\}$-type cluster

argued that this process of 'autocatalysis' – rather than the fortuitous formation of a molecule with the ability to replicate and evolve – led to life".

In this context the emergence of more complex molecular structures starting from simpler ones is the central problem. In this respect basic principles of induced cascade-type reactions such as that given in Scheme 2 are of special interest:

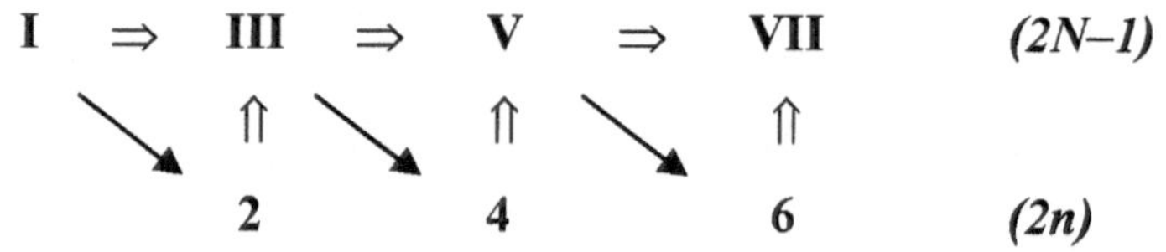

$$\textbf{I} \Rightarrow \textbf{III} \Rightarrow \textbf{V} \Rightarrow \textbf{VII} \qquad \textit{(2N–1)}$$

$$\textbf{2} \qquad \textbf{4} \qquad \textbf{6} \qquad \textit{(2n)}$$

Scheme 2.

The odd Roman numerals **2N – 1** signify maturation steps of a molecular system in growth or development and the Arabic numerals **2n** represent 'reagents' that react only with the special preliminary intermediate **2N – 1**. The 'species' **2n** can themselves be products of self-assembly processes, but can remarkably be generated template-driven by the corresponding intermediate **2N – 1**. *In the latter case each intermediate 2N – 1 carries information inducing the formation of the subsequent intermediate 2n (induced cascade-type*

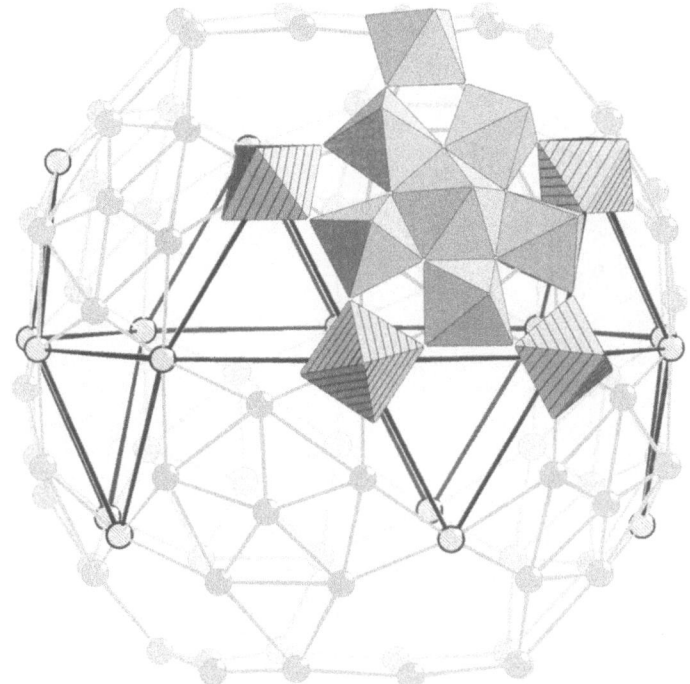

Fig. 23. Representation of the metal skeleton of the $\{Mo_{75}V_{20}\}$-type cluster with the emphasized 'magnetic' ring-shaped band formed by 10 V_3^{IV} triangles having common corners and one $\{(Mo)^0(Mo_5)^I(M_5^i)^{II}\}$ motif ($M_5^i = Mo^{VI} + 4\,V^{IV}$) in polyhedral presentation (vanadium centers: *hatched*)

reaction) with which it then reacts, thus demonstrating an interesting feedback effect.

The study of such reaction types will be a challenge for the future. In any case, one example for a growth process according to this model scheme has been discovered: A cluster anion containing 37 molybdenum atoms ($[H_{14}Mo_{37}O_{112}]^{14-}$) is formed via intermediates [27], one of which, the cluster of the type $\{Mo_{12}^V Mo_4^{VI}\}$ (containing an ε-Keggin-type nucleus capped by four electrophilic $\{Mo^{VI}O_3\}$ groups), can be isolated from the reaction medium [28a,b] (this $\{Mo_{16}\}$ fragment also resembles the core of other larger clusters, such as $[XH_nMo_{42}O_{109}\{(OCH_2)_3CR\}_7]^{m-}$ with $X = Na(H_2O)_3^+$: $n = 13$, $m = 9$; $n = 15$, $m = 7$ or $X = MoO_3$: $n = 14$, $m = 9$; $n = 13$, $m = 10$ [28c,d]). Interestingly, the next-step intermediate – formed after the relevant reduction of the four surface constituents of the type $\{Mo^{VI}O_3\} \rightarrow \{Mo^VO_3\}$ – acts both as nucleophile *and* template in directing the formation of further electrophilic molybdenum-oxide based $\{Mo_{10}\}$ and $\{Mo_{11}\}$ intermediates/fragments. These subsequently react with their template leading ultimately – in a type of symmetry-breaking step – to the target reaction product, the $\{Mo_{37}\}$-type

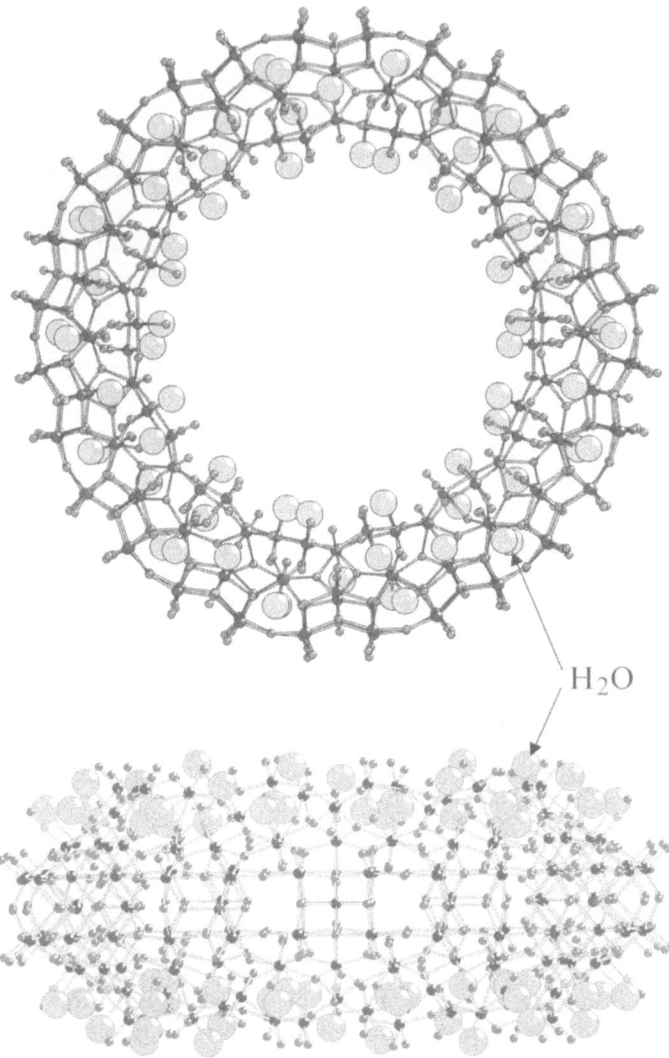

H₂O

Fig. 24. Ball-and-stick representation of a $\{Mo_{11}\}_{16}$ cluster anion showing its H_2O ligands as *light gray spheres*

cluster which remarkably has no (higher) symmetry element (Fig. 25). The most interesting aspect is that the former electrophilic $\{Mo^{VI}O_3\}$ groups are transformed/reduced into nucleophilic $\{Mo^{V}O_3\}$ ones after they have been attached to the cluster surface.

The reduction of units on the cluster surface is of relevance for the growth processes investigated here as it seems to be essential that the growing molecular system maintains its nucleophilic surface. On the other hand this is the entering point into the world of open (dissipative) systems, since the reduction of units (such as the $\{Mo^{VI}O_3\}$ groups) at the surface of the giant

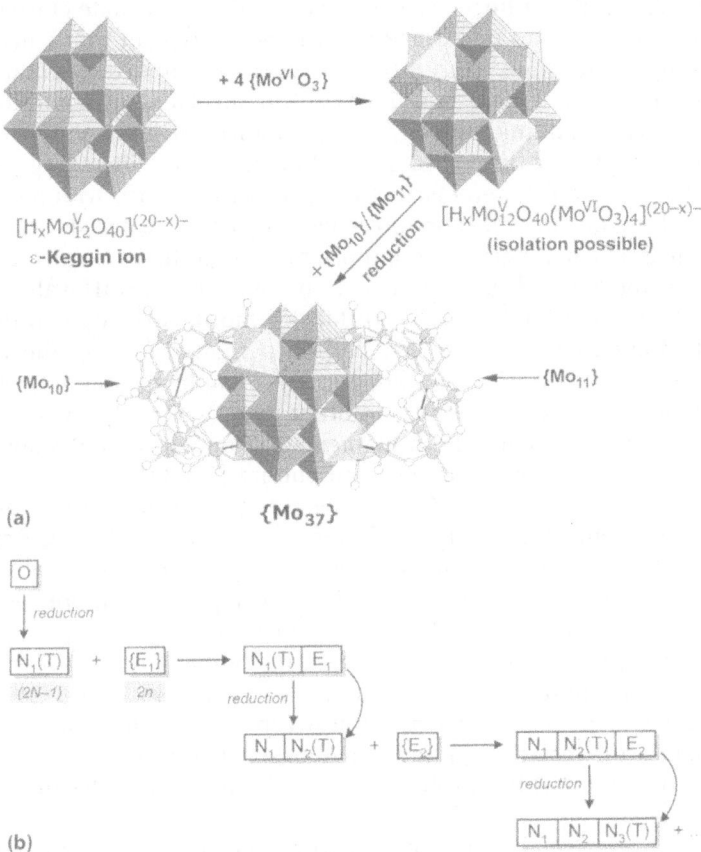

$[H_xMo_{12}^VO_{40}]^{(20-x)-}$
ε-**Keggin ion**

$+ 4 \{Mo^{VI}O_3\}$

$[H_xMo_{12}^VO_{40}(Mo^{VI}O_3)_4]^{(20-x)-}$
(isolation possible)

$+ \{Mo_{10}\}/\{Mo_{11}\}$
reduction

$\{Mo_{10}\} \longrightarrow$ $\longleftarrow \{Mo_{11}\}$

(a) $\{Mo_{37}\}$

O

reduction

$N_1(T)$ + $\{E_1\}$ $\longrightarrow$ $N_1(T)$ E_1
(2N–1)   2n   reduction

N_1 $N_2(T)$ + $\{E_2\}$ $\longrightarrow$ N_1 $N_2(T)$ E_2
                                                              reduction

N_1 N_2 $N_3(T)$ + ...

(b)

Fig. 25. a Schematic representation of a molecular growth process based on an induced cascade with molecular symmetry-breaking steps according to reaction Scheme 2 (see text); the resulting {Mo₃₇} cluster remarkably does not show any (higher) symmetry elements. **b** Reaction scheme corresponding to the growth process shown in (**a**), demonstrating principally the step-by-step procedure in which nucleophiles (such as the reduced ε-Keggin ion) attract electrophiles (such as {Mo^{VI}O₃}). Interestingly the latter can become nucleophiles ({Mo^VO₃} units) on the surface of the cluster upon reduction. O denotes species at the beginning of a growth process (formally the non-reduced ε-Keggin ion). $N_i(T)$ denotes a nucleophilic intermediate/fragment produced under reducing conditions (of the type **2N – 1** according to Scheme 2, e.g. the reduced ε-Keggin ion) with a potential template function for the generation of the electrophilic intermediate E_i (of the type **2n** according to Scheme 2, e.g. the {Mo^{VI}O₃} groups)

cluster can in principle be performed 'externally' by adding an appropriate reducing agent *at the right time*, a procedure which can therefore be repeated several times, thereby again and again initiating growth. (In this context it is important to note that these types of 'integrated' constituents are more easily reduced than the discrete ones.) This is, generally speaking, probably the door to a new type of chemistry leading to macroscopic structuring like in diatoms, but based on the linking of polyoxometalate units.

It can be shown that a huge variety of giant polyoxometalate clusters – and even different types – can be formed according to a type of unit construction under one-pot conditions (for different types of important giant cluster systems not considered here see [29–31]). On the basis of combinatorially linkable (virtual) building units (as disposition), such as the pentagonal $\{(Mo)^0(Mo_5)^I\}$ type, and different types of linkers/spacers including paramagnetic (open shell) centers, new fascinating aspects are opened with possibilities e.g. for structural chemistry, supramolecular chemistry, materials science, and especially magnetochemistry. It is important to note that we have a control parameter, namely the degree of reduction (but also the pH value), which influences the kind of the $\{(Mo)^0(Mo_5)^I(Mo_5)^{II}\}$ motifs formed, with the result that icosahedral structures (with C_5-type $\{Mo_{11}\}$ motifs) or tetra-/hexadecameric ring structures (with C_s-type $\{Mo_{11}\}$ groups) can be deliberately obtained. 'Playing' with pentagons is – besides in order to get several types of clusters – of interest because of their role in mathematics/topology, solid-state physics and perhaps in the history of science and culture [11, 32].

We are convinced that the $\{Mo_{11}\}_n$-type clusters will be the starting point for the development of a new type of supramolecular – including Keplerate-type – chemistry which is guided by topological or unit construction approaches. The spherical species are stable in aqueous solution and allow the encapsulation of different collections of guests in the cavity [33] and even the substitution of small guests through the abundant Mo_9O_9 ring-type apertures. A collection of the spherically shaped giant clusters provides a useful model system for special dynamical aspects of solid-state chemistry, since it enables for instance the investigation of nucleation processes (which are still not understood) but probably also of the first principles of structure-forming processes.

In the future it will be possible to investigate reactions between single molecules in the cavities of the giant cluster systems and especially to place molecules at distinct sites inside the cavity of the wheel-shaped clusters, which show a different reactivity as the discrete (isolated) ones.

A recent result obtained by us shows that the spherical $\{Mo_{11}\}_{12}$ cluster can be deliberately obtained by using the $Mo_2O_4^{2+}$ group (occurring in the compound $(NH_4)_2[Mo_2^VO_4Cl_2(OH)_2(H_2O)_2]$) as educt/template which initiates the formation of the $\{(Mo)Mo_5\}$ pentagons (virtual units) in "molybdate solutions"; correspondingly a cluster type with $\{(W)W_5\}_{12}$ pentagons can be obtained [34].

Acknowledgements. We are grateful to the Deutsche Forschungsgemeinschaft and the Fonds der Chemischen Industrie for financial support.

8
References

1. a) Müller A, Peters F, Pope MT, Gatteschi D (1998) Chem Rev 98: 239 b) Pope MT, Müller A (1991) Angew Chem Int Ed Engl 30: 34 c) Pope MT, Müller A (eds) (1994) Polyoxometalates: From Platonic Solids to Anti-Retroviral Activity. Kluwer, Dordrecht

d) Pope MT (1983) Heteropoly and Isopoly Oxometalates. Springer, Berlin Heidelberg New York

2. Voet D, Voet JG (1995) Biochemistry, 2nd edn. Wiley, New York, p 1076

3. First reports on the three-fragment {$Mo_{57}V_6$} cluster appeared in 1993, see a) Zhang S, Huang G, Shao M, Tang Y (1993) J Chem Soc Chem Commun 37 b) Müller A, Krickemeyer E, Dillinger S, Bögge H, Proust A, Plass W, Rohlfing R (1993) Naturwissenschaften 80: 560. Other related clusters (such as the two-fragment {Mo_{36}} cluster) were published later and errors concerning the formula in the first publications were corrected, see c) Müller A, Krickemeyer E, Dillinger S, Bögge H, Plass W, Proust A, Dloczik L, Menke C, Meyer J, Rohlfing R (1994) Z Anorg Allg Chem 620: 599 d) Müller A, Plass W, Krickemeyer E, Dillinger S, Bögge H, Armatage A, Beugholt C, Bergmann U (1994) Monatsh Chem 125: 525 e) Müller A, Plass W, Krickemeyer E, Dillinger S, Bögge H, Armatage A, Proust A, Beugholt C, Bergmann U (1994) Angew Chem Int Ed Engl 33: 849 f) Müller A, Bögge H, Krickemeyer E, Dillinger S (1994) Bull Pol Acad Sci Chem 42: 291 g) Müller A, Reuter H, Dillinger S (1995) Angew Chem Int Ed Engl 34: 2311. For details on the relevant basic {Mo_8} groups see also h) Müller A, Beugholt C (1996) Nature 383: 296. For the two-fragment isopolyoxomolybdate {Mo_{36}}-type cluster with [MoO] instead of [MoNO] groups see also i) Tytko K-H, Schönfeld B, Buss B, Glemser O (1973) Angew Chem Int Ed Engl 12: 330 and with the corrected formula j) Krebs B, Stiller S, Tytko K-H, Mehmke J (1991) Eur J Solid Inorg Chem 28: 883

4. a) Krätschmer W, Schuster H (eds) (1996) Von Fuller bis zu Fullerenen, Beispiele einer interdisziplinären Forschung. Vieweg, Braunschweig b) Coxeter HSM (1969) Introduction to Geometry. Wiley, New York

5. Chae HK, Klemperer WG, Marquart TA (1993) Coord Chem Rev 128: 209

6. Müller A, Krickemeyer E, Bögge H, Schmidtmann M, Peters F (1998) Angew Chem Int Ed 37: 3360

7. Liljas L, Strandberg B (1984) In: Jurnak FA, McPherson A (eds) Biological Macromolecules and Assemblies, vol 1, Virus Structures. Wiley, New York, p 97

8. a) Voet D, Voet JG (1995) Biochemistry, 2nd edn. Wiley, New York b) Alberts B, Bray D, Lewis J, Raff M, Roberts K, Watson JD (1994) Molecular Biology of the Cell, 3rd edn. Garland, New York c) Davies BD, Dulbecco R, Eisen HN, Ginsberg HS (1980) Microbiology, 3rd edn. Harper and Row, Philadelphia d) Harrison SC (1984) Trends Biochem Sci 9: 345

9. Müller A, Fedin VP, Kuhlmann C, Bögge H, Schmidtmann M (1999) Chem Commun 927

10. a) Kepler J (1596) Mysterium Cosmographicum; see also b) Kemp M (1998) Nature 393: 123

11. Müller A, Sarkar S, Bögge H, Schmidtmann M, Sarkar Sh, Kögerler P, Hauptfleisch B, Shah SQN, Trautwein A, Schünemann V (1999) Angew Chem Int Ed 38: 3238

12. Müller A, Krickemeyer E, Meyer J, Bögge H, Peters F, Plass W, Diemann E, Dillinger S, Nonnenbruch F, Randerath M, Menke C (1995) Angew Chem Int Ed Engl 34: 2122

13. Bradley D (1995) New Scientist 148 (2003): 18

14. a) Müller A, Krickemeyer E, Bögge H, Schmidtmann M, Beugholt C, Kögerler P, Lu C (1998) Angew Chem Int Ed 37: 1220 b) Jiang C, Wei Y, Liu Q, Zhang S, Shao M, Tang Y (1998) Chem Commun 1937

15. Müller A, Eimer W, Diemann E, Serain C, to be published

16. Müller A, Krickemeyer E, Bögge H, Schmidtmann M, Beugholt C, Das SK, Peters F (1999) Chem Eur J 5: 2114

17. Müller A, Koop M, Bögge H, Schmidtmann M, Beugholt C (1998) Chem Commun 1501

18. Müller A, Krickemeyer E, Bögge H, Schmidtmann M, Peters F, Menke C, Meyer J (1997) Angew Chem Int Ed Engl 36: 484

19. Müller A, Das SK, Bögge H, Beugholt C, Schmidtmann M (1999) Chem Commun 1035

20. Müller A, Das SK, Fedin VP, Krickemeyer E, Beugholt C, Bögge H, Schmidtmann M, Hauptfleisch B (1999) Z Anorg Allg Chem 625: 1187

21. a) Müller A, Shah SQN, Bögge H, Schmidtmann M (1999) Nature 397: 48 b) Ball P (1998) Nature 395: 745

22. a) Cotton FA, Wilkinson G (1980) Advanced Inorganic Chemistry, 4th ed. Wiley, New York, p 847 b) Kihlborg L (1963) Arkiv Kemi 21: 427 c) Müller A, Hauptfleisch B, in preparation
23. Müller A et al. (1999) in preparation
24. Müller A, Koop M, Bögge H, Schmidtmann M, Peters F, Kögerler P (1999) Chem Commun 1885
25. a) (1935) Gmelins Handbuch der Anorganischen Chemie, Molybdän. Verlag Chemil, Berlin, p 134 b) (1987) Gmelin Handbook of Inorganic Chemistry, Molybdenum Suppl., vol B3a, 8th edn. Springer, Berlin Heidelberg New York. See also c) Müller A, Meyer J, Krickemeyer E, Diemann E (1996) Angew Chem Int Ed Engl 35: 1206
26. Horgan J (1995) Scientific American, June Issue 74
27. Müller A, Meyer J, Krickemeyer E, Beugholt C, Bögge H, Peters F, Schmidtmann M, Kögerler P, Koop MJ (1998) Chem Eur J 4: 1000
28. a) Khan MI, Müller A, Dillinger S, Bögge H, Chen Q, Zubieta J (1993) Angew Chem Int Ed Engl 32: 1780 b) Müller A, Dillinger S, Krickemeyer E, Bögge H, Plass W, Stammler A, Haushalter RC (1997) Z Naturforsch 52b: 1301 c) Khan MI, Zubieta J (1992) J Am Chem Soc 114: 10058 d) Khan MI, Chen Q, Salta J, O'Connor CJ, Zubieta J (1996) Inorg Chem 35: 1880
29. a) Krautscheid H, Fenske D, Baum G, Semmelmann M (1993) Angew Chem Int Ed Engl 32: 1303 b) Fenske D, Zhu N, Langetepe T (1998) Angew Chem Int Ed 37: 2640
30. Ecker A, Weckert E, Schnöckel H (1997) Nature 387: 379
31. Wassermann K, Dickman MH, Pope MT (1997) Angew Chem Int Ed Engl 36: 1445
32. Mackay AL (1990) In: Hargittai I (ed) Quasicrystals, Networks, and Molecules of Fivefold Symmetry. VCH, Weinheim
33. Müller A, Polarz S, Das SK, Krickemeyer E, Bögge H, Schmidtmann M, Hauptfleisch B (1999) Angew Chem Int Ed 38: 3241
34. Müller A et al., to be published

Author Index Volumes 1-96

Baldwin AH, see Butler A (1997) *89*: 109–132

Balsenc LR (1980) Sulfur Interaction with Surfaces and Interfaces Studied by Auger Electron Spectrometry. *39*: 83–114

Banci L, Bencini A, Benelli C, Gatteschi D, Zanchini C (1982) Spectral-Structural Correlations in High-Spin Cobalt(II) Complexes. *52*: 37–86

Banci L, Bertini I, Luchinat C (1990) The ^{1}H NMR Parameters of Magnetically Coupled Dimers – The Fe_2S_2 Proteins as an Example. *72*: 113–136

Baran EJ, see Müller A (1976) *26*: 81–139

Bartolotti LJ (1987) Absolute Electronegativities as Determined from Kohn-Sham Theory. *66*: 27–40

Bates MA, Luckhurst GR (1999) Computer Simulation of Liquid Crystal Phases Formed by Gay-Berne Mesogens. *94*: 65–137

Bau RG, see Teller R (1981) *44*: 1–82

Baughan EC (1973) Structural Radii, Electron-cloud Radii, Ionic Radii and Solvation. *15*: 53–71

Bayer E, Schertzmann P (1967) Reversible Oxygenierung von Metallkomplexen. *2*: 181–250

Bearden AJ, Dunham WR (1970) Iron Electronic Configuration in Proteins: Studies by Mössbauer Spectroscopy. *8*: 1–52

Bencini A, see Banci L (1982) *52*: 37–86

Benedict U, see Manes L (1985) *59/60*: 75–125

Benelli C, see Banci L (1982) *52*: 37–86

Benefield RE, see Thiel RC (1993) *81*: 1–40

Bergmann D, Hinze J (1987) Electronegativity and Charge Distribution. *66*: 145–190

Berners-Price SJ, Sadler PJ (1988) Phosphines and Metal Phosphine Complexes Relationship of Chemistry to Anticancer and Other Biological Activity. *70*: 27–102

Bernt I, see Uller E (2000) *96*: 149–176

Bertini I, see Banci L (1990) *72*: 113–136

Bertini I, Ciurli S, Luchinat C (1995) The Electronic Structure of FeS Centers in Proteins and Models. A Contribution to the Understanding of Their Electron Transfer Properties. *83*: 1–54

Bertini I, Luchinat C, Scozzafava A (1982) Carbonic Anhydrase: An Insight into the Zinc Binding Site and into the Active Cavity Through Metal Substitution. *48*: 45–91

Bertrand P (1991) Application of Electron Transfer Theories to Biological Systems. *75*: 1–48

Bill E, see Trautwein AX (1991) *78*: 1–96

Bino A, see Ardon M (1987) *65*: 1–28

Blanchard M, see Linarès C (1977) *33*: 179–207

Blasse G, see Powell RC (1980) *42*: 43–96

Blasse G (1991) Optical Electron Transfer Between Metal Ions and its Consequences. *76*: 153–188

Blasse G (1976) The Influence of Charge-Transfer and Rydberg States on the Luminescence Properties of Lanthanides and Actinides. *26*: 43–79

Blasse G (1980) The Luminescence of Closed-Shell Transition Metal-Complexes. New Developments. *42*: 1–41

Blauer G (1974) Optical Activity of Conjugated Proteins. *18*: 69–129

Bleijenberg KC (1980) Luminescence Properties of Uranate Centres in Solids. *42*: 97–128

Boca R, Breza M, Pelikán P (1989) Vibronic Interactions in the Stereochemistry of Metal Complexes. *71*: 57–97

Boeyens JCA (1985) Molecular Mechanics and the Structure Hypothesis. *63*: 65–101

Bögge H, see Müller A (2000) *96*: 203–236

Böhm MC, see Sen KD (1987) *66*: 99–123

Bohra R, see Jain VK (1982) *52*: 147–196

Bollinger DM, see Orchin M (1975) *23*: 167–193

Bominaar EL, see Trautwein AX (1991) *78*: 1–96

Bonnelle C (1976) Band and Localized States in Metallic Thorium, Uranium and Plutonium, and in Some Compounds, Studied by X-ray Spectroscopy. *31*: 23–48